D0983583

Library

The Agrarian Roots of Pragmatism

The Vanderbilt Library of American Philosophy offers interpretive perspectives on the historical roots of American philosophy and on present innovative developments in American thought, including studies of values, naturalism, social philosophy, cultural criticism, and applied ethics.

Other titles in the series include

DEWEY'S EMPIRICAL THEORY OF KNOWLEDGE AND REALITY
John R. Shook

THINKING IN THE RUINS
Wittgenstein and Santayana on Contingency
Michael P. Hodges and John Lachs

PRAGMATIC BIOETHICS
Edited by Glenn McGee

TRANSFORMING EXPERIENCE
John Dewey's Cultural Instrumentalism
Michael Eldridge

The Agrarian Roots of Pragmatism

Edited by Paul B. Thompson
and
Thomas C. Hilde

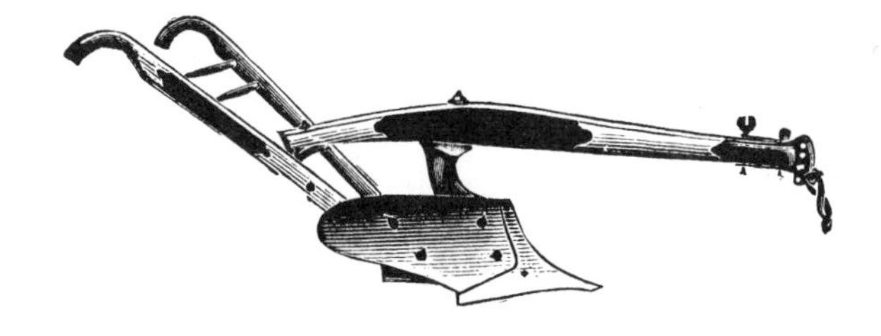

Vanderbilt University Press
Nashville

First Edition 2000

04 03 02 01 00 5 4 3 2 1

Library of Congress Cataloging-in-Publication Data

The agrarian roots of pragmatism / edited by Paul B. Thompson and Thomas C. Hilde.—
1st ed.
p. cm. — (The Vanderbilt library of American philosophy)
Includes bibliographical references and index.
ISBN 0-8265-1339-5
1. Pragmatism—History. 2. Agriculture—Philosophy—History.
I. Thompson, Paul B., 1951– II. Hilde, Thomas C. III. Title. IV. Series.
B832.A34 2000
144'.3'0973--dc21 00-008775

Published by Vanderbilt University Press
Printed in the United States of America

Contents

Acknowledgments vii
A Note on the Text ix

Introduction: Agrarianism and Pragmatism 1
THOMAS C. HILDE AND PAUL B. THOMPSON

Part I. Defining the Agrarian Tradition

Agrarianism as Philosophy 25
PAUL B. THOMPSON

American Agrarianism: The Living Tradition 51
JAMES A. MONTMARQUET

Land, Labor, and God in American Colonial Thought 77
CHARLES TALIAFERRO

Part II. Prepragmatist Agrarianism in America

Franklin Agrarius 101
JAMES CAMPBELL

Thomas Jefferson and Agrarian Philosophy 118
PAUL B. THOMPSON

Emerson and the Agricultural Midworld 140
ROBERT S. CORRINGTON

Wild Farming: Thoreau and Agrarian Life 153
DOUGLAS R. ANDERSON

Part III. Pragmatism and Agrarianism: Dewey and Royce

Provincialism, Displacement, and Royce's Idea of Community 167
THOMAS C. HILDE

Does Metaphysics Rest on an Agrarian Foundation? A Deweyan Answer and Critique 185
ARMEN MARSOOBIAN

The Edible Schoolyard: Agrarian Ideals and Our Industrial Milieu 195
LARRY A. HICKMAN

Part IV. Twentieth-Century Agrarian Thought in the Pragmatist Tradition

The Relevance of the Jeffersonian Dream Today 209
JOHN M. BREWSTER

Two on Jefferson's Agrarianism 254
GENE WUNDERLICH

Steinbeck and Agrarian Pragmatism 269
RICHARD E. HART

Coming Full Circle? Agrarian Ideals and Pragmatist Ethics in the Modern Land-Grant University 279
JEFFREY BURKHARDT

Notes 304
Bibliography 315
Contributors 329
Index 332

Acknowledgments

The seed for this book was planted at a meeting of the Society for the Advancement of American philosophy held in Philadelphia back in the mid-1980s. Among the eventual contributors to this book, Thompson, Campbell, Marsoobian, Corrington, and Hart (plus a few others, bringing the total to about a dozen) began over dinner to toss around ideas on the agrarian roots of pragmatism. Thompson, having just received a grant from the W. K. Kellogg Foundation to promote humanistic studies of agriculture, volunteered to pick up the check for the whole table (it came to about seventy-five bucks—not much even then), but insisted that in return everyone was obligated to write up their ideas. A first round of papers on this topic was published in a special issue of *Agriculture and Human Values*. Hilde appeared on the scene in about 1994, lamenting the fact that the location of these articles made them totally obscure for people working in the pragmatist tradition. Thompson and Hilde worked together off and on for six more years, adding papers and recruiting additional authors and then editing the volume. The collection before you is, thus, the eventual return on that Philadelphia fish dinner so long ago. Though most of this work was a labor of love for everyone involved, we would like to thank the W. K. Kellogg Foundation for the seventy-five dollars that was needed to get the project off the ground.

We would also like to thank our authors, who not only labored over their own contributions but served as commentators and reviewers (sometimes cheerleaders) for one anothers' chapters. We received a great deal of encouragement from John McDermott and Herman Saatkamp, for which we are grateful. David Danbom made very insightful suggestions about the entire project, as well as on Thompson's chapters. The book was conceived and the initial steps were taken when both Thompson and Hilde were associated with the philosophy department at Texas A&M University. We would like to thank our departmental colleagues for their tolerance of an oddball historical project. The clerical staff at Texas

A&M's Center for Biotechnology Policy and Ethics helped in with the correspondence that was essential for the early stages of the project. The College of Liberal Arts, the Institute for Biosciences and Technology, and the Office of the Vice President for Research (all at Texas A&M) contributed the funds that paid those staff salaries. Two individuals, John Borin and Daralyn Wallace, were particularly involved in the early stages.

Three individuals were also deeply involved in the final stages, taking charge of correspondence with authors, editing, indexing, and organizing the manuscript, and reading and reviewing the entire manuscript several times, with an eye to editorial and organizational questions. They are Todd Ferguson, Sarah Roberts, and Debra Jackson, all Ph.D. students in philosophy at Purdue University. Sarah, Debra, and Todd were each supported for a term by a research assistantship tied to the Joyce and Edward E. Brewer Professorship in Applied Ethics. It is very clear that this book might have been many more years in the making without the insight, diligence, and judgment that they brought to the final stages of the project.

Unfortunately, we have not taken enough advice from any of these people that we can displace from our own shoulders blame for errors of judgment or fact.

A Note on the Text

Citations listing author, date of publication, and page number are parenthetically inserted in the text. All parenthetical citations refer to the comprehensive bibliography that appears at the end of the volume. Notes may occasionally incorporate further citations, but always within the context of amplifying or commenting on the main text. Whenever practicable, we have edited previously published chapters to make use of canonical sources for principle figures. When the original reference (or the author's intent) was unclear, we have maintained the original citation. Benjamin Franklin's writings will be cited when possible from the Yale University Press critical edition: *The Papers of Benjamin Franklin,* ed. Leonard W. Labaree, William B. Willcox, and Barbara B. Oberg (New Haven: Yale University Press, 1959–), indicating volume and page number [for example, "(10:129)" for volume 10, page 129]. The *Autobiography,* ed. Leonard W. Labaree, Ralph L. Ketcham, Helen C. Boatfield, and Helene H. Fineman (New Haven: Yale University Press, 1964) will be cited using an "A" in the place of a volume number [for example, "(A:129)"]. Material from those years not yet reached in the new edition will be cited from the Albert Henry Smyth edition of *The Writings of Benjamin Franklin* (New York: Macmillan, 1907) cited as "W" followed by the volume and page number [for example, "(W10:129)"]. Thomas Jefferson's writings are cited from the two Library of America volumes: *Writings* (1984) and *Public and Private Papers* (1990). Except when otherwise noted, Josiah Royce's writings refer to John McDermott's two-volume *Basic Writings of Josiah Royce* (1969). Citations to John Dewey refer to Boydston's *The Collected Works of John Dewey* (1969–1991), published and referenced as *The Early Works* (EW), *The Middle Works* (MW), and *The Later Works* (LW).

Introduction

Agrarianism and Pragmatism

THOMAS C. HILDE AND PAUL B. THOMPSON

THE ESSAYS in this collection explore the relationship between agrarianism and classical American philosophy. We have selected a few previously published essays and recruited new chapters from distinguished authors who examine this relationship with respect to key figures, to formative periods in American political life, such as the Great Awakening and the Great Depression, and to contemporary issues in philosophy, pragmatism, and public policy for agriculture. We define classical American philosophy primarily with reference to the transcendentalists and the pragmatists, including both historical and contemporary themes that anticipate or draw upon the thought of Ralph Waldo Emerson, Henry David Thoreau, Charles Sanders Peirce, Josiah Royce, and John Dewey, the central figures that we are associating here with agrarian ideas.

Agrarianism designates different moral, social, and political, and even metaphysical philosophies that accord special roles to farms and the practice of farming. In its most common usage it indicates political movements for land reform. We will combine definitions offered by James Montmarquet and Paul B. Thompson, respectively, for our working model. *Agrarianism* is the belief that "agriculture and those whose occupation involves agriculture are especially important and valuable elements of society," and that "the practice of agriculture and farming establishes a privileged outlook upon fundamental questions of human conduct, and, sometimes, the nature of reality itself" (Montmarquet 1989, viii; Thompson 1990, 3). Agrarian philosophies stipulate and then develop a special value-relation between society, humans, and the land on which they practice agriculture. At these various intersections one also finds philosophical kinship with various themes in classical American philosophy. In the present volume, the authors examine the diversity of relations between agrarianism and the classical American philosophical tradition and, in particular, pragmatism and its progenitors. What follows is a general framework of some of the common ground, influences, conjunctions, and oppositions of pragmatism and agrarianism. Both of these traditions are rich in variety, and we do not wish to constrict the present authors to particular focuses or interpretations. The following therefore retains a fairly high level of generality.

Agrarianism

Some brief words on agrarianism are in order here, although in the first chapter, "Agrarianism as Philosophy," Paul Thompson explores in greater detail the historical sources of agrarian thought. In the twentieth century, agrarianism has been articulated, although not exclusively, as a romantic approach to nature and/or a romantic reaction to our modern techno-scientific era. It may arrive precariously close to political manifestations of the relation to the land as sacred: *our* land, the exclusive motherland of a nostalgic people; the particular and unique parochial land. The closest of relations between nature and humans is not merely the aesthetic appreciation of natural beauty or the simple extraction and appropriation of natural resources for human use. On this view our relation to the land can become politically instrumentalized as a means toward the propagandistic manipulation of human closeness to particular places for other ends. The land is idealized and the idea of the land is what is worked over. On the other hand, there is also a pragmatic aspect to agrarianism, less articulate as uniquely agrarian and perhaps more closely associated with questions of work, science, culture and tradition, democratic politics, our natural being, and the unique value of that special technological relationship with our environment: agriculture. If the manipulation of nature is what is essential to human ontology, then the association of soil and plants and animals and human labor and human relationships that makes up agriculture—the intimate nurturing, giving, taking, and renewing between humans and the land—brings us closest to nature in the fullest sense. In this vision lies the root of the values we as humans derive from our working relationship with the land and with each other on the land. Here is brute experience as pedagogical. Here are the metaphors, if it is not an exaggeration, of who we are and how we see ourselves. Agriculture becomes a labor not only of necessity but also of creativity, fulfillment, and loyalty.

Agrarian thought and sentiment includes a diversity of themes and directions. A cursory layout of central themes in agrarianism based roughly on Thomas Inge's (1969) distinctions involves at least the following:

1. Religious. A special spiritual good in the soil of agriculture and in farming reminds one of one's finitude in relation to God's infinite creation. In addition, God is the original husbandryman, and farming reflects God's own work.
2. Romantic. Modern technology and economy are corruptive and entail a loss of human dignity and independence. They are antithetical to the potential goodness of our immediate relationship with nature and the heartiness of work on the land.
3. Ontological. In contrast to the putative alienation and fragmentation of modern society, farmers have a sense of harmony and connectedness, a sense of place and of tradition, and a sense of identity.
4. Political-economic. Independence and self-sufficiency are basic characteristics of farming since, regardless of the performance of the rest of a national or regional economy, one still has the ability to produce for oneself. Farmers are virtuous citizens largely because their practice combines both private and public interest and good.
5. Social. Agricultural communities sustain cultural identity and values and provide a model for a society based on communal understanding and cooperation.

The essays in this volume range across each of these five types. We have not assigned our authors a particular type of agrarianism to examine or pursue. In several cases agrarianism has influenced American thinkers by representing the view that must be argued against, so we do not claim that any of the figures discussed (let alone the chapter authors) are agrarians in the sense of advocating one or more of Inge's themes. Furthermore, many of these themes appear to have little to do with farming or grazing as such. Agriculture is a source of metaphor, but whether material facts about the organization of the food system matter to any of these conceptions of agrarianism is an entirely open question.

It can be argued that a particular agrarianism arises in North America. In contrast to land ownership derived from the political authority of the aristocracy of Europe, the wilderness of America was at first inhospitable to European farming methods and anathema to traditional property relations despite attempts to force traditional paradigms onto it. Innovation, learning from Native Americans, and adaptation allowed European settlements to succeed, but it also engendered a robust appreciation of the need for flexibility and for independence from European authority. This in turn entailed a different formation of politics, economics, and religion: a different set of cultural, technological, and spiritual problems. As already noted, some of the themes that arise from this context reappear in classical American philosophy: pragmatism's naturalism, its emphases on experimental method, democratic community, and experience as the source and test of belief. Dewey's effort to collapse philosophical dualisms of all sorts, especially that between culture and nature, can be added to the list. Perhaps even the notion that experience is precarious, in Dewey's terms, may be seen most explicitly in the practice of farming as opposed to the concealing, dependency, and comfort of modern technological society. The pragmatists, however, appear to move away from agrarian thinking toward coming to terms with a modern technological world. This abdication may apply only to the romantic or metaphorical forms of agrarian rhetoric, however, since pragmatists are certainly not insensitive to the significance of tradition, context, and habit. Furthermore, Jefferson's agrarian/democratic vision and Emerson's romanticized nature are great influences on the pragmatic thinkers. Does their agrarianism make the translation as well? The crux of the matter will be the pragmatic question of how contextual embeddedness—here, we specifically mean an agricultural context—has substantively shaped and informed the directions, methods, and meanings of the American pragmatist tradition, whether positively or negatively. Pragmatism and agrarianism also find themselves intertwined in novel ways in the American agrarian revolts at the turn of the twentieth century. These bespoke the tensions between industrial society and agrarian concerns as fundamental to an evolutionary and functioning democracy. Agrarianism became a form of populism and/or progressivism—and often reactionism—engaged in a reexamination of the notions of democracy, individual freedom, and the cultural countenance of the coming America.

From the perspective of the twenty-first century, postindustrial American agrarianism sometimes appears to be an entirely romantic ideal—a contrived return to purifying and wholesome natural rhythms from a putative moral and spiritual degradation of American urban and suburban life, or from an insidious technoscientific order based on instrumental rationality. Wendell Berry and others, however, have continually and eloquently pleaded that we must have some benign relation to the land if we as a culture are to honestly discuss and attempt to live the meanings of our values, our communities, our ideal notions of work, and our relations to each other (see Berry 1977). The land, for these writers, entails not only self-reliance and self-determination, but also the basis for a free and virtuous society permeated with healthy personal, social, economic, and political relations and experience. Fundamental to the agrarianism of Berry, in particular, is an apparently unintentional affinity with John Dewey's attempt to collapse the dualism of nature and culture, expressed in Berry's view of agriculture as culture. The authors in the present volume explore how themes such as those above intersect and intertwine in the classical American philosophical tradition. For a more detailed and historical examination of the varieties of agrarianism, see "Agrarianism as Philosophy" by Paul B. Thompson.

Pragmatism and American Philosophy

The early history of the philosophical term *pragmatism* was one of misinterpretations and misguided extrapolations, as well as truly substantive philosophical disagreements. Representation of the philosophy behind the term was subject to aspersion by such illustrious philosophers as Bertrand Russell and Theodor Adorno, who had apparently given only cursory or secondary readings to the actual texts of Peirce, James, and Dewey. William James insisted that Charles S. Peirce was the inventor of the word *pragmatism* as the name for the unique system of philosophy the latter developed in the late nineteenth century. Seeing how the word had come to be equated in common parlance or by unsympathetic philosophers with a view of truth as "whatever works" (especially as a common interpretation of James's philosophy), and wishing to distinguish his own brand of philosophy from other "pragmatists," Peirce remarked that he would call his own brand *pragmaticism,* being such an ugly word that it would not inspire appropriation. The term *pragmatism* was most closely associated first with William James and then with John Dewey. Rather than referring to some central and precise doctrine, pragmatism represented a collection of original and not so original strains in modern American philosophy that were developed and explored in unique ways. As James wrote in 1906, "it is evident that the term applies itself conveniently to a number of tendencies that hitherto have lacked a collective name" (James 1977, 378).

Pragmatism eventually came to represent a diverse and rich school of American thought. Philosophers such as James Hayden Tufts and C. I. Lewis were thought to be pragmatists, while others such as Josiah Royce and George Herbert Mead were deeply influenced by pragmatism in different ways, even if purists were reluctant to admit them fully into the fold. At the millennium some philosophers debate who qualifies as a pragmatist. W. V. O. Quine and Richard Rorty are self-avowed pragmatists, but many scholars of the classical period of pragmatism find their work in many ways antithetical to that of Peirce, James, and Dewey (not to mention each other). Meanwhile pragmatism has spread, and the German philosophers Karl-Otto Apel and Jurgen Habermas are sometimes thought to be the current leaders in the ongoing development of pragmatism.

It is therefore folly to attempt a concise statement of pragmatism's core doctrines or to present a comprehensive overview of the diversity of positions and concerns of the pragmatists. Yet readers who come to this volume with little reading in either classical American pragmatism or the more recent debates are entitled to some orientation. One way to interpret pragmatism is to see it as a weaving of two broad strands. The first is the innovative antifoundationalism of Peirce that runs through the thought of later pragmatists. The second is the conception of experience that emerges from the Protestant settlers of New England, finding nineteenth-century expression in the transcendentalists. Whatever defects it might have, this way of framing pragmatism is useful in the present context because it allows us to place pragmatism's agrarian influences primarily in the second group. Pragmatism is not an agrarian philosophy, and the philosophical directions and concerns of Peirce are perhaps the reason why. His presence in the essays of this volume remains primarily implicit, although his influence on later pragmatist thought runs deep. For example, the importance of community as a philosophical theme in pragmatism grows out of Peirce's investigations into the nature of reality and, more particularly, his investigations into the distinction between nominalism and realism (see for example, Peirce 1992, 105). Yet, however much they may have been affected by Peirce's pragmatism and his logical or semiotic investigations, his fallibilism, and his theories of chance, continuity, and evolutionary cosmology, it is principally through the influence of the thought of Emerson and Thoreau that the philosophies of James and

Dewey may be said to have agrarian sources. The latter influences are prominent in the essays in this volume.

Given the undeniable presence of Peirce's thought in all of pragmatism, some words on the thinker are in order. Charles S. Peirce was a philosophical genius unappreciated by all but a few of his contemporaries. Only a fraction of the thousands of pages he produced over a lifetime dedicated to logic, mathematics, and philosophy were published before his death in 1914. Yet two short papers, "The Fixation of Belief" and "How to Make Our Ideas Clear," published in 1877 and 1878 respectively, introduce many of the key innovations for pragmatist philosophy. In the first essay, Peirce implicitly criticizes Cartesian philosophy, arguing that doubt and belief each play functional roles in learning and knowing in an individual's adaptation to and active transformation of its surrounding environment. People seek belief when plagued by doubt, and they arrive at belief when the unease that accompanies doubt is subdued. Why clutter one's philosophy with assertions of metaphysical certainties and structures resting on a spurious act of doubt? Such Cartesian maneuvers aim to confer certainty on mental states that will surely be abandoned in the face of a new problem, genuinely felt. As Peirce wrote, "let us not pretend to doubt in philosophy what we do not doubt in our hearts" (Peirce 1992, 29). In "How to Make Our Ideas Clear," Peirce proposes that we can best understand the meaning of a term or idea by examining the consequences of applying the idea in certain practical situations. So, to define "hard" we examine what happens when two things are rubbed against one another. The one that does the marking is hard. Furthermore, according to Peirce, the meaning of something consists in the conceivable practical effects for agents involved in transactions. Indeed, Peirce summarizes his pragmatic maxim as such: "consider what effects, which might conceivably have practical bearings, we conceive the object of our conception to have. Then, our conception of these effects is the whole of our conception of the object" (Peirce 1992, 132).

These two essays contain the seeds of pragmatism in at least two senses. First, they establish their philosophical claims by exhibiting a procedure that relies on and returns to practical experience. Practical experience is not equivalent to common sense. Practice here is engagement with the problem solving with which human beings find themselves occupied thousands of times per day. The simple structure of problem solving undergirds even the most complex forms of inquiry. Science and philosophy establish specific procedures for framing and solving problems, but these procedures are themselves the solution to a problem: how to gain a foothold for naturalistic inquiry in the face of revealed religion. This problem is in turn rooted in the eminently practical problem of survival throughout Europe's religious wars, reformations, restorations, and inquisitions. The European experience had created philosophies geared to religious tests that would measure a person through an inventory of beliefs. This social climate gave great utility to foundational philosophy—that is, to deductive systems of philosophy resting on logical truths and assorted metaphysical claims.

According to Charles Taliaferro (this volume) the impetus for a faith-based account of knowledge (or its opposite—a philosophy conceived as a defense against religious dogma) had already begun to give way among the early immigrants. By the nineteenth century, people in Peirce's intellectual circle saw science less as a competing doctrine contrary to religion and more as a refined intellectual method, exquisitely suited to the resolution of many practical problems. The great genius here is that Peirce concentrates on the *procedures* that yield belief, rather than the establishment of certain core truths. Peirce takes the idea of procedural philosophy where it had never gone before. Rather than specify indubitable and immutable criteria for belief or meaning, Peirce establishes a course which, if followed faithfully, would produce justified belief or grasp of meaning. He then argues that having possession of such procedures is adequate to the philosophical task; no foundations or infallible beliefs or methodological procedures are allowed. Peirce's procedures are public. They do not rely on private revelations; an observer can understand the result achieved. Peirce's procedure for establishing truth is social. Truth is the ideal limit at which would be arrived after an infinite amount of scientific disputation and experiment on a question. As Peirce writes, "the opinion which is fated to be ultimately agreed to by all who investigate, is what we mean by the truth, and the object represented in this opinion is the real" (Peirce 1992, 139). Everyone interested in convergence on the truth (and qualified) has a right and a responsibility to join in on the disputed matters. Even if we do not reach consensus, we can see that what we, as a "community of inquirers," are seeking is the consensus that we would reach given world and time enough. Though this is an impossible ideal, it is sufficient to provide an analysis of truth equal to the situations in which our idea of truth becomes problematic.

Peirce's two essays are also seeds of pragmatism in that they provided a springboard for all the later pragmatists, especially William James. James radicalized Peirce's problem-oriented philosophy, seeming initially to argue that the "will to believe" in the face of a problematic situation was an adequate basis for a general theory of truth. James's work was vulgarized by critics who accused him of saying that if you want it to be true badly enough, it is true. Whether or not this critique was fair, James (and then Dewey) developed pragmatism by laying stress on Peirce's implicit critique of foundationalism. Critique of traditional European philosophy became a salient element of the pragmatist approach. It was in this sense that pragmatism became a forerunner of deconstruction, the late twentieth-century program of showing how philosophical texts were held together as much or more by the social power relations that they tended to support as by their logic. Like recent continental theorists, pragmatists such as James, Dewey, and Peirce saw established theory as a vestige of past attempts to solve practical problems, but unlike deconstructionists, the pragmatists saw the detritus of deconstructed theory as a source of material for reconstruction.

Peirce elevated the philosophical status of practical intelligence and opposed the "spectator" theory of knowledge, the disengaged knower who merely observes. In this regard, Peirce anticipated Heidegger's distinction between "presence-at-hand"—a thing known by its observed properties—and "ready-to-hand"—a thing revealed through its use in work. Heidegger characterized these categories in ontological terms and went on to accord primacy to the ready-to-hand, which he saw rooted in the peasant farmer's way of life. For Heidegger, thinking must first recover and then proceed anew from this authentic starting point. There is little reason, however, to think that Peirce would have endorsed this turn in Heidegger's thought. It seems reasonable to conclude that Peirce incorporated one tenet of an agrarian philosophy—the emphasis on case-by-case trial-and-error problem solving—into his system. But stated as such, the tenet is not essentially agrarian. It is easy enough to make sense of it outside of an agrarian context.

Peirce's work was agrarian in only the weakest sense. He did not celebrate rural life, nor did he accord any special significance to farmers in any of his surviving writings, and it is difficult to believe that he would have done so. But his work is consistent with the second strand mentioned above—the pragmatist conception of experience. There is, however, little doubt that this influence, which is not *uniquely* American, was more potent in the thought of James and Dewey. For both, experience is central to their respective philosophies. Philosophical primacy was given to experience as transactional and to seeing both individual and social experience as an unbroken flow of not merely the traditional terms of stimulus (for example, problem) and response, but as stimulus in the midst of always/already doing. That is, human experience takes place in a field of dynamic and continuous relations in which stimuli simply redirect activity. Following the lead of many foundational philosophers, pragmatists presumed that experience was the limit concept for philosophy. But unlike the philosophers of the foundational tradition in which "impressions and ideas" constitute the building blocks for our conceptions of self, world, and reality itself, experience in the Jamesian terms of "radical empiricism" includes all phenomena affecting human consciousness, including the relations between things. James broadened the notion of experience to include all nervous and physiological responses that the human organism makes to its environment, not solely conscious experience. This move expanded the universe of possibilities dramatically and also deprived the old "impressions and ideas" epistemology of its authority and persuasiveness.

Furthermore, the pragmatist idea of experience is evolutionary with a strong dose of fallibilism. Responses that succeed in their immediate context tend to be repeated, to become habits. When they fail, the organism must innovate. This is fundamental to the human condition: "Life is interruptions and recoveries" (MW14:125). Innovation may be radical, but it is more likely to consist in a reassembly of discarded bits of discarded responses. For pragmatists (and especially Dewey) the process of evolution, adaptation, and transformation can be made much more intelligent with foresight and the application of scientific method. Of particular consternation are defunct responses to complex problems. Such responses tend to linger well beyond their usefulness and meaningfulness. To take a general example (and a major theme in Dewey's thought), rather than democracy being something final and achieved, a rote set of political habits, Dewey argued that democracy was rather an ongoing method and way of life toward better social, political, and economic arrangements. The former view belies the very principle on which it supposedly relies (see Stuhr 1993). Imprisoned by such "hypostasized" ideas (Dewey's word), the organism loses the capacity to innovate and to grow. Dewey aimed much of his brand of deconstruction at this problem and expected that innovative reconstruction could follow, given the application of increased intelligence. This rings of Darwinism. In fact, Dewey applied a broadly Darwinian approach throughout his philosophy. As Daniel Dennett writes recently, Dewey was indeed one of the first intellectuals (let alone philosophers) to appreciate the significance of Darwin's thought (Dennett 1995, 66).

Yet Darwin's theory was not a theory of experience, and the idea that experience is an evolutionary process of experimentation and adaptation predated Darwin by at least a hundred years, if it had not already been implicit in Aristotle. Hegel philosophized about a progressive evolution of spirit, for example, and drew from a large philosophical and historical literature on the origins of consciousness and civilization. In most cases, this "origins" literature gave pride of place to agriculture. The beginnings of settled agriculture coincide with the beginnings of civilization, and perhaps also with an expansion and refinement of the human capacity for conscious thought. This central theme binds together figures as diverse as Benjamin Franklin and Henry David Thoreau. It reaches its highest pitch in Dewey, who (like Hegel) sees both conscious and unconscious ideas as evolutionary (dialectical, for Hegel) products, rooted in complex material and intellectual relationships. Unlike Hegel, James and Dewey see experience as a continuously adaptive and innovative transactional process, one that is not designed, and which will never reach an apotheosis or completion. This is, arguably, the most radical move in pragmatist philosophy, the insight that separates pragmatists properly so-called from latter-day foundationalists, still seeking closure, as well as from deconstructionists, mired in a skepticism that disempowers the mind altogether.

The role of agriculture in the arguments of origins literature qualifies this mode of philosophizing as a form of agrarian philosophy. It may be expanded into philosophical theories of history, morals, and society. Such theories constitute the "agrarian roots" referred to in the title of this book. Many such theories were developed during the sixteenth and seventeenth centuries, and the first section of the book includes essays that examine some of them. To say that pragmatism has agrarian roots is thus to say that it was influenced by the Scottish Enlightenment, by the French natural philosophers who followed Buffon, and by the German Idealists (including Marx). This would be an exceptionally modest claim were it not for the fact that pragmatists systematically and radically transformed or even discarded much in these European philosophies while retaining elements that we might trace to agrarian roots. Pre-Darwinian evolutionary philosophy made much of agriculture, and in retaining and even exaggerating the relative importance of evolutionary experience the pragmatists brought implicit agrarian arguments along as well.

Finally, most of the thinkers discussed in this volume share a sense of meliorism. This meliorism is obvious in thinkers such as Franklin, but it was made philosophically explicit especially in the writings of James and Dewey. That is, they share the idea that the increased application of intelligence through critical inquiry and in light of habit and tradition can improve the human condition. Our postindustrial millennial days may very well give pause to a pragmatic meliorism, and they may breed a dark cynicism and irony of outlook. Yet, for the American thinkers and critics discussed in this volume, problematic situations are always an impulse toward further work to be done. We suggest to the readers of this volume that the thinkers presented here have much yet to teach, even in a melioristic spirit, and perhaps even more so in a largely nonagrarian society.

The Plan of the Book

The first section of this volume contains three essays, one each by Paul B. Thompson, James A. Montmarquet, and Charles Taliaferro. Together they serve as an overarching study of the historical development of agrarianism as well as the agrarian heritage in the contemporary American context. Thompson and Montmarquet make their own arguments for how we might view agrarianism at the millennium, while Taliaferro takes it back to the seventeenth century beginnings of European America. The first essay, "Agrarianism as Philosophy," by Thompson, provides a more in-depth overview than the present introduction of pre-twentieth-century philosophical themes and arguments in agrarianism. Thompson discusses three modes of agrarianism—its ideas, imagery and metaphors, and "mentality"—and agrarian thought by focusing on the Vanderbilt agrarians, property rights and subsistence rights, agrarianism in classical Greek philosophy, eighteenth-century Scottish and French natural philosophy, and the romantic strain of agrarianism. The forms of agrarianism connected to these views in some measure exemplify positions the authors in the present book argue against. Yet, Thompson maintains, the broader neglect in philosophy of the agrarian origins of even conventional philosophical arguments and positions attests to a neglect possibly of the roots of our own ideological tragedies and environmental crises. This is the case to the extent that these positions have been shaped by the mythologies and implicit philosophical claims in historically contextualized responses to problems of land, food production, and social-political and literary concerns.

Montmarquet's essay, "American Agrarianism: The Living Tradition," reprinted here from his 1989 book, attends to the nineteenth- and twentieth-century agrarian philosophies of Emerson and Thoreau, Populist agrarianism, Liberty Hyde Bailey, sage of the Country Life Movement, and the contemporary Wendell Berry. Montmarquet demonstrates that the views of Emerson and Thoreau are distinguished by a cleft in agrarian thought that also runs through the thought of more contemporary philosophers and agrarians. The belief that the domination of nature (through farming, among other means) is a natural and noble extension of the human will and spirit opposes the position of "small is better" when it comes to human transformation of nature and human productive processes. Bailey retains certain aspects of Emersonian and Thoreauvian views but attempts to navigate the implications of this split and the contradictions that arise in the practice of agriculture. Although sympathetic with his social critique, Montmarquet finds inadequate Wendell Berry's proposed solutions to the devastation of the land and small farming. But the solution found in Bailey's thought, one Montmarquet thinks crucial to the future of agrarianism, is a form of naturalism that unites experimentalism with the sentiment of the Thoreauvian "poetic farmer." This requires novel approaches to education, especially of the new stewards of the land, as well as the creation of economic opportunities which render more attractive the practice of farming.

"Land, Labor, and God in American Colonial Thought," by Charles Taliaferro, explores the colonial theistic tradition. Drawing mainly on the texts of early American Christian thinkers, Taliaferro notes that the view of the world as both creation and possession of God was interpreted in terms of a natural right of humans to the benefits of the land as well as the right of property. He argues, furthermore, that this theism and its displacement by Enlightenment thought served to provide a rough framework for the pragmatist attack on dualism through the emphasis on the primacy of will and reason throughout creation. This sets the stage for a series of chapters that examine the role of agrarian ideas in some of the central figures in American intellectual history.

The second section of this volume, "Prepragmatist Agrarianism in America," presents philosophical developments in early American thought, which become central themes or influences for the classical pragmatists. Benjamin Franklin is the subject of James Campbell's essay, "Franklin Agrarius." Campbell maintains here that Franklin was not merely the proponent of virtues of frugality and industriousness for individualistic reasons, but was arguing the case for social well-being in an agrarian society. Ideally, agriculture would be the central mode of producing in an independent land, according to Franklin, since it is the combination of labor and land that produces value—depending on the uses to which they are put—not commerce or trade. Indeed, Campbell contends, Franklin's writings express a moral primacy of agriculture. Agriculture, with its hard work yet relatively simple way of life, was the only honest way to acquire wealth, not for the sake of accumulation, but for the sake of conquering human need, as well as for happy, virtuous, and independent citizens living in democratic equality.

"Thomas Jefferson and Agrarian Philosophy," also by Paul B. Thompson, argues that the famous agrarian pronouncements by Jefferson in his *Notes on the State of Virginia* and other works and letters must be tempered by acknowledging the philosophical and historical contexts in which they were made. Thompson maintains that the agrarianism in Jefferson's writings is not all that original or uniquely American, despite its appropriation and interpretation by many contemporary agrarians (and others) as the canonical statement of the American agrarian heritage. This view, Thompson argues, is often the result of taking Jefferson's words expressing agrarian sentiment out of their political and rhetorical context. A reading of Jefferson in light of his philosophical contemporaries in America and Europe shows him to be adopting rather common positions and assumptions in regard to economy, land, and governmental institutions. One result of this broader reading is that Jefferson's agrarianism is not so much an influence on later American philosophers (who are, thus, drawing rather on eighteenth-century thinkers such as Adam Smith, Montesquieu, and the Physiocrats) as are his ideas regarding the primacy of nature, his democratic thought, and his educational and social ideas. These aspects of Jefferson's philosophy are agrarian only to the extent that they are sometimes viewed through the context of discussions of agriculture. Read together, Campbell and Thompson suggest that it was Franklin, not Jefferson, who was the important theorist of agrarian democracy. Why do we still remember Jefferson? Well, Franklin may have had the theory, but it was Jefferson who had the farm.

From the above colonial thinkers, we turn to nineteenth-century American philosophers in the next two chapters. The first of these, "Emerson and the Agricultural Midworld," by Robert S. Corrington, traces Emerson's thought on the relation between human being and activity and nature as constituted by the "midworld," the clearing in which the truth of nature is made accessible to human beings. Emerson's early view was that the poet has a unique status in that he or she submits nature to his or her own conscious transformations of it, revealing at the same time the potencies of nature. It is the mind of the poet that translates or, further, shapes the world through the poet's symbolic language. This "linguistic idealism" undergoes a change in Emerson's thought from the optimistic and heroic view of human imagination toward a more Stoic view of nature as a force that demands obeisance, and ultimately toward a darker view of nature as inaccessible and human selves as essentially lost within its flux. This late position becomes tempered by Emerson's realization that it is not the poet who best expresses the midworld, but those who obey the laws of nature while harnessing its precarious forces in practice for human use. This is the role of the humble farmer, who conforms to nature for human advantage by studying and learning its laws and learning to instrumentalize them. This practical midworld replaces the earlier hubristic conception. Corrington writes, "The pine fence replaces or augments the poem as a fitting symbol of the midworld." Corrington maintains that in Emerson's philosophy it is thus the farmer who sustains the relation to nature through praxis, as well as an understanding of both human limitations and possibilities of human transcendence.

Douglas R. Anderson's essay, "Wild Farming: Thoreau and the Agrarian Life," explores Thoreau's version of farming as a midworld. In this case, it is the "borderland" between the wild and the tamed. Anderson argues that Thoreau uses the farm metaphorically as a way of expressing the necessity of retaining a wilderness for the sake of both the environment itself and for human beings. The midworld of the farm looks toward both civilization and wilderness. Thoreau had believed famously that civilizing tendencies—even among farmers—could engender dull mediocrity and conformity. For human culture to grow, in terms of both selves and communities, they need something to grow into, and wilderness represents the sphere of possibility of such growth. "Wild farming" indicates the effort to navigate between the wild and the cultivated, the latter already being inescapable to a large extent, as Thoreau had discovered at Walden. Self-cultivation requires its space of freedom, and herein lies Thoreau's qualified agrarianism. For wild farming demands an ethic of preservation of wilderness instrumental to human cultivation and freedom. This, of course, requires a culture awakened from the vapid safety of its consumerist slumber, and Thoreau, Anderson argues, is sounding a clarion call through metaphor, argument, and his lived example.

The third section of this volume, "Pragmatism and Agrarianism: Dewey and Royce," is devoted to two central figures in American pragmatic philosophy. In "Provincialism, Displacement, and Royce's Idea of Community," Thomas C. Hilde approaches Josiah Royce from the angle of Royce's lived experience and the parallel development of his philosophy of community and loyalty. Maintaining that geographical and cultural displacements and losses in Royce's life may be viewed as key to understanding his idea of community, Hilde argues that Royce's idea of community skirts problems that are central, if only implicit, in agrarian notions of community as an ideal. Royce attempts to come to affective terms with the transition from an agrarian-based society, and this appears in novel ways within his intellectual development. Ultimately, Royce's conception of community finds a vital rather than reactionary role for the displacement and loss of industrializing society. This entails a conception of community larger than that of the "placed community" cherished by many contemporary communitarians, agrarian and otherwise, and it requires the ethic of what Royce calls "loyalty to loyalty."

Two contributors write on Dewey in this volume. Armen Marsoobian is the author of "Does Metaphysics Rest on an Agrarian Foundation? A Deweyan Answer and Critique," and Larry A. Hickman contributes "The Edible Schoolyard: Agrarian Ideals and Our Industrial Milieu." Marsoobian elicits from Dewey's work an argument for why the question in the paper's title makes sense. Dewey had argued that the *philosophical fallacy*—taking a product of inquiry or experience as something antecedently existent—on which he thinks Western metaphysics is grounded appears at a point in which hunting society shifted toward agricultural society. The precarious, contingent nature of "the hunt" in which goods are gained through active, instrumental practice came to be viewed in symbolic terms and thus also in terms of gifts of "grace." When the drama of the hunt was converted into symbolic narratives the precarious was rendered in terms of control—the ability to re-create the drama of the hunt. The "account," as it is later put in ancient Greek philosophy, becomes the selective interpretation of fundamentally problematic experience. Therefore, the scientific methods that arose out of agricultural practice already contended with a mentality that viewed human significance in terms of "grace," of unity and ends, rather than the activity of doing and undergoing in human experience. Marsoobian maintains that these reflections provide Dewey with his imperative toward reconstruction of essentially conservative, static ways of understanding the world, especially as embodied in Western metaphysics.

In "The Edible Schoolyard," Larry A. Hickman discusses Dewey's emphasis on the growth of intelligence and democratic life through educative processes of experimentation and cultivation. This paradigm for intelligence exists also in agricultural production. That is, the cultivation, processing, and preparation involved in food production requires judgments of practice based in experiment rather than merely judgments of description. In his education theory, Dewey sought to promote the development of the experimental attitude in children so that methods could be applied by students to other realms of their lives beyond the classroom. Moreover, this attitude comprises the Jeffersonian idea of democracy. Although Jeffersonian means may be outdated and ineffective in contemporary society, Jeffersonian democratic ends retain their ideal significance. Human beings educated in experimental methods will ultimately be the best democrats since democracy itself is an ongoing, evolutionary process. The paradigms of agricultural practices remain educative toward democratic ideals and the growth of intelligence even in a society that is no longer agrarian.

The final section of the volume is "Twentieth-Century Agrarian Thought in the Pragmatist Tradition." These essays explore twentieth-century agrarians and agricultural practice in the context of a radically altered, relatively nonagrarian society. The essays collectively argue the case for a renewed inquiry into agrarian thought in the pragmatic spirit of understanding the value in agrarian experience. Such inquiry is ultimately important for pragmatic social reconstruction. In the 1940s, a popular song asked, "How are you going to keep 'em down on the farm, after they've seen Pair-ee?" This was, in fact, a serious question for students of American policy and rural development. Wendell Berry's novel *A Place on Earth* is set in a farming hamlet during World War II. The characters, whose joy and heartbreak are deeply interwoven, do not realize that they are on a precipice. Even if their sons return from Europe, they will not return to this place on earth. Wartime technology (nitrogen production and DDT) will transform farming, even if the sight of Paris does not. In 1948, A. Whitney Griswold published *Farming and Democracy*, a study dedicated to the proposition that while Jeffersonian agrarianism *had been* the reason for the success of the American republic, it was now (in 1948) not only possible but necessary to discard the agrarian foundation. This severance precipitated several decades of agricultural and rural development policy based on the belief that agriculture, like any other sector of the general economy, should be organized so as to maximize economic efficiency in the use of land, labor, and technology inputs to produce salable food and fiber. Gone was the idea that agriculture's purpose is, in part, to produce healthy democratic communities.

The late John Brewster's essay "The Relevance of the Jeffersonian Dream Today" was written originally as a response to Griswold's *Farming and Democracy*. Brewster (1905–1965) received his Ph.D. in philosophy from Columbia University in 1938. He had been a student of George Herbert Mead at the University of Chicago and transferred to Columbia expecting to complete his studies with Mead, who died before assuming the Columbia position. Brewster spent his entire professional career with various economic research branches of the U.S. Department of Agriculture. He published a series of articles on the economics and politics of agriculture under conditions of industrialization, and his research articles won awards from the American Agricultural Economics Association.

Brewster never occupied an academic position in philosophy, but he enjoyed a lifelong friendship with David Miller, former head of the Department of Philosophy at the University of Texas. Miller was also Mead's student and arguably the leading Mead scholar of his generation. Miller and Brewster had known each other in Oklahoma before either went to the University of Chicago to study philosophy. Miller mentioned these shared experiences and shared philosophical interests in a 1983 conversation, recalling some details from Brewster's dissertation on the problem of universals. Though at best vaguely familiar with Brewster's writings on farming, Miller was of the opinion that Brewster's thought was compatible with his own, which is to say a natural extension of Mead. He clearly believed that Brewster's writings would be a valuable place to look for the connections between pragmatism and agrarian thought.

The present essay was published in a 1963 book, *Land Use Policy and Problems in the United States*, edited by Howard W. Ottoson. The book includes essays by eighteen land-use specialists, representing the disciplines of geography, history, economics, and (if we count Brewster) philosophy. Carl Sauer, Marion Clawson, and Gene Wunderlich, the author of "Two on Jefferson's Agrarianism" in this volume, are among the contributors. Ottoson's introduction quotes Griswold in noting that the "Jeffersonian ideal . . . which suggested a causal relationship between family farming and the political system of democracy," has been an important source of guidance for U.S. land policy. The chapter authors were challenged to update this suggestion by examining how family farming had changed since Jefferson's day and to note how technology, nonfarm job opportunity, farm credit, and agricultural research had supplemented the family farm's contribution to the development of agriculture (Ottoson 1963, vi).

Brewster enjoyed playing the role of "philosopher among economists," and some who knew him report that this act was too often combined with an imperious and prickly manner that was off-putting. Not surprisingly, the reception given to Brewster's philosophy from farm economists was mixed. Some found him a useful and refreshing voice, while others saw him wedded to philosophizing with little relevance to problem-solving inquiry—hardly conduct befitting a pragmatist. Was Brewster a pragmatist, a significant theorist of American political culture, and a neglected figure in the recent history of American philosophy? Brewster himself expressed admiration for pragmatism, yet is said to have denied holding a pragmatist view of truth. Along with the 1963 essay "The Cultural Crisis of Our Time," the essay reprinted here is the best representation of Brewster's mature thought. Readers may form their own judgment of him based on the following essay, and from reading Wunderlich's critical review.

Wunderlich discusses both Brewster and Griswold in "Two on Jefferson's Agrarianism." Wunderlich shows that both played free and easy with Jefferson's texts. Griswold presented Jefferson as endorsing a spiritual relation between farmer and land—something he did not do. Brewster inserts John Locke's thought on property to fill gaps in Jeffersonian agrarianism. Both make Jefferson less a naturalist and less pragmatic than a close attention to his texts would warrant. Wunderlich argues that Brewster's and Griswold's selective reading of Jefferson has concealed the relevance of Jefferson for pragmatism and for contemporary agriculture. Jefferson's agrarianism, as distinct from "keep 'em down on the farm" agrarianism, prepares the way for environmental awareness and stewardship, for making agriculture a linchpin in an environmental pragmatism.

Richard Hart continues this theme in "Steinbeck and Agrarian Pragmatism." Immediately prior to World War II and amidst the lingering Great Depression, John Steinbeck wrote *The Grapes of Wrath,* a journalistic novel that portrays crisis in rural American life. Bound to each other and to place by years of work and dedication, farm families evicted from Dust Bowl lands founder economically and morally in a world of *caveat emptor.* Unable to function in a world ill-suited to their morality of personal loyalty and mutual assistance, the Joad family shatters under the weight. Steinbeck's novel is one of the first and most powerful documents noting the transition in farming, under way in the 1930s but gaining steam after World War II. Hart emphasizes not *The Grapes of Wrath,* but *The Sea of Cortez* and *Of Mice and Men.* In these works, Steinbeck's experimental naturalism is evident, as is a deep appreciation of humanity's embeddedness in the biotic community. Hart produces a reading of Steinbeck that extends Wunderlich's review of Jefferson and points us toward a pragmatic environmental ethics.

Jeffrey Burkhardt continues with this theme in the final essay, "Coming Full Circle? Agrarian Ideals and Pragmatist Ethics in the Modern Land-Grant University." Burkhardt surveys the history of the modern land-grant university and its failures to operate according to the democratic agrarian ideals on which it was founded. He notes the role of "neo-agrarianism" in the future of these institutions. The neo-agrarian trend (including members as diverse as Jim Hightower, Wendell Berry, and Marty Strange) is one whose ethic comprises agricultural sustainability, alternative technologies, ethics education, and an attack on technological and scientistic fetishes and their practical consequences in agricultural education and research. Neo-agrarianism, according to Burkhardt, is fundamentally—although unconsciously—Deweyan in ethical, social, and political orientation because of its stress on contextualism, human community, and democratization.

Conclusion

Collectively, the essays in this volume constitute both a historical and a philosophical treatment of American thought from the earliest days of European settlement to the present. Historically, agrarian ideas emerge from being unquestioned assumptions to central tenets of transcendentalist thinking. They are continually altered and criticized throughout the twentieth century, though they influence figures such as Dewey and Steinbeck. Chapter authors disagree over the relative importance of agrarian ideas for the key figures in American intellectual history, and on the relative influence of Thomas Jefferson, the man most closely associated with agrarian thought. There is, however, a thematic unity to these essays. Agrarian naturalism—the belief that culture and conduct are conditioned by nature because they are of a piece with nature—becomes pragmatic naturalism, giving way to a new set of puzzles as to how we are to understand the rural landscape and our responsibilities for its use.

During the twentieth century, whatever is agrarian in pragmatism seems increasingly less important. Just when it is least expected, however, the agrarian theme reemerges in connection with the rise of environmental pragmatism, a philosophy that strives to bring naturalism, experimentalism, amelioration, and pedagogy to our understanding of the natural environment. Environmental pragmatism (see Katz and Light 1996) draws upon the agrarian recognition that human beings are most thoroughly integrated into nature in their work. Any adequate ethic of the environment must begin with the way that human beings produce themselves, their culture, and indeed the nature/culture of the future in the way that they bring forth their sustenance through the productive use of soil, water, sunshine, and other natural resources.

In this respect, the essays in this volume point toward the future rather than the past. We will never again be "an agrarian nation" in the sense understood by Franklin and Jefferson. We may, with effort, recover the wisdom of farming and husbandry: to see our lives as fitted to a place on earth, to see our work as cultivation of that place (hence, consistent with its demands), and as husbanding spiritual and natural resources for those yet to come. Yet we must resist the blood-and-soil myths that have twisted the lessons of husbandry into some of the twentieth century's most destructive ideologies. That is why environmental ethics and neo-agrarianism, to use Jeffrey Burkhardt's term, must be pragmatist, rather than foundational philosophy. The essays in this volume are an attempt to extend the conversation between pragmatist philosophy and agrarian thought into the twenty-first century.

Part I

DEFINING THE AGRARIAN TRADITION

Agrarianism as Philosophy

PAUL B. THOMPSON

IS THERE such a thing as agrarian philosophy? Any survey of recent work by academic philosophers would suggest not. So far from being thought essential to philosophy, for at least 150 years few philosophers have seen reason to dedicate an essay or even a paragraph to agriculture. Yet, as we shall see, eighteenth-century European philosophy was permeated by agrarian ideas not only in political economy, but also in epistemology and theory of science. How did philosophical analysis of agriculture fall so far and so fast? Is there any point in revisiting eighteenth-century agrarianism? Is there any potential for an agrarianism of the third millennium?

This chapter provides an abbreviated overview of agrarian themes, beginning with the southern New Critics who posited the conception of agrarianism most familiar to many contemporary American intellectuals. From there the search for agrarian philosophy turns to seventeenth-century English debates on land use, then moves back all the way to the ancient Greeks. Going forward again, eighteenth-century political economy incorporates (indeed, presupposes) agrarian ideas consistent with those attributed to the Greeks. Excepting the southern New Critics, this chapter leaves the nineteenth and twentieth centuries to the other authors in this book. In truth, it is impossible to summarize agrarian themes concisely. *Agrarianism* has at various times meant very different (but nevertheless related) things. I do not aim to lay down the gauntlet by defining agrarianism this way or that, as if I could examine various thinkers or intellectual traditions to see whether they match up. Nevertheless, to say that a theory, style of thought, or form of argument has agrarian roots, it is necessary to have *some* sort of conception of what agrarianism is.

The essays in this book center on a conception of agrarianism framed by three central ideas. First, the essays follow an implicit historical narrative that begins with early American religious thought and becomes something like a national secular religion personified by Thomas Jefferson, defender of liberty and master of Monticello. The story continues with the transcendentalists and Dewey, ending with postwar attempts to apply progressive principles to the politics of agriculture and rural development. The combined essays in the book assemble a case for a modestly agrarian interpretation of pragmatist moral and political writings, and a less modestly pragmatist approach to agricultural politics and environmental ethics. They place Dewey's work between strongly agrarian progenitors and neo-agrarian environmental pragmatists. Jefferson, Emerson, and Dewey are milestone figures in anchoring the particular strand of agrarianism that is of most interest to our authors. It is in this sense that the agrarianism of interest is tied to pragmatism.

Second, the essays in this book emphasize agriculture. They interpret agriculture as the cultivated, material transformation of soil, water, and sunlight into the living tissue of plants and animals for use by human beings. Although there will be little discussion of actual farming practices, these practices do matter for the conception of agrarianism under development. If an important claim were shown to be inconsistent with actual agricultural practice, that would be an important consideration against it. Another way to put this is to note that for a pragmatist, all inquiry begins in response to a problem. Here we are interested in problems that originate with agriculture, or for which agriculture would be an expected response.

In many contexts, agriculture implies settled civilization. The essays here are consistent with that assumption, but a complete agrarian philosophy would theorize the transition *to* farming from whatever went before, as well as *from* farming to whatever comes afterward. American thinkers' concern for that latter transition is a key point for this book. Of course, many scholarly studies of agrarian transition are little concerned with agriculture as such. For example, Paul Bové's 1986 study of the Agrarian New Critics affiliated with Vanderbilt University in the late 1920s through about 1935 is entitled "Agriculture and Academe," but Bové mentions agriculture only to chide John Crowe Ransom, one of the group's leading figures, for knowing little about southern agriculture. Bové both critiques and exemplifies "agrarian" writing that ignores any concern for actual crop farming or animal husbandry. Emphasizing the material performance of food and fiber production is one way of introducing rigor into the discussion of agrarianism, and of evaluating whether a given claim or idea is agrarian in the relevant sense.

The essays in this book also concentrate on agrarian *philosophy*. Some attempt to explain what this means will be made below. For now it is useful to make a three-way distinction. First, there are ideas that attribute special powers or status to farming. Second, there is agrarian imagery in art and literature. Finally there are studies of agrarian mentality. Only the first of these lead naturally to agrarian philosophy. Agrarian ideas are explicit statements of how basic production practices relate to the formation of personality and social institutions, or how they figure in history. Such ideas emphasize farming and pastoralism because agriculture was the core subsistence mode for European civilizations undergoing modernization. However, agrarian ideas might as easily emphasize fishing or, with an advanced technology, the biochemical transmutation of sludge into edible proteins. Whether farming enjoys any moral or political advantage is, of course, one of the contested themes within agrarian philosophy. Agrarian imagery can be found in paintings, drama, fiction, poetry, biography, and memoirs that praise rurality. These works convey the quality and character of rural experience to their audience, though they may take a critical stance toward that experience. Such literature and art might rest on philosophical principles, but it might instead intend only to evoke a sense of agrarian mentality. *Mentality* refers to the personality and thinking habits of a person or group. Those who possess a given mentality need not be particularly mindful of having it. An agrarian mentality is, then, simply the mind-set of someone living in or particularly well suited to an agricultural society. As will be discussed below, some twentieth-century historians and social psychologists attempted to characterize the mind-set of the farm family as a rural ideal type. That project may, at bottom, rely on agrarian ideas (as they are defined here), but it too often smacks of armchair sociology, devoid of method and obscure in purpose. Many who praise agrarianism have articulated the power and virtue of agrarian mentality in mystical terms, leaving agrarian ideas, as we use the term, entirely implicit. Having a collective mentality is tantamount to having a culture, but neither personality nor culture requires explicit and critical evaluation of their basic tenets. Agrarian mentality is important for agrarian philosophy only to the extent that the implicit becomes explicit and meets philosophical demands for critical assessment.

The Vanderbilt Agrarians

As already noted, the word *agrarianism* is irretrievably linked to a group of historians, social theorists, and literary critics associated with Vanderbilt

University in the 1920s and 1930s. It would hardly be fitting for a book published by that university's press to ignore them entirely. This group, which includes Allen Tate and Robert Penn Warren among its most distinguished members, devised and promoted New Criticism in literary theory and, with William Faulkner, were also among the leading figures in what was called the southern Renaissance. Two volumes of essays—*I'll Take My Stand: The South and the Agrarian Tradition,* originally published in 1930, and *Who Owns America?* published in 1936—linked the New Critics and southern intellectual history to the agrarian tradition. According to John L. Stewart's *The Burden of Time,* generally regarded as the standard work on the Vanderbilt agrarians, the core doctrine of the group was expressed in Ransom's book *God Without Thunder* (1930).

> Farm labor and an agrarian culture established upon small subsistence farms . . . permitted the proper exercise of the sensibility and enabled the whole man to function as an entity whose total experience can best be represented in myths and the arts. But under an unchecked industrialism the sensibility withered, and the reason, now unopposed, dominated all thought and conduct, producing the dehumanized abstractions and the drab life that characterized the northern technological culture. (Stewart 1965, 140)

As far as it goes, this paragraph is consistent with the Emersonian agrarianism that Robert Corrington exposits in his contribution for this volume. However, Stewart also characterized the Vanderbilt agrarians as "anti-progressive, anti-rationalist, and anti-humanist . . . insist[ing] on the irreducible mystery in life, the all pervasiveness of evil in human affairs, and the limitations of man's capacity to understand and control his environment and his own nature" (quoted in Bové 1988, 113).

None of the figures discussed in the succeeding essays can be described this way, and the characterization would certainly not apply to Dewey, who was progressive, rationalist, *and* humanist. Indeed, the Vanderbilt agrarians themselves harbored a conscious antipathy toward Dewey's thought. Allen Tate, for example, rebukes Edmund Wilson's advocacy of social planning by writing, "his Planned Economy, seen through the whiskers of Professor [John] Dewey, is the most fantastic piece of wish-thinking I've ever seen" (quoted in Bové 1988, 118). Dewey is also singled out for rough treatment in Lyle H. Lanier's contribution to *I'll Take My Stand* (Twelve Southerners 1930, 136–149). Any mutually supportive relationship between Vanderbilt agrarianism and Dewey's pragmatism would not have occurred to followers of either camp, though Ransom later became a grudging admirer of Dewey's work (Stewart 1965, 189). If the Vanderbilt agrarians influenced Dewey, it was probably in encouraging him to minimize his use of agrarian rhetoric, lest he be misunderstood. Yet the focus on Dewey's pragmatism is justified in this context because it promises first to uncover a significant agrarian dimension of pragmatist philosophy and second to reveal a rich but neglected vein of agrarian philosophy. The Vanderbilt agrarians were not thinking in a Deweyan or pragmatist way. Ironically, our interest in a narrative framed by Jefferson, Emerson, and Dewey puts Ransom and his cohort on the margin. It would be interesting to compare the Vanderbilt agrarians with the pragmatists and to interpret both groups as competing for Jefferson's legacy. However, such a comparison presupposes the analysis being developed in this book.

As already noted, Vanderbilt agrarians have been criticized for expounding on rural life while knowing comparatively little about agriculture. This also puts them in violation of our second parameter. Ransom's unsigned introduction to *I'll Take My Stand* describes *agrarian* as being in no particular need of definition, then goes on to provide one.

> Technically, perhaps, an agrarian society is one in which agriculture is the leading vocation, whether for wealth, for pleasure, or for prestige—a form of labor that is pursued with intelligence and leisure, and that becomes the model to which the other forms approach as well as they may. But an agrarian regime will be secured readily enough where the superfluous industries are not allowed to rise against it. The theory of agrarianism is that the culture of the soil is the best and most sensitive of vocations, and that therefore it should have the economic preference and enlist the maximum number of workers. (Ransom et al. 1930, xix)

This passage represents the whole of the eleven-page "Statement of Principles" given to agriculture, the balance being a polemic against industrialism and progressive ideologies. While it might have served well enough as the first word on an agrarian philosophy, the passage receives little explication in *I'll Take My Stand*. The phrase "agrarian society" is used interchangeably with "the South." Here and especially in *Who Owns America?* the argument tends heavily toward critique of the industrial

North. Agrarian society comes to be defined largely as an "other," something that is not the hated northern regime.

Stewart writes that despite their personal origins in farming communities, the Vanderbilt agrarians knew little about farming or farm life. This failing was manifest in their inability to "provide a substantial and believable image of life on a southern farm that was capable of satisfying to the degree they claimed for it the spirit and aspirations of man" (Stewart 1965, 131). In fact, the life of a southern farmer was exceptionally hard, largely so because southern soils were poor and southern climes were hospitable to bird and insect pests. Southern yeomen were at a significant disadvantage when their farms are compared to the fertile soils and temperate climates of New England, the Midwest, and, once irrigated, the Central Valley of California. Even after the end of cotton, southern farms were often successful to the extent that they approximated a plantation rather than agrarian model, with armies of cheap labor and production practices geared to produce entirely for nonlocal markets.

Scholars such as Bové are kinder to the Vanderbilt agrarian's style of argument, but this may be because they are even less inclined to believe that agriculture could be important for any kind of critical or intellectual study. Recent works of literary criticism stress how Ransom, Tate, and the rest used agrarian discourse to affect the balance of power within intellectual politics, the milieu with which they were most personally involved. These recent studies move even farther from farming and food production than the original writings. In some cases, critics express admiration for the Vanderbilt agrarians critique of industrial capitalism, while others see their movement as primarily significant for the emergence of southern literature. Feminists such as Susan Donaldson see the Vanderbilt agrarians as just another brick in the southern power structure. She finds their arguments aimed primarily at protecting the southern white male hierarchy.[1] Yet this trend of scholarship only serves to move the discussion even farther from the material practice of agriculture.

This is not to say that critics have treated the Vanderbilt agrarians unfairly. If we measure their work in terms of agrarian ideas, agrarian imagery, and agrarian mentality, relatively little of Vanderbilt agrarianism falls into the realm of ideas. Ransom did produce philosophical essays on agrarian themes, but as a group the Vanderbilt agrarians were more interested in promoting a new and intellectually rich interpretation of southern culture than in examining agrarianism as such. This task required an attack on the northern, European, and Old South assumption that culture must be brought *to* the South, and a positive statement of southern cultural forms. Agrarian philosophy might have lent itself to such a task for any number of reasons, but the Vanderbilt agrarians were more inclined toward literary works in which philosophical principles were masked or studies of mentality in which substantive agrarian ideas were left implicit. In part this is because Ransom's aesthetic philosophy made strong claims for nonrational, virtually mystical, intuitive knowledge. Such intuition flees rationality or science, he believed, hence the best rendering of intuitive wisdom is through work and art.

Not surprisingly, then, scholars of the Vanderbilt agrarians tend to note novels and poetry when reviewing their work. Stewart states that Robert Penn Warren's *All the King's Men* is the most thoroughly agrarian of the entire group's total output. Bové argues that personality studies (two, *Stonewall Jackson* and *Jefferson Davis,* by Allen Tate, and one, *John Brown,* by Warren) are the most important works for advancing the critique that the Vanderbilt agrarians wished to mount. These three books rehabilitate the image of the South by exhibiting virtues and vices within the context of a lived life. Tate's book on Jackson praises the fit between the soldier/teacher's mind and its object, the pursuit of war. Davis and Brown are moral failures due to pernicious (that is, northern) intellectual influences. The emphasis on conveyance by mood, color, and implication was consistent with Ransom's idea that scientific rigor and discursive thinking would destroy what could be grasped immediately through the arts.

More broadly, emphasis on mentality was common coin in the southern Renaissance. Faulkner's novels contrast the mind-set of a fading but gentile elite with the rising Snopes's, petty bourgeois entrepreneurs, and scalawags unburdened by any sense of common morality, much less noblesse oblige. W. J. Cash (no agrarian himself) offered a compendium of personality types in *The Mind of the South* in 1941, a decade after *I'll Take My Stand.* Cash's book became required reading for southern intellectuals well into the 1970s. These literary and historical renderings made the case for and the critique of southern culture by conjuring personalities that rang true to southerners' experience (at least for white males). The studies of mentality permitted judgments of moral worth that were unmediated by the principles of moral philosophy. As such, they might become part of a systematic philosophy, but only if there are arguments that link mentality to character and conduct, and that articulate how and why certain mentalities come about. The southern studies of mentality provide little philosophical underpinning, however appropriate they might have been for the purposes of the Vanderbilt agrarians and their contemporaries (see O'Brien 1988).

Philosophers should take the Vanderbilt agrarians seriously, but this brief review of Ransom and the Vanderbilt agrarians is, I hope, sufficient to show that a philosophical agrarianism running through Dewey is a substantially different kind of intellectual animal. A complete program of research on philosophical agrarianism would have to interpret and critique the philosophical commitments in the writings of the Vanderbilt agrarians. It would usefully go on to make a comparative study of twentieth-century agrarian thought, taking up not only the Vanderbilt agrarians and the line of thought followed in this volume, but also the darker variety that fascinated Heidegger. Yet such a study presupposes a philosophy of history that considers how farming and husbandry are linked to the formation and transmission of ideals and culture. It should include a normative account of why the loss or decline of these mechanisms can be seen as social problems. This volume is dedicated, thus, to a philosophical explication of the pragmatist line of thinking on problem formation and subsistence production. Without presupposing or implying any particular judgment on the Vanderbilt agrarians' philosophical merit, these essays take up logically distinct and philosophically prior questions.

Agrarianism, Property, and Rights

If the Vanderbilt agrarians represent a poor entrée to agrarian philosophy, where should we turn? James Montmarquet's *The Idea of Agrarianism* is the only recent study of agrarian thought by a philosopher. Montmarquet provides a detailed discussion of the evolution of agrarian thought from the ancient Greeks through its involvement with European movements of enclosure and constitutional reform. As he traces it, the agrarian ideal begins with the Greek belief that the farming society provides a life that harmoniously promotes virtue, and it evolves into a political philosophy committed to a universal right to land and a limit on any individual's right to the accumulation of wealth. By the time of Locke and until the time of Jefferson, agrarian philosophy was caught in a deep ambiguity. On the one hand, it was a continuation of a medieval natural law tradition with a strong anticommercial bias. On the other hand, it was evolving into a libertarian philosophy of absolute noninterference in the individual's rights to control the use of land (Montmarquet 1989).

Montmarquet examines agrarian ideas as they contribute to notions of nobility, yeomanry, and the dignity of work. In this literature, a materialist orientation to food and farming ultimately gives rise to agrarian social thought through the analysis of rights. At the same time, materialist agrarian philosophy has always been accompanied by literary and aesthetic soliloquy. Montmarquet discusses agrarian poetry and novels in a long chapter on romantic agrarianism, but in truth literary forms of agrarianism hold little interest for him. For Montmarquet, the central, materialist strand of agrarian argument splits along two lines, with the English debate over enclosure. Enclosure refers to the practice of fencing or walling off (that is, literally enclosing) lands that had traditionally been used as a "commons." Commoners were, in fact, so named because they shared the use of open-access lands where they could pasture animals and raise subsistence crops, though with no title or right to sell the land itself. With enclosure, commoners were driven off these lands, and their right of access was denied. But at the same time, using more efficient farming methods on enclosed land led to an increase in the general availability of food.

Are the rights of access claimed by commoners legitimate? John Locke's *2nd Treatise of Government* contains what are arguably the most influential thirty pages ever written on property rights. Locke explicitly endorsed enclosure when he wrote:

> [He] that encloses land, and has a greater plenty of the conveniences of life from ten acres, than he could have from an hundred left to nature, may truly be said to give ninety acres to mankind: For his labour now supplies him with provisions out of ten acres, which were but the product of an hundred lying in common. (Locke 1690, 23–24)

According to C. B. MacPherson's celebrated study, this pattern of thinking ties Locke to "possessive individualism," a libertarian vision of society that stresses property rights. The commoners, on the other hand, were calling for land reform. Montmarquet's sympathies are clearly with the latter group. He describes the views and activism of Gerrard Winstanley and the Diggers, George Ogilve and Henry George. For these revolutionary thinkers, the crucial fact about farming is that farmers are weak, impoverished, and up against the owners of property. Claims for land reform became the focus of left-leaning agrarian movements through the twentieth century. Like MacPherson, Montmarquet interprets Locke's *2nd*

Treatise as an argument for unconstrained individual accumulation of property and for noninterference in the individual's absolute right to dispose of property however he or she pleases. Montmarquet's history is indispensable to the study of agrarian philosophy, and it is impossible to do justice to his analysis here. Yet it is still possible to question whether he has identified a peculiarly agrarian philosophy in this typology. In particular, the two threads of Montmarquet's social agrarianism break down along philosophical lines that require no reference to farming or food production. The libertarian possessive individualist insists on limiting the power of the state to the enforcement of noninterference in individual liberties, including property rights. The egalitarian radical insists on rights that secure basic opportunities and human needs, even when satisfying these rights requires a redistribution of landholdings and other forms of wealth. Agriculture figured historically in the evolution of these two views on justice because fencing common lands to introduce more intensive crop and animal production was a flash point for political debate. The radical egalitarian view of universal subsistence rights opposes enclosure, while the libertarian doctrine of noninterference supports it.

Of course an adequate philosophical analysis of enclosure is subtler, but what is crucial here is merely to see that agriculture is primarily important as a case study for the conflict between libertarian and egalitarian conceptions of rights. It is a particularly important case study from a historical perspective, for the two notions of rights arise out of an agricultural controversy. Yet contemporary rights theorists are fully capable of expositing arguments for both libertarian and egalitarian theories that make no reference to agriculture or to the material basis of society. It is true that the subsistence right claimed by radical egalitarians must involve access to food. To the extent that agriculture is the means to producing food, agriculture is inextricably (though not conceptually) involved in subsistence rights. Henry Shue has argued that such a right is among the most basic of universal human rights, and that securing this right demands reform in the world's production and distribution of food as a matter of justice, not charity (Shue 1980). If we cannot foresee a situation in which subsistence rights would not entail significant duties for agricultural producers, we are compelled to associate this argument with a genuinely agrarian idea. We all have rights to what we must have to live, and agriculture produces food, which we must all have to live. However, in market economies the right shifts from a claim against agricultural producers to a claim against the structure of society as a whole. Hence agriculture and the food system must be organized in a manner that does not frustrate each individual's right to food, but subsistence rights might be satisfied by food stamps or welfare payments, rather than as direct claims against individual farmers.

The same cannot be said, however, for libertarian property rights. At one time in European history, access to land for use in cultivation of crops and pasturing animals was the difference between survival and starvation for the vast majority. In such a world, a property right in land is equivalent to a subsistence right. But libertarian property rights are held in many goods that do not produce food, and most readers of this book are more able to secure their subsistence through exercising their property rights in automobiles or computers than through tilling the soil. The libertarian property right, thus, does not appear to suppose any unique dependence on or relevance for agriculture. The standard libertarian analysis of property tells us why *anyone* might hold property rights dear. It does not tell us why agrarians would find property in land especially dear. The theory's conception of property may have evolved by considering land-use disputes, but it does not derive its conception of property from distinctly agricultural land-use patterns, or from the problems of farmers attempting to protect their claim on land. In this sense, agrarian ideas are not at the root of the libertarian conception of property.

In summary, libertarian rights are conceptually independent from agricultural land use, and there is only the weak sense in which even subsistence rights depend on agriculture. The same subsistence needs that reformers cite to establish rights to food also establish rights to shelter, heating oil, and, in modern societies, transport. Thus both libertarian and egalitarian rights are important to the practice of agriculture only to the extent that they are universally applicable to any human practice. One does not need a philosophy of agriculture to explain why such rights claims are made. Montmarquet identified egalitarian claims for subsistence rights with radical agrarianism in a 1985 paper, "Philosophical Foundations of Agrarianism," and this theme receives extensive treatment in his 1989 book. However, by 1989, Montmarquet seems to have concluded that there is more to agrarianism than the radical's call for land reform. Though he sees much of the writing on agrarianism following either the aesthetic or the social strand, American transcendentalists and social reformers (such as Liberty Hyde Bailey) are an exception to the trend, advocating a moral vision that reconstitutes and integrates elements of an agrarian ideal that began with the Greeks. Montmarquet's chapter in this volume reprints that section of his seminal analysis.

Agrarian Philosophy in Ancient Greece

Victor Davis Hanson is a classicist who has written extensively on both Greek and contemporary agrarianism. Hanson believes the ambiguity between romantic and social agrarianism began in the ancient world. Lyrical agrarianism consists in pastoral elegies and praise of rusticity. It can be traced to Virgil's *Georgics*. Hanson, however, has even less interest in idle praise of farming than Montmarquet. Instead, he notes how Greeks believed that farming produces neither sympathetic nor sensitive personalities, though it does produce a hard-bitten individual who is master of his fate. Hanson also offers this line of descent for the agrarian ideal:

> With Hesiod's world begins the entire notion of agrarianism that was soon to become the foundation of the Greek city-state, and later to be enshrined in the West as the exemplar of a democratic society: a culture of small, independent yeomen on the land, who make their own laws, fight their own battles, and create a community of tough like-minded individuals (Hanson 1996, x)

Hanson's agrarians have little interest in anticommercial conceits. Any farmer knows that the farm must pay.

Hanson's agrarians *are* deeply committed to property rights, however, for that is what secures their independence. Nonetheless, in Hanson's treatment agrarian property rights are not equivalent to libertarian rights of noninterference. *Property* for the agrarian means land, and a property right establishes a right to use the land and to retain control over foodstuffs and other commodities produced on the land. Agrarian property rights may (but need not) include a right to transfer the land to another user by sale or gift. Saxon freeholders, for example, retained secure rights to control common lands during their period of actual occupancy. Upon quitting occupancy, however, lands reverted to the commons, available for anyone to take up farming or grazing anew (Hargrove 1980). With access to land, agrarians may bring forth the sustenance of life largely through their own endeavors. Without land, the agrarian is at the mercy of the broader economy. But owning transferable title to land is of secondary importance, at best.

Here marks the beginnings of a philosophy that is truly agrarian. Libertarian property rights apply equally to land, labor, capital, or ideas. Libertarians are fond of broad claims that link liberty to the protection of private property (see Hospers 1971). Yet in what sense is a person (such as most of us) who must sell services in a labor market truly independent? We are assuredly dependent on the structure and functioning of that market. We cannot bring forth sustenance without this social support network. The liberty secured by rights of noninterference is a highly constrained liberty. A farmer or pastoralist can secure a life with very little immediate dependence on a broader network of economic relationships. In this sense, agrarian property rights are more deeply linked to liberty than contemporary libertarian property rights.[2]

Hanson derives his conception of agrarian philosophy jointly from his study of the ancient Greeks, and from his experience as a family farmer near Selma, California. He sees Greece as a farmer would see it. First, the soil and climate of Greece favor a system of small-holding farmers who have the freedom to adapt their farming practice to variations in soil or slope that change literally from acre to acre. The soils were rocky, lending them to hand work with pick and hoe, but rendering them less suitable for the cultivation of vast acreage. That fact, plus the unavailability of water to supply large-scale irrigation gave the independent yeoman an advantage over the centrally managed plantation systems of the Mycenaean age. Lacking the massive workforce of the plantation system, Greek farmers planted a mix of crops that would mature at different seasons of the year and limit the potential for catastrophic risk. The former practice allows a small workforce to be continuously at work harvesting first one crop then the next. Crop diversity limits risk because it averages out the damage caused by weather and blight: when apricots are bad, pears could be counted on to be good, for example; so the chance of a total crop failure is reduced substantially. Given Greek soil and climate conditions, tree (fruit and olives) and vine crops were the logical choice to supplement cereal production. This, in turn, creates the environment for evolution in both mentality and social life:

> Any farmer who plants trees and vines, unlike the pastoralist or even the grain grower, invests his labor and capital in a particular locale *for the duration of his life*. In this interdependent relationship, the cultivator's presence and commitment to a stationary residence ensure that the young orchard and vineyard will be

> cared for and become permanent fixtures on the landscape. People who choose this form of agriculture have confidence that they can and will stay put, that they can and will keep the countryside populated, prosperous and peaceful. They are not just a different sort of farmer, but a different sort of person as well (Hanson 1995, 42)

Thus is personality affected indirectly by the geographical features that favor one kind of farming over another. Hanson writes, "From isolated rural residence the very independence of the Greek character must have originated (1995, 59).

To make a long story short, more or less, broad distribution of landholdings eventually gives rise to self-governance, if not democracy in the full sense of the word. Independent agrarians realize that they must provide a common defense, a fact that binds farmers and herders from a particular place with a common interest. The geographical proximity and military vulnerability of these lands lay down the basic structure of the polis. Farmer-comrades experienced in war see the self-interest in developing a state structure that will be constrained by the interests of farmers who are neither rich nor poor, Aristotle's *hoi mesoi* (Hanson 1995, 116). Hanson argues that Athens's sea power opened the way for new markets, new imported goods, and new wealth from trade, benefiting all, at first. But the new wealth was not geographically tied to Athens. Hanson documents a change in the definition of the *dêmos* that corresponds to a change in the practice of democracy.

> So under the original notion of agrarian oligarchy/timocracy/democracy/polity, the term *dêmos* actually meant the native-born resident citizenry who owned land and formed the popular government. *Dêmos* was in some contexts nearly the same term as *polis* itself. . . . The later reinterpretation of *dêmos* as *everyone* born into the *polis,* landed and landless alike, was the creation of fifth-century Athenian democracy. (Hanson 1995, 206)

While we might regard this as progress from one perspective, Hanson (and if he is right, Aristotle) believes that it caused the undoing of Athenian society. Agrarians were stuck with their land, planted in vines and trees. They were committed to the future stability of Athens as a specific place, geographically inclusive of their farms. Traders could move from port to port, seeking the most favorable terms. As such the trader's self-interest did not necessarily coincide with the goal of making Athens into a regionally based polis with a sustainable system of governance.

Hanson argues that Athens was experiencing a shift from relatively agrarian foundations to an economy based on trade during the lifetime of Socrates, Plato, and Aristotle. Though each of these philosophers differs dramatically from the others, Hanson sees all of them as attempting to influence this transition. Crucially, each is committed to what we might call "place prosperity," as distinct from individual wealth. Each philosophizes in service to the ideal of creating and maintaining a way of life at a particular place, though the possibility of succeeding in this ideal was certainly more robust for Socrates than for Aristotle. If, in contrast, the point of life is simply individual wealth accumulation and quality of life, why build a polis when one can simply pick up and move when the good times are gone? All this, perhaps, provides a context for Aristotle's praise of farming: "Indeed, the best common people is the agricultural population (*geôrgkios dêmos*), so that it is possible to introduce democracy wherever the multitude lives by agriculture or by pasturing cattle" (quoted in Hanson 1995, 116–117). Perhaps the agrarian assumptions implied by the polis were so obvious to the Greeks that there was no need to articulate them. Such assumptions are not obvious today, hence we tend to read these philosophers as offering universalistic philosophy rather than highly contextualized arguments.

Hanson's history is suggestive and fascinating, but in this context little stands or falls on whether Socrates, Plato, and Aristotle were agrarian philosophers in any deep sense. We should, however, pay attention to the argument that links production at a place, the interests of the producing class, the social goal of place prosperity, and the creation of democratic institutions for governance. The crucial agrarian claim is that land-based production establishes the basis for a political community in a way that manufacturing or trade cannot. The agrarian must stick with the land to make a life; the craftsman or merchant may move from place to place. Hanson claims that democracy (and indeed philosophy) arises naturally from communities committed to place prosperity, and that dissolution and decay are the rewards of individual wealth maximization. We may read *agrarian* broadly enough to include some fishing communities in this argument, as well as nomadic groups that had an established route and territory that constituted their place. But the argument cannot be read broadly enough to include plantation or industrial systems of agriculture whose material practices emphasize the extraction of wealth from the land, and accumulation by the owners of land and capital.

Agrarian Natural Philosophy in the Eighteenth Century

Something like Hanson's argument in *The Other Greeks* was a commonplace of eighteenth-century political philosophy. In particular, the philosophers of the Scottish Enlightenment theorized the transition from informal moral sanctions to state institutions by arguing that (a) the production of material sustenance is essential and as such it establishes parameters for norms and sentiments; and (b) governments and social conventions are the evolutionary product of change in or adaptation to those parameters. These agrarian tenets were not the primary object of Scottish philosophy. They were a by-product developed to support the Scots' larger moral and political theory. There is little evidence that any contributor to the Scottish Enlightenment thought the first tenet controversial, though they knew that the second would contradict popular views on the social contract. One of Hume's best-known essays, "On the Social Contract," makes a strong case for the evolutionary origins of the state. The Scots wrote when the transition from a population with 80 percent involved in primary agricultural production to the present 1½ percent was only beginning, but it was already clear where it was headed.
Agricultural technology and the food system were undergoing significant changes. Not the least was the death of "moral economy," in which access rights to food and land were based on informal relationships among lifelong neighbors who were well known to one another. In its place was a new "political economy" that demanded formal institutions for definition and enforcement of access rights (Thompson 1971)). Theorists such as Barrington Moore analyze the rise of social contract theory as a response to the declining influence of village morality (Moore 1966). Scottish political philosophy eschewed the hypothetical contract of arguments associated with Hobbes and Locke. They turned instead to an evolutionary theory of social relations, asking how hunting or agricultural societies would favor one set of moral practices and how emerging commercial societies would differ (Robertson 1987, 468–471). Francis Hutcheson, Adam Ferguson, and Adam Smith each worked this theme, though each did so to suit his own purposes (and without attribution to the others). Smith's version of the argument is summarized in Thompson's chapter on Jefferson.

The French also had a version of the view that material circumstances establish parameters for the evolution of culture and personality. Georges-Louis Leclerc Buffon presided over a school of natural historians who collected data on plants, animals and human cultures from around the world. These data were organized in a theory which stipulated that soil type and climate determined whether humans would hunt or cultivate, and it fixed limits on what they could do with respect to the latter. Climate, mediated by the pattern of production, was then thought to influence culture and character. Thus, it stereotyped people from southern climes as being naturally lazy due to the ease of bringing forth sustenance and aversion to exertion in hot weather. French natural philosophy of this sort lavished praise on those places where climate and soils had fostered agriculture, producing a people who prized the virtue of industriousness. Indeed, Buffon and the philosophes devoted considerable energy to speculating as to whether regions lacking the soils and temperate climate of Europe could ever produce a truly great civilization. Such speculation reflected poorly on the southern colonies of the Americas, a fact that prompted Thomas Jefferson's famous agrarian manifesto in his *Notes on the State of Virginia* (Wills 1979).

To see where the French school differed from the Scottish, we may consult David Hume. Hume did not think that a strong agriculture would produce virtue. His essay "On Commerce," concluded that any agrarian society will soon produce a surplus of labor. When surplus labor is not employed in manufacturing, "A habit of indolence naturally prevails" (Hume 1993, 160). The essay "On National Characters" reviewed the possible ways of explaining the observed tendency of individuals from national groups to share habits of industry, thrift, or indolence. Hume offered a refutation of the view that topography, soil, or climate were of singular importance in determining these characters, further distancing himself from Buffon and the French. Yet what may be of most interest here is his account of the two possible forms of influence on national character.

> By *moral* causes, I mean all circumstances which are fitted to work on the mind as motives or reasons, and which render a peculiar set of manners habitual to us. Of this kind are, the nature of the government, the revolutions of public affairs, the plenty of penury in which the people live, the situation of the nation with regard to its neighbors, and such like circumstances. By *physical* causes, I mean those qualities of the air and climate which are

> supposed to work insensibly on the temper, by altering the tone and habit of the body, and giving a particular complexion, which, though reflection and reason may sometimes overcome it, will yet prevail among the generality of mankind, and have an influence on their manners (Hume 1993, 113–114)

Although Hume does not oppose a "nurture" view by appealing to nature in the form of blood inheritance, neither does he suggest that personality and moral character rest on holding or rejecting certain beliefs. Moral character derives from the fit among person, work, and society, and physical causes play a role. His is, to this extent, still a materialist view consistent with agrarian presumptions.

Both the Scots and the French, thus, give precedence to material influence over the formation of personality and social institutions. The Scots emphasize moral causes, meaning what is contingent about any human population's way of organizing labor to bring forth the sustenance of life. The French emphasize what is given to agriculture by nature—that is, soil and climate. Other seventeenth- and eighteenth-century political economists brought agriculture to prominence in different ways. James Harrington's *Oceana* (1656), a treatise on geopolitics and sea power, argued for a wide distribution of landholdings among the population as the basis for secure military power. François Quesnay and the marquis de Mirabeau proposed the Physiocrat economic theory, built on the premise that only agriculture could produce a true increase in a nation's wealth. All of these views were premised on a method of inquiry that stressed the material basis of society, and that placed agriculture at the center of the analysis. None of them were idealistic, sentimental, or romantic.

Agrarian Romanticism

What then *did* become of the romantic agrarian ideal, the praise of farming and the pleasure of work amidst nature? It is a theme that has seldom lacked advocates. Montmarquet places the English poet Oliver Goldsmith (1728–74) at the head of the list. Goldsmith's long poem *The Deserted Village* articulates poignant and picturesque dimensions of traditional rural life. These rural charms are seen as threatened by the intensification of agricultural production, manifested in Goldsmith's time in the enclosure movement. Commercialization and intensification are the enemies of the themes that Goldsmith wants to celebrate, hence the poem's most cited lines:

> Ill fares the land, to hastening ills a prey
> Where wealth accumulates and men decay:
> Princes and lords may flourish, or may fade:
> A breath can make them, as a breath has made;
> But a bold peasantry, their country's pride,
> Whence once destroy'd, can never be supplied.
> (Goldsmith 1902)

Is *The Deserted Village* simply nostalgia for a time that never was? Certainly Goldsmith's critics have thought so. Agrarian rhetoric and imagery have been especially attractive to literary romantics who invest the past with almost mystical significance. Ransom and the Vanderbilt agrarians, for example, fall into this tradition. Yet, with the partial exception of Heidegger, romantic philosophers do not give such pride of place to agriculture. Montmarquet argues that eighteenth-century romantics such as Goldsmith point us toward two enduring insights. One might apply to any romantic imagery: "The power of the romantic critique of [our] civilization retains a kind of force, even as the possibility of a truly agrarian alternative belongs to a long buried past" (Montmarquet 1989, 216). The other is more specific to agriculture. Montmarquet believes that whatever credibility *The Deserted Village* lacked in eighteenth-century Europe, the transformation of agriculture since that time has indeed obliterated the way of life celebrated by romantic authors. He concludes his discussion of Goldsmith by writing, "Perhaps what Goldsmith claimed to see, he actually foresaw" (Montmarquet 1989, 194).

If we turn specifically to eighteenth-century America, we find J. Hector St. John de Crèvecoeur. Crèvecoeur was an unusually literate immigrant with royalist sympathies. He maintained a correspondence with Jefferson, published *Letters from an American Farmer* (one of the first works of American literature), and for a time farmed successfully in New York State. According to Jan Wojcik, Crèvecoeur imagines subsistence farming as a grand occupation, an American philosopher's stone. To Crèvecoeur America sings, "If thou wilt work, I have bread for thee; if thou but be honest, sober and industrious, I have greater rewards to confer upon thee . . . ease and independence" (Crèvecoeur 1986, 89). Wojcik summarizes Crèvecoeur's moral wisdom, saying,

> The farm in the right place is Crèvecoeur's metaphor for a moral life. The mind of a good farmer is a kind of genius loci, a spirit of good place, properly distanced from court and wilderness, near others of its kind, and within driving distance of good markets and honest civilization. (Wojcik 1984, 32)

On the one hand, Crèvecoeur's idyllic praise of farming might represent an alternative agrarianism, one that prefigures romantic philosophy and the nature aesthetics of the nineteenth century. That would make him the forerunner of American transcendentalists. On the other hand, the *Letters* might rest on the less idyllic foundations being laid by the political economists. On this view, we read Crèvecoeur as engaging in speculative psychology that extends an argument initially made on materialist grounds. As Montmarquet reads Goldsmith, Wojcik reads Crèvecoeur as having more relevance to late-twentieth-century changes in agriculture than to his own time (see Wojcik 1989).

The notion of a "right place" resonates with the agrarian's interest in place prosperity. It may be the most politically potent theme that emerges from romantic agrarianism. Crèvecoeur asks the rhetorical question, "[W]here is that station which can confer a more substantial system of felicity than that of an American farmer, possessing freedom of action, freedom of thought, and ruled by a mode of government which requires but little from us?" (Crèvecoeur 1986, 52). As Simon Schama shows in *Landscape and Memory*, most European cultures developed art and literature that invested mystical significance in the power and grace of native soil (Schama 1995). It was the soil itself that particularized otherwise general philosophical claims about memory and tradition, claims that would not afford any moral or aesthetic advantage to any single national culture. This feature of agrarian rhetoric trades on the seeming permanence of land, especially when seen in comparison to an endless parade of princes, regimes, and forms of social organization. Though English and German artists might choose similar subject matter, the fact that it was English and not German land that was being celebrated (or vice versa) provided the principle of difference that was essential to nationalist mentality. It was the land and not the regime that established unique national identities. Paintings and poetry with rural subjects paved the way for a form of nationalism that saw agrarian rhetoric contributing to the cultural integrity of a particular nation. This, in turn, laid the basis for a turn of thought that allowed each national people to see itself as exceptional (and generally superior) when compared to the history and culture of all other nations.

It is as if a romantic agrarian who surveys the arts finds poetry like Goldsmith's and paintings like Brueghel's. "Yes," she says to herself, "these themes are common to all cultures. But our own national culture is the exception, because only *our* culture is grounded on *our* native soil." Of course, this is not a respectable philosophical argument. Crèvecoeur himself seems to locate American exceptionalism more reasonably in political freedoms, though there are plenty of passages in the *Letters* extolling American soil and climate. Countless others would move immediately to the view that America was special because American land was singularly blessed by God. For evidence, drive through the American heartland and count the billboards declaring "This is God's country."

"And *your* country *isn't*!" they say to outsiders in a hoarse stage whisper. The claim of divine blessing on land points us toward some of agrarianism's least savory dimensions, one that pits communities and ethnic groups against one another in a bizarre holy war. The appearance of "God's country" billboards on every other county line suggests that God's grace has taken a singularly local turn in North America.

In our day, Thomas Jefferson is the eighteenth-century author most likely to be linked to agrarian romanticism. Jefferson famously writes, "Those who labour in the earth are the chosen people of God, if ever he had a chosen people, in whose breasts he has made his peculiar deposit of substantial and genuine virtue." On the one hand, this may tie Jefferson to a mystic and nonrational preference for country over city. The entire *Notes on the State of Virginia* may be read as a treatise in the native soil tradition of romantic nationalism. On the other hand, as "Thomas Jefferson and Agrarian Philosophy" of this volume argues, Jefferson may be engaged in an intellectual project more in line with that of the French philosophes. His terse but evocative agrarian prose may simply assume that his readers already know the basic agrarian argument, so that he may gesture to it without providing amplification.

There are three points that must be made here. First, at the time of Crèvecoeur and Jefferson, belief in the power of agrarian practices to control the development of both formal and informal social institutions was virtually ubiquitous. This is not to say that intellectuals were uncritical of this idea, but the agrarian thinking behind much of what Crèvecoeur wrote in the *Letters* or what Jefferson recounted in his *Notes* was part of what every schoolchild in the late eighteenth century knew to be true.

This knowledge was based on folklore, common sense, and the fact that even a lifelong urbanite had direct experience of agrarian transition, of the dramatic social changes from feudal to yeoman to industrial agriculture. As such, most systematic philosophy took agrarian ideas seriously, and especially so when the aim was to modify or correct implicit agrarian beliefs. In other words, Crèvecoeur and Jefferson did not come up with anything particularly novel, though they may have offered particularly eloquent expressions of this common knowledge. The burden of proof was on those, such as Adam Smith and David Hume, who anticipated the end of agrarian society and manners, and who were already engaged in the search for something new. As the essays in this volume attest, the burden of proof has now been reversed, and we need an explicit and convincing statement of what everyone once knew to be true. Implicit knowledge of an agrarian paradigm made the poetry and art meaningful, but the very implicitness of this knowledge made it likely to be underspecified and mythologized, rather than described and theorized.

Second, perhaps Crèvecoeur did contribute something new in the idea that working in the presence of nature permits realization of an authentic, natural democratic sensibility, but if so this was a minority view in the eighteenth century. The Europeans (like Hanson's Greeks) believed in the stability of agrarian democracy, but not because agrarians had a rarefied sensibility. Agrarian society melded self and public interest in a way of which no other material organization of society seemed capable. Agrarian society may select for certain personality types, but one need not claim that it *molds* people with certain moral traits in order to make the political economists' argument. Of course, there are obvious ways to connect the environmental determinism developed by French natural philosophers and the native soil arguments of the romantics. The most expansive versions of this linking argument were made a hundred years after Crèvecoeur and Jefferson wrote. For example, Brazilian Euclides da Cunha's historical novel *Os sertões (Rebellion in the Backlands)* is a war story proceeded by 150 pages of native-soil rhetoric. In Cunha's story, the people of the Brazilian backlands develop a physiognomy and a mentality that makes them particularly well suited to survive in its rugged conditions and climatic extremes. They are of mixed racial background and have been shaped by harsh backland conditions. As a result, they are not properly descended from either Portuguese or Indian blood, but can truly be said to be a new race. They have survived, perfecting a form of ranching unique to the climate and topography.

Their habits, complexion, and even the sinewy strength of their small bodies reflect the environmental conditions in which they have evolved.

Cunha posits a force within soil and water acting on the media of human physiology and personality. The people of the backlands are less intellectual and more immediate in their response to circumstance. In this respect, Cunha's admiration for the backlanders mimics Emerson's praise of farming. It is easy to see why this kind of literature would appeal to New World people struggling to separate from the European culture and searching desperately for an authentic culture with which to replace it. The book remains popular in Brazil to this day. To my taste, *Os sertões* takes the picture painted by Goldsmith, Crèvecoeur, Emerson, Ransom, Tate, and the others to an extreme conclusion. Here, environmental determinism (and the national exceptionalism that goes with it) operates as genetic selection, resulting in a biologically determined human temperament suited to soil and climate. There is more than a trace of cultural bias (even racism) in this mode of thinking. When it succeeds too well, agrarian rhetoric deployed to establish the autonomy and identity of an oppressed group itself becomes repressive. Recent critics of the Vanderbilt agrarians have used postmodern methods to unmask racial interests. The essays in this volume are intended to exposit a better use of agrarian ideas.

Finally, as already noted, many national cultures adopt a landscape myth as the root notion of difference for their particular conception of national identity. Soil, water, and climate are determinative in these myths, and the human beings that adapt, over time and generations, to a particular place develop habits of observation, patterns of response, and social norms of collaboration that make them better able to cope with the special challenges of a particular landscape. In some landscape myths, such as Cunha's, the human body is itself transformed through a form of subspecies evolution. In other myths, a people is said to be of the land or even owned by the land. In most of these myths, the peculiar adaptation of a people to the land is thought to give them a special moral claim to inhabit and cultivate the land. But do these myths articulate anything of enduring philosophical value?

Following Montmarquet, it seems prudent to conclude the discussion of romantic agrarianism simply by recognizing that these myths have continuing appeal, something they should not have if they were truly as ludicrous as rational critique makes them seem. Romantic agrarian themes enjoyed poetic expression going back to Virgil. They frame important questions in aesthetics, art, and art criticism. Is cultivated nature the proper object for aesthetic contemplation? Does the aesthetic experience coincide with a recovery of sensibilities lost with the retreat from nature? It is impossible to answer the question of whether this early romantic conception of agrarianism *competes* with the materialist ideas of the political theorists, or if it instead simply presumes that analysis as its starting point. For that matter, even Vanderbilt agrarianism, which seems clearly to be grounded in a critical aesthetic project, may presume, if vaguely, a materialist argument. For philosophical naturalists, the interpretation that sees relatively romantic agrarianism as an offshoot of materialist arguments makes the romantic texts far more interesting, but it is a question that must be left open here.

Agrarian Philosophy at the Millennium

One could argue that full-blown agrarian philosophy ends in the eighteenth century, if, indeed, it had not already ended long before. Yet the philosophically rich nineteenth century has many figures—Hegel, Marx, Mill, Spencer, Emerson, and Spengler, to name a few—who take *some* note of agriculture, if sometimes only to argue against agrarian political economy. Even theories that oppose key agrarian claims can be said to have agrarian roots in the way they are configured as a result of that opposition. By 1925 agrarian philosophy was clearly in retreat, though agrarian imagery and models of agrarian mentality would continue to influence literature, criticism, and historiography for another forty years. By midcentury "what every schoolchild knows" about the agrarian basis of society was, in fact, known by few in industrial Europe and North America. Heidegger's agrarian orientation went unnoticed by the majority of his readers. The background knowledge that a Smith or Buffon (or even Marx and Spencer) would have taken for granted no longer existed. In this sense, their texts are an incomplete representation of their thought, and much of what is missing is ineluctably practical.

Why such a precipitous decline in agrarian thinking? That question takes us well beyond the scope of this volume. Surely there are many causes. One may simply be the fact that the mass of population (and especially elites) was spatially and spiritually farther from food production than ever before. Having no practical experience beyond ordering from a menu or purchasing brightly colored packages in a supermarket, why should an ordinary person ever have cause to think about the material organization of the food system? This ignorance was augmented with a clear antipathy toward farming and farmers that has roots in urbanite beliefs that agrarians are rude, crude, and somewhat stupid. Antirural prejudice goes back a long way (see Williams 1973), but the twentieth century has seen an especially virulent version arise among American intellectuals, and especially among philosophers. While sociology, economics, history, and literature tolerate rural studies, philosophy does not. Farming is like farting in most philosophical circles: one avoids mentioning it as assiduously as one avoids doing it.

The fact that blood-and-soil nationalism gave rise to repressive social movements and two world wars provides a more serious reason for rejecting agrarian ideas. Native-soil ideologies contributed to the outbreak of World War I, and German National Socialists expressed allegiance to agrarian ideals explicitly. It was arguably these agrarian claims that seduced many intellectuals—Martin Heidegger, Ezra Pound, Allen Tate—into expressions of support for the Nazis. Much earlier, Darwinian ideas had been grafted onto agrarian roots, producing dubious claims for the superior fitness of Europeans and European civilization. With the development of genetics, the old argument about the determining influence of material production shifted from soil to blood and became the basis for the eugenics movement, as well as Nazi genocide. After the Holocaust it has become easier to see why these turns in agrarian theory were morally flawed, but it is more difficult to see how they drew on, yet differed from, the agrarian-materialist arguments that went before.

Unfortunately but not surprisingly, the reaction to these monstrous perversions of theory has been to shun all agrarian rhetoric entirely. Saying anything that does not connote immediate disapproval of agrarian themes is quickly taken to constitute evidence of racism and insensitivity on the part of the speaker. Yet clear strands of agrarian thought are making inroads in evolutionary and ecological economics, where once again economists see institutions as products of selection pressure exerted by material forces. Historically, these forces have their root in food production, so farming and fishing societies are especially useful case studies. As Smith (and Dewey) argued, agrarian institutions can become dysfunctional, creating turmoil and strife for years before social evolution replaces them. Agrarian mores and informal rules can also regulate human activity within environmental constraints, however. Theorists such as Daniel Bromley (1989) and Elinor Ostrom (1990) have begun to integrate classically agrarian themes into rational choice theory.

A pattern of thinking characteristic of eighteenth-century agrarian political economy also lives within that community of historians and social scientists who try to explain the seemingly unique pattern of European development. This literature is complex, but scholars such as Immanuel Wallerstein (1974), E. L. Jones (1981), Alfred Crosby (1986), and Charles Tilley (1992) have again reversed the balance, giving more credence to soil than blood. Jared Diamond's Pulitzer Prize–winning *Guns, Germs and Steel* (1997) makes a particularly strong case for the view that geographical parameters including soil, climate, topography, plant diversity, and populations of insects and other animals are the basis for the eventual evolution of European civilization. In this new literature, Europeans are not superior, but merely lucky. Furthermore, their good fortune may have made them less well prepared to face the crises to come. Philosophy has a role to play in these new intellectual developments, but eschewing the philosophical study of agrarian ideas has left philosophers more ignorant of both the power *and* the problems in agrarian argument. Is our society perhaps more vulnerable to repressive ideologies or environmental crisis as a result?

American Agrarianism

The Living Tradition

JAMES A. MONTMARQUET

I HAVE SUGGESTED that agrarianism may now be doomed to a perpetual rear guard status, its champions with nothing more significant to do than sniping occasionally at a triumphant industrialism (see Montmarquet 1989). I would like to make the case, however, that agrarianism may enjoy a significant future as well as past. I would like to make the case that there is more that lives, or could live, in this ancient tradition than what I have called, somewhat disparagingly, "romantic agrarianism." It will be, in fact, my purpose in the last section of this chapter to explain, in as much detail as my clouded crystal ball allows, just what this future is likely to be—but first some historical preliminaries.

Actually, we have not yet taken American agrarianism much beyond Jefferson and his farmer contemporary Crèvecoeur. We have, it is true, explored the distinctly southern tradition of Taylor and his twentieth-century descendants, the Vanderbilt agrarians, and we have traced the main lines of Steinbeck's isolated genius. Alongside the southern agrarian tradition there is also, however, a distinctly Yankee lineage. Crèvecoeur belongs to this, as do Thoreau and Emerson; and in this century there are the interesting figures of Liberty Hyde Bailey and the leading agrarian writer of our own times, Wendell Berry. Where the southern tradition is courtly, conservative, and faintly suspicious of democracy, this northern tradition is idealistic, democratic, and strongly committed to individual freedom. This idealism extends not only to the social and political aspects of farming, but sometimes even to its technical side. There is a receptivity—though with important exceptions both past and present—to science and its applications for agriculture.

I intend here, however, no general history of this strain in American thought and still less a general discussion of agrarian movements, like Populism, which do not fall neatly under either the heading of "northern" or "southern." Instead, I want to develop some themes to be found in Emerson, Thoreau, Bailey, and Wendell Berry; I want briefly to contrast these with some leading concerns of the nineteenth-century Populists and the Grange and in general to set the stage for our discussion of the future of agrarianism in this country.

Emerson and Thoreau

Let us begin with Ralph Waldo Emerson's most direct ruminations on the subject of agriculture: his essay entitled "Farming" (1858). Here Emerson opens on a classical note by recalling humanity's ancient respect for this even more ancient way of life:

> Men do not like hard work, but every man has an exceptional respect for tillage, and a feeling it is the original calling of the race, that he himself is only excused from it by some circumstance which made him delegate it for a time to other hands. (749)

But the significance of farming, for Emerson, was not determined entirely by its place in the economic nexus uniting man and nature, important as that place surely was. He continues: "All men keep the farm in reserve as an asylum where, in case of mischance, to hide their poverty—or a solitude, if they do not succeed in society" (749).

The farm, thus, is also a kind of healing place—to which the tired or the failed urban dweller's children may repair, to be "cured by that which should have been [their] nursery, and now shall be their hospital" (749). The farm had this healing power through its connection to the highest good of Emersonian philosophy: nature. The farmer's slow ways were nature's themselves. His work, in drawing upon nature's bounty and utilizing nature's powers, created a permanent benefit to the farmer and, if all went well, a permanent benefit for everyone. "The man who works at home," says Emerson speaking of the farmer, "helps society at large with somewhat more certainty than he who devotes himself to charity"—adding, significantly in terms of any comparison one might want to draw with his southern contemporary, John Taylor, that the truest abolitionist is the free farmer who "stands all day in the field, investing his labor in the land, and making a product with which no forced labor can compete" (751).

But it is nature more than freedom, indeed more than farming, which is Emerson's true subject in this essay. Insofar as Emerson wishes to celebrate the farmer, it is as Nature's manager; it is farmers who, by working natural improvements, become "confuters of Malthus and Ricardo" (and their essentially pessimistic outlook on political economy). In fact, Emerson argues (contra Malthus) that succeeding generations who work the land actually tend to increase its yields. Waxing the optimistic Yankee, Emerson says:

> The last lands are the best lands. It needs science and great numbers to cultivate the best lands, and in the best manner. Thus true political economy is not mean, but liberal, and on the pattern of the sun and the sky. Population increases in the ratio of morality; credit exists in the ratio of morality. (757)

One gets a deeper and more substantial insight into Emersonian philosophy of nature and the role it accords agriculture in the ultimate scheme of things, from Emerson's very first book, *Nature* (1836). Here, Emerson distinguishes several aspects of nature, which he entitles "Commodity," "Beauty," "Language," and "Discipline"—one of which, the first, directly concerns agriculture.

From a spiritual standpoint, Commodity is the lowest form or usage of nature. But it is also, he says, the only one which all human beings are capable of appreciating. It is "all those advantages which our senses owe to nature"—even though these benefits are "only temporary and mediate" (7). They are, unlike the higher services of nature to the human soul, "not ultimate." Still, despite this limitation, Emerson is not one to disparage the merely practical: Commodity remains a vital part of his view of nature. Commodity is a token of Emerson's confident sense that nature is there, in part, to serve and to be used by human beings. Today we have come to expect indifference, and sometimes even hostility, toward the practical, agricultural uses of nature from those who, like Emerson, are mainly concerned with the appreciation of its more spiritual qualities. But this is not at all Emerson's view of the matter. In this regard, much more like Jefferson or Quesnay than Rousseau, he writes:

> The field is at once his [humankind's] floor, his work-yard, his play-ground, his garden and his bed. . . . Nature, in its ministry to

> man, is not only the material, but the process and result. All the parts work incessantly into each other's hands for the profit of man. (8)

Emerson's enthusiasm for the practical utilization of nature is not, however, limited to agriculture and rural activities. On the contrary, all of the contrivances by which people put nature to use for their own benefit—industrial as well as agricultural; urban as well as rural—excite his interest and approval. For Emerson, there is no sharp, Physiocratic line to be drawn between the genuinely productive (agricultural) and the sterile (industrial) uses of nature.

As we ascend through the remaining forms by which nature can be known and felt by the human observer, however, a philosophical reversal takes place. The commodiousness of nature to man now is understood in the context of a larger alienation. Thus, in the section on spirit he writes of the relation between ourselves and the world:

> We are as much strangers in nature as we are aliens from God. We do not understand the notes of birds. . . . We do not understand the uses of more than a few plants. . . . Is not the landscape, every glimpse of which hath a grandeur, a face of him? Yet this may show us what discord there is between man and nature, for you cannot freely admire a noble landscape if laborers are digging in the field hard by. The poet finds something ridiculous in his delight until he is out of the sight of men. (36)

As man's spiritual powers are augmented, there is a growing sense of alienation between nature and spirit—an alienation which includes the gulf we find here separating the poet and the planter. This separation, however, this inward "denial of our sympathy with nature" (39) is not final. It is overcome ultimately through the deeper recognition that all reality is spiritual—that is, through recognition of the core of truth in philosophical idealism. With the insight that all reality is spiritual in nature, Emerson supposes, man can reassert his total hegemony over a nature now understood to be nothing other than his own creation. He closes, characteristically, on a note of optimism:

> Build therefore your own world. As fast as you conform your life to the pure idea in your mind, that will unfold its great proportions.

> [The advancing spirit] shall draw beautiful faces, warm hearts, wise discourse and heroic acts around its way, until evil no more is seen. (42)

Now we can make somewhat clearer the sense of the mystic proclamation with which "Farming" ended. Political economy, recall, is the triumph of the moral will over the apparent Malthusian limitations of nature. The greater the natural obstacle, the more complete the revelation of spirit when this obstacle is overcome. And, thus, the true subject of political economy is not Malthusian natural limitations, but the means to overcoming them through the human mind and will.[1]

So, ultimately in Emerson's philosophy, the agrarian idea of the special character of farming is overcome in the general rout of Nature by Spirit. In contrast, Emerson's Concord, Massachusetts, neighbor, Henry David Thoreau, managed to remain more or less an agrarian by consigning Spirit to a much less aggressive spot in his philosophical universe. If Emersonian man soars ever upward on the beating wings of Spirit, Thoreau's man only gazes upward while staying quite rooted to the earth. If Emerson embraces the political economy of an expanding and increasingly interdependent nation, Thoreau's concern remains much more one of self-sufficiency, of taking time to do the essential things of life well and not rushing through the important things so as to have time for the unimportant.

Certainly agriculture, or the simple kind of agricultural life which Thoreau tried to lead at Walden Pond, fit well with his essential concerns—though he was under no illusions that farmers always led simple lives, close to nature and the like. Most of us know Thoreau's observation about the majority of people leading lives of "quiet desperation," but how many recall its continuation? "From the desperate city," he said, "you go into the desperate country." And he said, also in the first chapter of *Walden:* "The farmer is endeavoring to solve the problem of a livelihood by a formula more complicated than the problem itself. To get his shoe strings he speculates in herds of cattle" (Thoreau 1937, 29). In contrast to this rural version of urban madness, Thoreau offered the idea of the "poetical farmer," who "does nothing with haste and drudgery, but everything as if he loved it" and who looked forward not to the income his crops will bring, nor to any particular future satisfaction, so much as the enjoyment he receives in the present from labor. "He is never in a hurry to get his garden planted, and yet it is always planted soon enough."[2]

Thoreau himself was a kind of poetical farmer—not fully one, since he was not entirely a farmer—yet surely he partook of this spirit. At Walden Pond he planted beans—like the Diggers, to whom he bears a number of similarities. Thoreau, however, was more about himself, about observing and improving himself than about the massive social schemes of a Winstanley. He has worked much over his bean crop, Thoreau says—but in the future he will work more to sow different seeds: of virtue, of "sincerity, truth, simplicity, faith, innocence, and the like" (148). The poetry of farming, for Thoreau, was not mainly the enjoyment of the physical exertions or even the practical skills it demands. It was primarily a matter of finding meanings which go beyond the physical. The ancients, Thoreau observes, treated agriculture as a sacred art, had festivals and celebrations of agriculture. Today, unfortunately, the farmer

> sacrifices not to Ceres and Terrestrial Jove, but to the infernal Plutus rather. By avarice and selfishness, and a grovelling habit, from which none of us is free, of regarding the soil as property, or the means of acquiring property chiefly, the landscape is deformed, husbandry is degraded, and the farmer leads the meanest of lives. He knows Nature but as a robber. (149–150)

Farming should be as much an occasion for profound reflection as physical labor, a becoming aware of the heavens as well as the earth. There is virtue to be found in farming, Thoreau agrees with Jefferson. But here I find Thoreau more realistic and more to my liking. These virtues require, as we say, "cultivation"—as much on the farm as in the city. Virtue is certainly not the inevitable outcome of the efforts of farming; it requires special effort and attention in its own right. Farmers are not "blessed"—they have, at most, the opportunity of making themselves so; but even this will necessitate rejection of many of the ways of other farmers.

Reflecting on these two great American thinkers, we must remark that, whatever their philosophical and political affinities, it is quite possible to see Thoreau and Emerson as originating, or standing very near the origins of, two entirely different kinds of thinking about agriculture. Let me briefly explore each in turn; we shall come back to them at the end of this chapter.

Thoreau, obviously, is the spiritual godfather of those today for whom "small is beautiful" and (to coin a phrase) "less is better than more." Thoreau is also the spiritual ancestor of those for whom agriculture must be understood as coexisting with—rather than as destroying or superseding—those other forms of vegetation and animal life we find on our planet. (Thoreau even wrote that "the sun looks upon our cultivated fields and on the prairies and forests without distinction" and that "in his [the sun's] view, all the earth is equally cultivated like a garden" (150). So from Thoreau there comes a regard not only of ecology narrowly conceived as involving issues like proper soil care, but also much more broadly conceived, in the mode of Aldo Leopold and John Muir, as a concern for all life—no, all things—on this earth.

It is possible to see Emerson as heading up quite a different tradition, where "big is beautiful" (even if small can be so as well) and where more—more of what people want—is better than less. Here science and technology are the natural extensions of our knowledge and will. If these are the source of problems, they are the best and sometimes our only solutions as well. In this world made commodious to the human spirit, human creations will have a special place under the sun—a sun which is, after all, itself an expression of a larger "spirit" emanating from the human man. Here, the practical consequence of philosophical idealism is a kind of scientific, technological, and economic pride (not to say, hubris).

Liberty Hyde Bailey and the Country Life Movement

Emerson's and Thoreau's Concord enjoyed a social and an economic stability which contrasts sharply with the lot of the American farmer in the decades following the Civil War. American agriculture during that period came, and with a vengeance, under the influence of larger social forces and interests. This was a time of industrial expansion and, at least as far as farm prices and income were concerned, agricultural contraction. And if farming was still the dominant occupational category in the United States, this did not seem reflected in the degree of political and economic power actually held by farmers. Out of this troubled period came two important rural movements, requiring some consideration at this point in our discussion: the Grange (the "Order of Patrons of Husbandry," as it was officially known), founded by Oliver Hudson Kelley in 1867,[3] and the Populist Party, founded by Gen. James Weaver in 1891.

The Grange, with its secrets, its rituals, and its ranks based in Greek and Roman mythology, offered some of the ceremony whose absence Thoreau perceived in American farming. (Whether old Thoreau, who died five years before its founding, would have appreciated the small-town, Masonic atmospherics of the Grange may be doubted, however!) The Grange was a cultural, educational, social, and, on local levels, an economic alliance of farmers. According to their Ten Commandments, the Grangers were bidden to "choke monopolies, break up rings, vote for honest men and make money" (not necessarily in that order of importance). Members purchased life insurance, attended adult education classes, enjoyed social intercourse with other farm families, belonged to cooperatives—all based around the Grange Hall.

Despite considerable pressures to the contrary in those decades of agrarian dissent, the political involvement of the Grange was limited to pressuring both political parties to promote farmers' interests; it neither ran nor officially endorsed candidates for public office. Leading members of the Grange—perhaps most notably Ignatius Donnelly (on whose literary career see Montmarquet 1989, chap. 5 n. 2)—were to test this rule; but the Grange remained, to a fairly considerable degree, outside of partisan politics. This was true even when the Populist Party in 1892 ran a national campaign and candidate explicitly appealing to agrarian interests.

The Populists, to be sure, did not make a narrow appeal just to agrarian concerns, but saw instead farmers and all other "productive citizens" (a category which included virtually everyone with the exception of monopolists, stock swindlers, lawyers, and Jews)[4] as naturally allied with each other against the enemies just mentioned. Farmer support was to be enlisted in this moral struggle between "the robbers and the robbed,"[5] but so was that of city laborers, businessmen, and professionals. Perhaps because their appeal was so diffuse, perhaps because farmer interests differed regionally, support for the Populists in 1892 and William Jennings Bryan, whom they endorsed in 1896, was regional and not based along farmer/nonfarmer lines.

The Populists and Bryan freely invoked the kind of agrarian ideal on which the ideology of the Grange was based. Thus in his famous "cross of gold" speech, Bryan intoned that

> the great cities rest upon our broad and fertile prairies. Burn down your cities and leave our farms; and your cities will spring up again as if by magic; but destroy our farms and grass will grow in the streets of every city in the country. (McKenna 1974, 131–139)

But neither Populist candidates nor Bryan got anything like the kind of nearly unanimous farmer support they would have needed for election. Instead, by the start of the twentieth century, a case could be made that the position of the farmer in American society and the basic Jeffersonian faith of many of our fathers were considerably imperiled. Now for the first time America was no longer primarily an agricultural nation; the 1910 census showed less than one-third of the population worked the land. Despite the efforts of the Grange movement, those who remained on the farm were perceived as suffering from considerable cultural, educational, technological, and economic impoverishment—relative to their urban counterparts. America clearly was a young nation surging ahead in those years of the Roughriders, the Panama Canal, and the unabashed nationalism of Teddy Roosevelt. Was not agriculture in great danger of being left behind, in a perpetual dark closet of rural ignorance and boredom?

The Country Life Movement was meant to address precisely these concerns. It was the response of the Roosevelt era to this supposed dark closet. It meant to educate, to uplift, to do many of the things for the farmer which Kelley's Grange had intended. But this was to be—what the Grange was not and never was meant to be—a national response to a national issue. President Roosevelt himself gave original impetus and direction to the movement when he created a Country Life Commission made up of leading scholars, educators, and journalists, charging them to make a comprehensive report on the wide range of economic and cultural problems perceived as afflicting agriculture in 1908.

Although short on specific solutions and fallen upon an unresponsive Congress (which failed even to appropriate the $25,000 needed to print it), the Report of the Country Life Commission nonetheless served to focus attention on rural problems and gave impetus as well as a widely recognized name to the Country Life Movement. The report came out under Roosevelt's successor, but the broader movement which it fostered certainly bore the stamp of TR's approval and enthusiasm. Roosevelt's vaunted love of nature and feeling for the outdoors would have made him sensitive to such rural problems as the need for land conservation (which, with the closing of the frontier, was an important new concern). There was, moreover, an obvious affinity between the qualities of rugged individualism traditionally attributed to farmers and those valued by the president himself.

Seen, though, as a grand attempt to improve the life of the American farmer, the Country Life Movement had an ambivalent quality which is worth remarking on from the start. In its underlying concern for rural life, its faith in the potential for what this life might be, this movement clearly belongs to the Jeffersonian tradition in American thought.[6] Indeed, one important wing of this movement—termed the *urban agrarians*—emphasized the need to transform *urban life,* allowing city folk to enjoy more of the pleasures and the healthfulness of country living. Still, this was mainly a movement bent on reforming, of improving the conditions of life in the countryside; this was a response to the perceived *deficiencies* of country life. As such, it was concerned with the often neglected, darker side of the Jeffersonian agrarian ideal. Behind the ideal of a simple, ruggedly independent, hard-working farming life, it saw in many instances drudgery, isolation, and cultural and educational backwardness.

A similar ambivalence underlies the economics of this movement. The first two decades of the twentieth century were marked by a considerable economic turnabout for farmers. A greatly expanding, demand for farm products, together with a relatively constant supply, meant steadily rising prices and an unprecedented agricultural prosperity. (These decades are often termed the "golden age" of American agriculture.) For city dwellers, however, these same rising prices and stagnant levels of production were seen otherwise—as major problems. An important current within the Country Life Movement, and a prime indicator of the largely nonfarming basis of this movement, sought to remedy this situation by introducing industrial efficiencies and technological innovations in the countryside. Even though an avowed goal of the Country Life Movement was to improve the economic position of the farmer, the effect of the reforms it advocated was predictably to lower prices and, likely as well, the income of the average farmer.[7]

The Country Life Movement, then, was not a rural movement so much as a movement on behalf of rural people.[8] Its aim, insofar as it was specifically rural, was more to lessen the burden of farm labor to improve farm income, more to raise the general level of farm culture than the level of farm prices. Like the Populists, it wished to see farming allied with a variety of other social interests, but without the latter's absolute dualism between the plutocrats and the good, hard-working members of society. (In fact, many of these "plutocrats"—the railroad magnate James J. Hill, for one—belonged to the Country Life Movement.)[9]

Ultimately, it was not a plutocrat or Teddy Roosevelt, but a Cornell University professor and the co-chairman of the Country Life Commission, the colorfully named Liberty Hyde Bailey, who emerged as the leading intellect of the Country Life Movement. Bailey's views qualify him in certain respects as one of those urban agrarians who wished to reform city life,[10] incorporating older, agrarian values into the metropolis; but, as we shall see, he was equally an advocate of reform in the countryside. His views were broad enough to encompass both wings of the Country Life Movement.

Bailey's vision for agriculture, first of all, harked back, in its combined attention to farming and naturalism, to Jefferson and especially to Crèvecoeur, to a time when in a more wooded America the farmer was more apt to be a naturalist. "Good farmers," said Bailey, "are good naturalists." In contrast to our present view of the successful farmer as a kind of applied chemist and applied agricultural economist, Bailey's naturalist view stressed a more "romantic," a less rigidly "scientific," approach to nature (see Bailey 1911, 30). Such an understanding was to be based on biological knowledge—of course, Bailey was a scientist and a professor. But more importantly, it was to involve a level of emotional involvement and appreciation of the natural surroundings of the farm. It was this involvement and appreciation which represented, for Bailey, the distinctive charm of the agrarian life. And it was this interest, he thought, which represented a solution to the often very real boredom and drudgery of farm life. "Only as we love the country is the country life worth living," he said (quoted in Bowers 1974, 56). This naturalistic spirit was not, however, a mere possibility to be recovered, but in many instances a continuing reality. He wrote:

> The country living is essentially an outlook to nature, and the farmer is a naturalist. In proportion as he is a good naturalist he is a good farmer. The farmer, woodsman, hunter, explorer, knows as much about the things in the out-of-doors than you can find in any book. . . . The best naturalists do not write. (Bailey 1911, 54)

This reawakening to nature on the farm, in fact, was to be part—Bailey hoped—of a larger national reawakening to the value of nature. In this regard Bailey belonged, as I have suggested, to the urban agrarians. Compared with farming life (with all its problems and deficiencies), he found life in the city distinctly inadequate. He was fond of pointing out (here, of course, echoing Bryan) that the city was economically "parasitic" on the country.[11] But, more than that, he saw city life causing a steady loss of those vital qualities of naturalness, spontaneity, and a vivid appreciation of what he referred to as "the commonplace." Urban man quests anxiously after "the new, the strange and the eccentric" (Bailey 1911, 5); but his soul cannot find peace or satisfaction because he has lost connection to the purer vitality of nature. Thus, Bailey speaks here of a longing, felt by every sensitive soul, for "something that is elemental in the midst of the voluminous and intricate, something free and natural that shall lie close to the heart and really satisfy his best desires" (1911, 54). Among the evils of the city there was its hedonism, its materialism, but, above all else, its phenomenal waste of effort and energy. "I marvel," he wrote of city life, "at the enormous waste of human effort, and at the insincerity and indirection; and I wonder what might be the state of civilization were half of this energy and shrewd ingenuity applied to effort that would make for usefulness" (quoted in Bowers 1974, 59).

Bailey's political views were built upon this perception of the differences between urban and rural life. In the city, there was the fundamental conflict between the "laboring class" and the "corporate and monopolized interests." Between these poles, each tending "to go to extremes," there was agriculture, the "natural balance-force or middle-wheel of society": conservative where the others were each in their own way radical; law-abiding where the others were each in their own way lawless; individualistic where the others were each tending toward group conformity (Bailey 1913, 16ff.). In keeping with this role, farmers, in Bailey's scheme, were not to become simply another "pressure group" agitating for higher prices, lower interest rates, and the like. Instead, they were to work mostly on an individual level and for ends which were primarily noneconomic.[12]

This conservative function, however, was by no means guaranteed simply by living and working on the land. Like Thoreau, he realized that greed, laziness, and a lack of regard for nature could be rural as well as urban evils. The most typical and probably most dangerous expression of these, in Bailey's view, was the failure to use the land conservatively. If Bailey was normally concerned about promoting traditional agrarian individualism, even this value was checked for him by a regard for the land and specifically for soil conservation. This regard for conservation even conditioned what would be seen today as very much a contrary impulse: the desire to "conquer" nature through such projects as the Panama Canal or the kind of western irrigation projects that were just then beginning. For Bailey, as much as for Roosevelt himself, such projects represented the ultimate test of a nation's military spirit, the test of whether it had maintained its "fighting edge." But the conquest of nature, for Bailey, never meant (as seemingly it has often come to mean in our own day) the kind of technological overreach which ignores nature's laws and limits. On the contrary, the foundation of the irrigation communities he foresaw was a workable conservation policy. He wrote, prophetically:

> Society has a right to ask that we be careful of our irrigated valleys. They are abounding in riches. It is easy to harvest this wealth by the simple magic of water. We will be tempted to waste these riches, and the time will come quickly when we will be conscious of their decline. (1913, 49–50)

Speaking more generally of questions of conservation under the heading "No man has the right to plunder the soil," Bailey observed that the farmer "owes a real obligation to his fellowmen for the use that he makes of his land" (188).

Our discussion to this point has emphasized the agrarian, the Jeffersonian side of Bailey's philosophy, but we must not forget that Bailey authored a report which was, in effect, highly critical of traditional agriculture and keen on reforming its perceived evils. The ignorance of proper soil conservation, of course, would have been one of these evils—but there were important others as well. Bailey believed in agrarian independence and individualism, but he was not insensitive to problems of rural isolation (see "Community Life" chapter in Bailey 1913). He believed in simplicity and hard work, but he was quite aware of the need to lessen the drudgery of housework, to make rural life more appealing to the farmer's wife as well as the farmer; as an educator, he could only be highly critical of the traditional rural school; and, above all, as a scientist, he was convinced that further improvement in farm life required the spread of technical knowledge. This was, of course, the idea behind agricultural extension work—and one of the primary thrusts of the Country Life Report.

In Bailey's thought, and to a great extent in the Country Life Movement generally, we find expressed some of the fundamental contradictions facing the analyst of agrarian problems: how to reconcile rural individualism and independence with the need for concerted actions and reform; how to reconcile science with tradition and the feeling for nature; how to reconcile the excitement of large-scale development with a proper conservative regard for traditional ways; and, perhaps most perplexingly, how to reconcile the simplicity and naturalness of the traditional rural life with the need to make its tasks lighter and more acceptable to the average person. The way Bailey saw around apparent conflicts and contradictions was *education*, specifically rural education. In answer to the "fundamental question" of how to make country life all that it was capable of being, he submitted that "the fundamental need is to place effectively educated men and women into the open country. All else depends on this" (1913, 61). Properly educated, rural men and women could appreciate what is elemental and valuable in rural life, yet use the best that science and technology had to offer.

So we can find, then, strands of Thoreau's naturalism and primitivism alongside Emerson's more robust enthusiasm for knowledge and the moral will of humanity in Bailey's philosophy. Uniting them, as we have seen, is a belief, based on his faith as an educator, in the consistency of rigorous science and a more emotionally based "feeling" for nature. If Saint Thomas could speak of divine grace as "perfecting nature," Bailey might have said that the science of nature perfects natural feeling. And with this perfection, this harmony of reason and emotion, Bailey was convinced there could be an agriculture which both respects nature and her limits—and which is highly productive and able to use the advances of science.

Wendell Berry: Contemporary Agrarian

With the publication of his singular work, *The Unsettling of America: Culture and Agriculture* (1977), Wendell Berry has emerged among the leading agrarian thinkers in contemporary America. Although his main themes are by no means unique—he is alternately reminiscent of Jefferson and Thoreau—the particular forcefulness with which he develops these is very much his own. Absolutely central to Berry's thought is the contrast he makes in chapter 1 of this work between the "exploitative" and the "nurturing" mentalities. The exploitative mentality was the one Europeans brought with them to the New World: "The first and greatest American revolution, which has never been superseded," Berry remarks, "was the coming of a people who did not look upon the land as a home land" (4). Exposure to this new mentality destroyed those Indian cultures to which it was alien—but more than that, such exposure has threatened and continues to threaten the existence of subcultural formations, such as many farm communities, within the dominant society. More generally, of the difference between the exploitative and the nurturing mentalities, Berry writes:

> I conceive the strip-miner to be a model exploiter and as a model nurturer I take the old-fashioned idea of a farmer. The exploiter is a specialist, an expert; the nurturer is not. The standard of the exploiter is efficiency; the standard of the nurturer is care. The exploiter's goal is money, profit; the nurturer's goal is health—his land's health, his own, his family's, his community's, his country's. (7)

The exploiter, Berry goes on to remark (citing Confucius), is ultimately victimized by his own internal disorder, which he seeks to remedy by imposing an external order, only to be locked into unavoidable contradiction:

> There is nothing more absurd . . . than the millions who want to live in luxury and idleness and yet be slender and good looking. We have millions, too, whose livelihoods, amusements and comforts are all destructive, who nevertheless wish to live in a healthy environment. (12)

Bound up with this fragmentation and disorder (and "the disease of the modern character") is the baneful role of the "specialists": "people who are elaborately and expensively trained *to do one thing*" (19). With specialization comes a kind of personal irresponsibility for the consequences, especially the environmental consequences of what one is doing. For one's responsibility under this regime of specialization extends to doing some one thing—for whose consequences a multitude of others, other specialists, will be responsible. Where once the average person had to do many things well, now we have "professionals" trained in universities which are themselves collections of specialists (compare 44). Under specialization

> The community disintegrates because it loses the necessary forms, understandings and enactments of the relations among materials and processes, principles and actions, ideals and realities, past and present, present and future, men and women, body and spirit, city and country, civilization and wilderness, growth and decay, life and death—just as the individual character loses

> the sense of responsible involvement in these relations. No longer does human life rise from the earth like a pyramid, broadly and considerately founded upon its sources. Now it scatters itself out in a reckless horizontal sprawl, like a disorderly city whose suburbs and pavements destroy the fields. (21)

Like Cobbett and most agriculturalists, however, Berry's environmental philosophy is one of correct and "kindly use" (compare chap. 3), not mere contemplation of nature's beauties.[13] Such use will be a mixture of old and newer technologies; it is open to the use of horse-drawn plows, for instance, but does not contemplate anything like a total reversion to these (compare 200–201). More importantly, it will be a usage based on *skill* developed over years of experience (65). Or, as Berry says of the "good farmer," he is

> a cultural product; he is made by a sort of training, certainly in that his time imposes or demands, but he is also made by generations of experience. This experience can only be accumulated, tested, preserved, handed down in settled households, friendships and communities that are deliberately and carefully native to their own ground, in which the past has prepared the present and the present safeguards the future. (45)

Such cultural products Berry sees well exemplified in farms which rely heavily on older methods, shunning (like the Amish) or using only very selectively chemical herbicides and tractors. But the impetus for a significant move to these types of farm, Berry admits, will be hard to come by. Of the present "orthodoxy," he says, "it would rather change than die, and may change only by dying" (173).

Change in this situation will only come, suggests Berry, from the margins. He explains what he means with an analogy of those who provide an alternative to "encrusted" orthodox religions.

> It is changed by one who goes alone into the wilderness, where he fasts and prays, and returns with a cleansed vision. In going alone, he goes independent of institutions, forswearing orthodoxy. . . . In going to the wilderness he goes to the margin, where he is surrounded by possibilities—by no means all good—that orthodoxy has excluded. (174)

Certainly in rejecting as he did the academic establishment and adopting a farming life, Berry has done something rather like the religious prophets of which he speaks. Like Winstanley and the Diggers, he has attempted to start the social process with a personal commitment.

As Berry sees it, there are grounds for hope, despite the power of the current establishment. For the agricultural establishment, in the very sweep of its national domination, has widened the range of possibly effective, marginal alternatives. These alternatives, he believes, may supply diversity in place of monoculture, a respect for carrying capacity in place of environmental irresponsibility, an adequate labor force in place of depopulation, a self-sustaining economy in place of economic specialization, and, by way of summing up, health rather than profit as a goal (compare 182–183).

In the last chapter of *The Unsettling of America* there is a memorable description of an encounter between the author and an old farming couple living near his home in Henry County, Kentucky. We read such passages as these, which call to mind Wordsworth's *Michael:*

> House, yard and barn always showed a resident pride. . . . The pastures were mowed every summer. . . . More than anything else those little timber bridges bespoke the old man's care; the usual thing would have been to drive regardlessly across such shallow drains and so wear the banks away. (189–190)

Suddenly and sadly, though, the farm begins to change noticeably. "Disorder" begins to appear everywhere. The old man has died; his widow has moved to the city, and the farm has been rented "to people who, though technically they had become its residents, clearly did not live there" (19). To agricultural experts and specialists, Berry remarks with some bitterness, this man and his farm were simply anachronisms. Yet,

> [t]he curious thing is that many agriculture specialists and "agribusinessmen" see themselves as conservatives. They look with contempt upon governmental "indulgence" of those who have no "moral fibre" than to accept "handouts" from the public treasury—but they look with contempt upon the most traditional and appropriate means of independence. What do such conservatives want to conserve? (191)

Berry offers still one more dichotomy—between production and reproduction. Production, he suggests, "is the male principle in isolation from the female: it aims to do everything at once, to exert itself absolutely, to break every record" (217). John Crowe Ransom, it may be recalled, had offered a similar characterization of the "masculine principle." But Berry's understanding of the "feminine" differs totally from Ransom's association of it with a kind of useless "bleeding heart" liberalism. Reproduction, for Berry, would involve the male principle in harmony with the female. It is patient, "respectful of the nature of things" and aiming not at an all-time best, but on doing well over and over again: "At their best, farmers have always had this ancient purpose of reproduction. Without it, they make their art as sterile as mining" (217).

The thought here, in a way, is reminiscent of Quesnay and his struggle with the question of whether only agriculture (or also fishing, mining, and the like) is genuinely productive of wealth. Says Berry, in effect: mining may be productive, but it is not *reproductive*. What distinguishes agriculture from other activities which extract nature's products is precisely the related attitudes of nurturing and reproduction. With these a kind of mental discipline is imposed which is different from such purely and intrinsically exploitative activities as mining. Thus, as opposed to the Physiocrats, Berry argues fundamentally for the psychological—rather than the economic—superiority of agriculture over commercial and industrial activities. Agriculture is ultimately valuable for the mental order it requires and imposes on the farmer—if it is to be done well.

Berry's philosophy may strike some as an attempt to apply a basically Jeffersonian outlook to our seemingly very non-Jeffersonian world.[14] In time even Jefferson, it could be said, gave up his agrarian purism. One worry could be put in this way: if laws and police are required because of certain persistent human tendencies, which we would gladly eradicate but apparently cannot, is not our present civilization (with its television, its labor-saving devices, and so on) equally a testament to other, arguably undesirable but not very easily eradicable human traits? It is fine to criticize, as Berry does, these apparently all-too-human foibles; but can we seriously base a social philosophy on the mere hype that all of us someday will live in a less slovenly, less present-oriented, more socially conscious way? Of course we should approach life in these ways. But how in a relatively free, relatively democratic society is this going to happen very quickly? And what are we to do in the meantime? For instruction on these points, we can but turn to Berry's specific public policy recommendations (218–222).

After a very rich plate of social criticism, one must be content with some rather thin gruel, as far as specific recommendations are concerned. From the start, Berry eschews "big solutions" and crash programs involving massive governmental, scientific, or industrial intervention in agriculture, because these run the obvious risk of hastening the demise of the small family farm. Government can, however, intervene to protect "the small and the weak from the great and the powerful" through policies of taxation which achieve an equitable distribution of property. Government can also—and this proposal I find a bit surprising and disconcerting, given the sorry history of intervention in this area—institute a system of price and production controls "to adjust production both to need and to the carrying capacities of farms." Local self-sufficiency, moreover, should be promoted and the corporate orientation of land-grant colleges curbed.[15]

These remedies, however, as I think Berry would admit, will be effectual only in the context of a renewed individual commitment—ultimately to what Berry sees as the overriding value to be achieved by this or any society: the value of *health,* not merely in the hygienic sense of personal health, but the health, the wholeness, finally the holiness, of Creation, of which our personal health is only a share (222).

The Future of Agrarianism

The old divisions remain with us today. Berry's and Thoreau's minimalism versus Emersonian scientific optimism; the Country Lifers' feeling for traditional rural life versus their desire to introduce industrial efficiencies into agriculture; Bailey's ecological concerns versus his desire to put science to use for agricultural production; Bailey's desire to preserve what is genuine and valuable in rural people's emotional response to nature versus his confidence that educated outsiders could aid significantly in the reconstruction of rural life. Underlying these several dualisms, however, is a more considerable, a more pervasive one. I first noted its existence back in the original discussion of the nature of agriculture and the transition from hunter-gatherer to agricultural societies (Montmarquet 1989). The more complicated account of the politics, sociology, history, philosophy, and literature of agrarianism carried us at times away from it—but now we return to face it more squarely. This is a dualism not just of different approaches to agriculture, but of two entirely different approaches to life.

First there is the tradition of big—or better, since it implies a process—expansionist agriculture. This is the tradition of Virgil's typically

Roman embrace of "civilizing" technology, of Cato's and Cicero's also typically Roman love of "gain," and of the Physiocrats' combination of laissez-faire economics and a belief in the distinctive productivity of agriculture. It is a legacy of the classical, aristocratic tradition insofar as the latter is a tradition of overseership and the management of large-scale estates. But, interestingly, it is also the tradition of Karl Marx, Thomas More, and that strain of radical socialism which sees science and large-scale social investments in agricultural projects as the key to ensuring an abundant food supply. Finally, this is the culmination of an entire tradition of aggressive, expansive, "yeoman" agriculture. It is the natural expression of an agriculture which is free to expand, to utilize its formidable energies and capacities. If the philosophical underpinnings of this view lie in the intricacies of John Locke's theory of property, its natural expression lies in the work and activity of much less complicated men, typified by that "yeoman of Kent" caught "sitting on his penny rent."

This, however, is a tradition beset with the consequences of its own success. We can see ample evidence for this today in the many "deserted villages" which are the consequences of the expanding size and productivity of individual farms. But we also know, especially thinking of the history of enclosure, that this is no new development. Arthur Young, a leading eighteenth-century advocate of large-scale agricultural investment and increased efficiency, put it this way shortly before his death: "I had rather that all the commons of England were sunk in the sea, than that the poor should in future be treated on enclosing as they have been hitherto."[16] The social problems wrought by agricultural expansion, however, go beyond the condition of the poor. As Goldsmith's haunting term, *the deserted village,* clearly suggests, it is not just the poor who are at risk; it is virtually all who have lived, or still do live, or will live, on the land.

In the introduction to *The Idea of Agrarianism* (Montmarquet 1989), I described agriculture as a precarious balance between aggression and passivity: between an aggressive attempt to wring whatever one can get from nature and a recognition that there are natural limits to such assaults, lest agriculture cease to be itself. The agrarian tradition under present scrutiny is just this first side of agriculture carried to a level of philosophical and literary expression. And, of course, to the extent that this first side, carried to its ultimate real-life conclusion, implies the sort of "postagricultural" future I discussed in the introduction to my book, it implies an end of agrarianism as well (see 1989, 18–21). The agrarianism of the old "yeoman of Kent" gives way to his distant descendant, a biotechnologist working in a field become laboratory. Such agrarianism truly carries the seeds of its own destruction.

Second, then, there is that other side of agriculture's precarious balance, also raised to the level of philosophy and literature. This, one might say, is the tradition of *limited* agriculture. In our own century, in such authors as Wendell Berry, this tradition has manifested itself as a concern with the "sustainable" use of farming resources, with the environmental consequences of intensive use of chemicals, but also with those qualities of personal character which reside in farming well—as a concern with the craft as opposed to the applied science of agriculture.

Politically, this is the tradition that emphasizes, with Jefferson, the larger democratic, the larger social purposes to be served by agriculture—if it remains sufficiently small scale. Philosophically, this agrarianism draws on a different aspect of the classical tradition. It is not especially concerned with the virtues of a well-managed estate, so much as with the ethical limits of wealth thus gained. Thus it draws upon the classical, Aristotelian view that unlimited acquisitiveness is wrong and that the proper bounds of wealth are set by the needs of the household and what is genuinely good, genuinely healthy, for human beings. This tradition thereby stands opposed to that tendency, qualified in Locke but essentially unqualified in Adam Smith and Quesnay, to favor a right of property ownership, based not on the limited needs of household but on the potentially unlimited desires of the individual.

We can go on. In terms of the dualism treated in the third chapter of *The Idea of Agrarianism*—between a Benedictine regard for the inherent dignity of simple farm labor and a Promethean rebellion against the boredom and inherent limits of any such life—this view will opt for dignity and simplicity—even at the risk of boredom. In this tradition, farmers may be many things, but they are not—not *primarily*—managers, soils scientists, engineers, or economists. The farmer may be all of these things at one time or another—but it is his role as physical laborer, in physical contact with the soil, that all such other activities are brought into a higher and more valuable unity. Here Steinbeck's almost mystical faith in the powers of this bond brings out, surely, what is implicit in this tradition.

In its rejection of violence and of violent, sweeping land redistributive schemes and its acceptance of the virtues of a life bereft of many or most material comforts, this second agrarian tradition may be seen as conservative, even reactionary; but, as we noted in connection with William Cobbett, this is a kind of reaction which, with the encroachments of agrarian capitalism, becomes equally a kind of radicalism. Thus, it is steadfast in its defense of the traditional rights of land occupants (be they in a technical, legal sense "owners" or not); and this implies a conservatism with respect to any such idea as "land reform"—but becomes a radicalism when traditional rights are themselves swept under in a rising commercial tide. (Happily, in the United States, where rich stocks of unowned lands existed, the vacillation between these extremes gave way to a more stable Jeffersonianism.)

The problem with this tradition, unlike its rival, is not whether it will remain genuinely agrarian so much as whether it will remain at all. I have observed (1989) the increasingly romantic absorption in the past of much agrarian thought during the last two centuries—in effect, since the Industrial Revolution. And I have wondered, also in effect, just what sort of a future this or any other type of agrarian could possibly enjoy in a civilization which seems bent on wiping out the remaining bases of its social and economic life. But the problem is not just with that amorphous enemy "civilization." With the Country Lifers, let us be honest enough to acknowledge that there have been deficiencies endemic to the rural life. And let us also acknowledge that loose talk of the virtues of a "simple life" and the "dignity of farm labor" has often served to obscure these very problems. For the ordinary person, such dignity has usually meant drudgery, and such simplicity has often been just a label to describe the condition of doing without or having less than others. Clearly, there *has* been something lacking in many of these "simple lives," and, equally clearly, this has been one force in pulling many of the most talented, intelligent, and energetic young people off the farm.

So I praise Liberty Hyde Bailey for recognizing these problems and chide those writers today, like Wendell Berry, who uphold the traditional virtues of agricultural life without much acknowledging its deficiencies. In fact, I go further. I submit that if there is a single "solution" to be found to the problem of formulating a viable agrarian philosophy today, its main lines are to be found in Liberty Hyde Bailey's writings and philosophy. True, as the previous discussion attempted to bring out, there are unresolved tensions in this philosophy and in these writings. I prefer to see these, however, as the unresolved and presently unresolvable tensions of agrarian survival in today's nonagrarian world.

Here, the key term which underlies virtually all of Bailey's concerns, is *naturalism*. Naturalism, in Bailey's sense, involves the kind of delicate balance we have been identifying with agriculture itself. Naturalism does not reject science or the scientific method outright; but, in comparison with the natural scientist, the naturalist is more concerned simply to observe nature and to limit experimentation to the kinds of relatively informal investigations a farmer might actually employ in the course of making a livelihood. The naturalist will use only in a very selective and cautious way the results of more rigorous experimental techniques, knowing that the highly artificial conditions under which these results have been established are not at all the same as the specific ones with which he or she is faced. Bailey's naturalism involves a second balance, equally delicate: the naturalist is inspired by feelings, by what may be an instinctive emotional response of human beings to nature in a farm setting; but these feelings must be informed by a level of knowledge and practical experience which probably exceeds that of Thoreau's poetic farmer. As I put it earlier, knowledge does not replace so much as perfect this primitive feeling for nature.

The importance of such naturalism is not just that it makes one a better farmer; it is—as Bailey pointed out—that it makes the farming life more interesting, more worth living, and more worth choosing to live. We recall here Bailey's words: "Only as we love the country is the country life worth living"—and may wish to add as a corollary: "Only as we are interested in the natural surroundings of the countryside will life in the countryside be interesting to us." For that is the whole point in Bailey's scheme, even of the intervention of trained educators from the outside: to make rural life more interesting by heightening our emotional and intellectual responses to its possibilities. Ideally, the farmer not only learns more about nature in order to become a more successful farmer, but also farms as a way of learning more about nature.

Today we have largely realized one part of Bailey's program. The farmer today is inundated with outside information. If there is a shortage today, it is of farmers, not extension agents. And if there is room for significant improvement, it is not in the quality, and certainly not in the quantity, of information available to farmers so much as in the underlying attitudes, prejudices, and sensibilities of those receiving this information.

Let me try to get to the bottom of the matter, as I understand it. The farmer, it is often said by way of replying to the "agrarian myth," is no better than the rest of us. The farmer's values, it is said, simply reflect those of the larger culture. This is true, let us admit, and becoming truer all the time. But if anything like Bailey's program is to be realized in the modern world, I would urge that the farmer must be in many respects better than the rest of us. Better how? Essentially in those qualities embraced by Wendell Berry's notion of a nurturer. We are, frankly, a society of exploiters. If the farmer is to be a nurturer, then he or she will have to be, in many respects, a better man or woman than most of us.

How is this to be accomplished? How are we to have "better people" doing farming? Here, as I have been suggesting, I do not think that Bailey's program fits the current situation. Rather than educating the sort of people who happen today to be farming, fundamentally what we need to do is to attract a different and, in the ways I have suggested, a better sort of individual to farming. We need to attract educated young people from both rural and urban backgrounds, young people with a capacity to learn farming but not necessarily with a farm background, young people whose idealism and whose naturalistic interests could bring about a genuine renewal of rural life and culture in the United States.

Where are these people and how can they be encouraged to go into agriculture? This could be a subject for a book in its own right, but here I shall be brief. The problem is not one of finding people meeting this description; there are countless highly educated young people with these qualities, many of whom would like under the right conditions to go into farming. They abound in the graduate programs of the various agricultural colleges, among younger members of the Extension Services—and they certainly may be found, here and there, in less obviously promising locales. The problem is mainly one of economic opportunity. At present, there is a vicious dual problem facing anyone who would enter farming: a combination of still very high land costs and low expected earnings. In theory, of course, costs of economic assets like land should drop proportionally as expected income goes down. But farmland, owing in part to the very same noneconomic values which underlie the agrarian tradition we are talking about, often does not function in this regard entirely like other economic assets. When the value of silver drops sharply, silver mining land may become almost totally valueless. But not so with respect to farm commodities and farmland.

How are such economic opportunities, then, to be afforded qualified people? There is, of course, the obvious answer: some form of government subsidy. I am convinced, though, that the carrot of government assistance must be coupled with the stick of much stricter curbs on existing ecologically unsound agricultural practices. This would hasten the departure from agriculture of those who, for whatever reason, are unable to farm in a socially responsible fashion, thereby increasing the opportunities for those who can and will farm in this way. Such curbs, properly enforced, will help to ensure that today's young idealists brought into agriculture with public assistance do not become tomorrow's ecological reprobates.

I could go on at some length as to how the sort of individual I have alluded to might be encouraged to enter agriculture. But it is much more to the point of the present discussion to make explicit how any such accomplishment might relate to the future of agrarianism. Consider that farmer-naturalist, Crèvecoeur. Crèvecoeur is, first of all, a very good model of the sort of individual who today needs to be brought into agriculture. Here was a man whose economic interest in such activities as beekeeping was hardly distinguishable from his naturalist's interest in bees—and in every part of the natural environment of the farm. It was the general absence of such individuals as this which led Richard Hofstadter to decry America's lack of any genuine "rural culture." Of course, a pessimist, or a staunchly anti-agrarian thinker like Hofstadter, might contend that Crèvecoeur was just America's Hesiod: a man who wrote before "the steamroller of the city" overwhelmed an essentially rural culture. I see no reason, though, to go that far. Even today, with all the cultural deficiencies of rural life, there continues to exist not just the rural culture celebrated by Andrew Lytle—the "low culture" of traditional farming beliefs, rituals, songs, and dances—but a more refined, knowledgeable, and self-critical cultural tradition to which Hesiod, Crèvecoeur, Berry, and, arguably, Lytle himself belong. The hope of agrarianism is in part the hope that this culture may continue to exist—and grow. I suggest nothing more optimistic than the prescription that there are people who, if they could be encouraged to come into agriculture, might provide a basis for continuing and expanding such a culture.

Now, such a rural culture has two aspects which need to be distinguished. On the one hand, a rural culture means a way of farming and a knowledge of ways of farming which is passed along from farmer to farmer, from farming parent to farming child. This, obviously, is a mostly oral and not, per se, a literary tradition—even though it may often give rise to technical manuals on agriculture. On the other hand, a rural culture means a specifically literary or philosophical expression of the values inherent in rural culture (in the previous sense). This, of course, would be a typically literary culture. And although it will be based on the broader experiences of a wide number of people sharing a common tradition, it will be in a direct sense the product of a distinct minority within a minority. Today the writings of a Wendell Berry or a Wes Jackson issue, as they would probably admit, from a very narrow cultural base. They issue, to use Berry's term, from the *margins.* All society can do, hopefully, if it shares the kinds of motivations and adopts the kinds of measures I have been discussing, is to expand slightly the base of this highly literate minority.

"Man," I noted earlier, "is a hunter and forager who for the most part hunts and forages no longer" (1989, 2). Over seven thousand years this hunter-forager acquired the status of a grower and over the next thousand, with increasing rapidity, lost this status. The literature of agrarianism gives slight record of the first change and is a great record of the second. If agrarianism is to survive, as I believe it can, it will serve as a different sort of a record: not of mass changes in the lives of the many, but of the persistent counterpressure of those few who choose to go in a direction of their own.

Land, Labor, and God in American Colonial Thought

CHARLES TALIAFERRO

THIS ESSAY provides a backdrop to the overall project of the collection *The Agrarian Roots of Pragmatism* by highlighting some main currents in early European American thought prior to the emergence of pragmatism as a distinct philosophical movement. More specifically, my aim is to sketch the pre-Revolutionary, colonial approach to land and labor in order to provide a context and a contrast with the agriculture and philosophy of the nineteenth and twentieth centuries that enabled pragmatism to take root in the United States of America. How does agrarian pragmatism (or pragmatist agrarianism) look over and against the colonial heritage? To frame the question more specifically, let us consider one of the hallmarks of pragmatism.

When pragmatism emerged as a philosophical movement, it was hailed for overcoming a series of dualities. These dualities have been enumerated across a wide range: the duality between God and world, mind and body, fact and value, evidence and truth, self and nature, individual and community, eternity and time (see, for example, Langsdorf and Smith 1995, 5). These pairs have been thought of as philosophical disasters, plaguing our metaphysics, ethics, and epistemology. Take, as an example, the commonplace complaints against a duality between the mind and body. If the two are distinct, in which the body is physical and the mind is nonphysical, how can the two interact? At least initially it appears that to suppose a nonphysical thing can have an effect on some physical thing is either nonscientific or a conceptual absurdity, like supposing that there could be a square circle. And assuming that our character or moral identity is somehow lodged principally in our choices, emotion and intelligence (and thus our minds), are we not then led to an intolerably negative approach to the body as a mere tool or habitation for the mind? Moreover, on the assumption of a mind-body split, how are we to form a coherent, comprehensive account of how we know about other people's minds? I might see the bodies of other people, but do I thereby see their minds? On the face of it, then, a dualism of mind and body puts us in a hostile philosophical predicament in which our claims to know about others is undermined. John Dewey railed against such a dualism and took as one of his chief aims the overcoming of all such dualist polarity. He contended that these dualities were intertwined and recalcitrant. Some efforts to redress these dualities did not go far enough: "The older dualism between sensation and idea is repeated in the current dualism of peripheral and central structures and functions; the older dualism of body and soul finds a distinct echo in the current dualism of stimulus and response (Dewey 1960, 233). Dewey held that any philosophy that admits irreducible dualities is doomed. By his lights, the creation of these dualities is a mark of modernity and overcoming them a distinguishing mark of American pragmatism. "The cleavage that has resulted between theology and positive science, between the mundane and the heavenly, between temporal interests and those called eternal has created the special divisions which in the form of 'dualisms' have determined the chief problems of philosophers that are 'modern' in the historical sense" (Dewey 1968, 6).

Were there such invidious dualities in colonial time? And if so, how did they fare? Was there a destructive dualism between land and labor, self and nature, or a dualism at the center of human nature between mind and body that afflicted early American approaches to agriculture or that impeded integrating agriculture into American life?

Colonial Theology on Land and Labor

The earliest British colonial settlements were largely preoccupied by the practical matter of bare survival, a task that tested these communities of laborers, clerics, soldiers, merchants, seamen, and adventurers. But this effort did not so much eclipse theological and political reflection as much as it underscored its importance. There was an urgent need for the colonialists to have a reliable theological and political framework to uphold the settlements through the hardships of illness, epidemic, crop failure, collaboration, and conflict with Native Americans. Because European agricultural practices did not transplant well to the New World, colonialists had to change radically, and this compelled them to develop a theology of human nature, language, and culture that allowed them to adopt new ways of living. The colonialists needed to incorporate Indian (and thus, at the time, non-Christian) practices of raising crops and to rely on Indian stores of food in crisis. Without some overall guiding framework, bare economic exchange and agriculture would not have been intelligible or feasible. It was because of the colonialists' need to understand themselves in this new context that theology and politics—along with agriculture, hunting, fishing, trading, woodwork, building, and military activity—were very much a part of the practical activities of the colonies.

The Puritan migration to the New World was largely articulated at the time as part of the drive to live with integrity in relation to God and in communion with one another. Many of the contemporary Puritan sermons, diaries, and tracts in the seventeenth and eighteenth centuries underscore the necessity of integrating right living and right habitation. To fail to live righteously in a land is to fail to achieve the entitlement to govern in a land. The presumed hypocrisy and the abuse of power in the Old World displayed to many Puritans the illegitimacy of the prevailing church and state. In Great Britain, Puritans and other nonconformists and dissenters opposed the episcopacy in the 1630s and then, in the 1640s, opposed both the Church of England in the person of Bishop Laud and the Crown in the person of Charles I. The condensed, rapid sequence of events that unfolded dramatically confirmed to the Puritans the political illegitimacy of remaining in Britain: The Church of England was abolished in 1644; Charles I was executed in 1649, the same year that the English parliament resolved to support the Puritan settlements in New England; and Cromwell was made lord protector in 1653; but within only seven years Cromwell is dead, the monarchy is restored, and the Church of England reestablished. Not all colonialists supported Cromwell, and there was serious colonial support for Charles I and for the restoration. (Bushman 1992 offers an insightful overview of the period.) Still, according to a prevailing Puritan theology of state and history, legitimate rule required justice and righteousness. The link between right living, right habitation, and government is especially clear among Puritans who opposed church and Crown. One can see this in Thomas Hooker's (1586–1647) *The Danger of Desertion,* and in John Davenport's (1597–1670) *The Saint's Anchor-Hold.* Davenport upheld the right to hide those who fled Britain at the Restoration in order to escape prosecution. A morally corrupt sovereign power does not have the entitlement to prosecute those who oppose it.

A large part of the earliest American Puritan teaching was in the great covenental tradition that was forged in the European Reformation on the basis of biblical narratives and teaching. According to this tradition, human beings are related to God in a twofold fashion, first by creation and then by regeneration. The appeal to creation is foundational and deserves the lengthier treatment here, for the philosophy of creation had a vital role in colonial reflection on land and labor.

According to the Christian theism current in the seventeenth- and eighteenth-century British colonies, God is believed to be the supreme Creator of the cosmos. Unlike this contingent and temporal world, God exists necessarily and eternally. God is omnipotent, omniscient, omnipresent, and all good. The dependency of the cosmos is radical. The world cannot somehow outgrow its dependence upon God any more than there could be a number so great that there could not be a greater one. By its very nature, the existence of the cosmos at any given instant is partly constituted by God's creativity. Our very being, then, is derived from God, and we are endowed with the powers we possess due to the generous, creative, and sustained will of an all-good Creator. Some American theologians disagreed about whether God sustains the world by a continuous creative act or whether God continually re-creates the world at each point in its existence, a view that implies that the cosmos is successively created from nothing at each instant. But on either account, the cosmos is believed to be profoundly dependent upon God such that if God's creative power were withdrawn, the cosmos would cease to be.

The dependence of the cosmos upon the creativity of a sovereign, provident God had a dramatic ethical and political implication. Such an account of creation recognizes in God a supreme power and was assumed to imply that the whole cosmos belongs to God. We are God's. Indeed, the whole cosmos may be thought of as God's property. On the grounds of this theistic creation and possession, the cosmos is delineated as something that is owned. The cosmos is not something radically evil or alien to omnibenevolence, wisdom, and virtue; it is, rather, the creation of the being in whom omnibenevolence, wisdom, and perfection are unsurpassed. Property rights, then, are not human inventions; they begin with God (Tall 1992; Avila 1983; Grace 1953). On this front, New World theology was very much in keeping with the greater Christian tradition as well as Judaism and Islam (See the Hebrew Bible/Christian Old

Testament: Pss. 24:1, 50:12, Ezek. 18:4, 1 Chron. 29:11–18; Qur'an, "The Light").

In *God's Promise to His Plantations,* John Cotton (1585–1652) sets forth a strong version of the thesis that all creation belongs to God. It is because of God's supreme authority as the one to whom all things owe their being that God may give land to those whom God chooses.

> This placing of people in this or that country is from God's sovereignty over all the earth, and the inhabitants thereof, as in Psalms 24:1, *The Earth is the Lord's, and the fullness thereof.* Therefore it is meet he should provide a place for all nations to inhabit, and have all the earth replenished. (in Heimert and Delbanco 1985, 77)

This appeal to scriptural authority very much fits the reformed reading of early biblical narratives frequently cited in covenental, colonial sermons, the most prominent being the divine promise and guidance to Abraham (Gen. 15 and 17), Noah (Gen. 8–17), Moses and the children of Israel in their exodus from Egypt and their journey to the land of promise (first six books of the Bible), and to the disciples of Christ in the New Covenant (for example, Luke 22:20).

The authority of divine power and sovereign entitlement has, of course, been used for a wide range of ends, including support for monarchical government, the feudal system tout court, slavery, certain otherwise cruel treatments of animals, and the dispelling and destruction of indigenous people. But at the very outset, the appeal to divine ownership was used in the colonies first and foremost to establish a commons, an arena that was to be regarded as belonging to a group of people. Due to God's bounteous gift to people as part of creation, God was also believed to give to individuals and groups (for example, parishes) land in virtue of their promise to cultivate the land and flourish.

In accord with mainstream colonial theology, persons are believed to be made in the image of God. As such, persons have powers that are God-given, which, if used properly, may be employed to contribute to their own and others' welfare to the glory of God. In the exercise of one's powers whereby one cultivates and improves something in the commons, one thereby comes to take possession of it in a way that protects it from the use of others. Human exercise of power is on a very different footing than God's exercise of power in creation, but rightful human endeavor may still be seen as a remote reflection of God's bounteous creativity. This understanding of the rise of the civil right to property was often articulated with biblical examples and held in check by appeal to biblical teaching. Thus, the right to own property was not a gift from God in order for the owner to treat it capriciously. Ownership and cultivation must not be unjust.

Consider, for example, John Winthrop's (1588–1649) *Reasons To Be Considered for Justifying the Intended Plantation in New England and for Encouraging Such Whose Hearts God Shall Move to Join with Them in It.*

> The whole earth is the Lord's garden and he hath given it to the sons of men, with a general condition, Genesis 1:28: *Increase and multiply, replenish the earth and subdue it,* which was again renewed to Noah. The end is double moral and natural, that man might enjoy the fruits of the earth and God might have his due glory from the creature. . . . That which lies common and hath never been replenished or subdued is free to any that will possess and improve it, for God hath given to the sons of men a double right to the earth: there is a natural and a civil right; the first right was natural when men held the earth in common, every man sowing and feeding where he pleased, and then as men and the cattle increased, they appropriated certain parcels of ground by enclosing and peculiar manurance and this in time gave them a civil right. Such was the right which Ephron the Hittite had in the field of Machpelah, wherein Abraham could not bury a dead corpse without leave, though for the out parts of the country which lay common, he dwelt upon them and took the fruit of them at his pleasure. (In Heimert and Delbanco 1985, 72, 73)

The natural right to benefit from the earth, then, is part of the general bequest of God's bounty, and the civil right to own demarcated parts of the earth stems from the divine invitation to flourish. On these salient points Winthrop was striking a theme that would receive further articulation by John Locke (1632–1704) in his *Second Treatise on Civil Government.*

Contemporary scholars diverge in their estimation of John Locke's influence on early American thought, with some giving prominence to Lockean economic individualism (Appleby 1984; Boorstin 1964; Hartz 1955), while others resist seeing Locke as the dominant voice in early America (Dunn 1969; Pocock 1975). Wherever one places Locke in

American history, there is a sense in which Locke was simply one voice in recognizing the ownership of all things by God and God's consequent underwriting of property claims to individuals and groups in virtue of their exercise of God-given powers. Like Winthrop, Locke restricted the use of civil property on grounds of the common good (there had to be enough land for the private use of others; there could not be an ownership right that leaves others destitute) and rightful cultivation (Taliaferro 1992). Locke also sought to refine the application of the appeal to divine ownership; God did not give supreme, unchecked authority to the monarchy over subjects, parents over children (contra Sir Robert Filmer), or a person over their own bodies (thus, Locke objected to suicide on the grounds that this involved destroying something owed to God).

This conception of natural and civil ownership was not limited to Christian theists, but upheld by deists such as Thomas Paine (1737–1809) as well.

> Though every man, as an inhabitant of the earth, is a joint proprietor of it in its natural state, it does not follow that he is a joint proprietor of cultivated earth. The additional value made by cultivation, after the system was admitted, became the property of those who did it, or who inherited it from them, or who purchased it. . . . Cultivation is, at least, one of the greatest natural improvements ever made by human invention. It has given to created earth a ten-fold value. (Paine 1995, 399, 400)

So, the second way of acquiring goods through labor was subsequent to the more fundamental entitlement that all people have to the earth by virtue of their creation.

The early theological treatments of divine ownership make much of two features of the philosophy of God that deserve underlining: monotheism and divine omnipresence. There are no other gods, and thus there is no other recourse to God's overarching, primary claim to all creation. The ocean, forests, the settlers as well as Native Americans, were all God's. Moreover, as an omnipresent reality, there is no place in creation where one might achieve any absolute isolation from God. The harshness of frontier, pioneer life, with its blend of hunting, farming, and self-defense, was thus understood within a broader context of a greater frontier with God. There was no supremely private sector *(res privato)* in creation where one may flee from God's presence. To God, everything is public *(res publica)*.

While the appeal to creation established and provided a theoretical foundation for respect and community, the second aspect of colonial, covenant theology set up a community that was called to live in fellowship with Christ. According to much early American theology, Puritan as well as Anglican and Roman Catholic, human beings have fallen from grace and are in need of a relation with Christ, corporate and individual, to effect atonement or atonement with God. Puritans in Geneva and New England may have had a more radical view of the effects of the fall than, say, non-Puritan Christians in Rome or Canterbury, but all alike understood the need for divine grace. And the Puritans made this need a crucial point in forging their covenental theology of regeneration and the rites and expectations of such renewal.

Perhaps one of the most compelling statements of this theology of creation and regeneration can be found in John Winthrop's famous sermon in 1630, *A Model of Christian Charity,* delivered aboard the ship *Arbella* prior to its landing in Massachusetts Bay. God owns all things and calls on all of us to restrain greed and to remain . . .

> knit more nearly together in the bonds of brotherly affection. . . . There are two rules whereby we are to walk one towards another; justice and mercy. . . . We must delight in each other, make others' conditions our own, rejoice together, mourn together, labor and suffer together, always having before our eyes our commission and community in the work, our community as members of the same body. So shall we keep the unity of the spirit in the bond of peace. (In Heimert and Delbanco 1985, 83, 91)

In light of this covenental promise, our individual lives are to be animated and partly defined by the welfare of others. This interwoven identity had affective as well as practical dimensions. According to Winthrop, economic welfare needs to be understood in terms of a greater divine economy on which we all depend, an economy which is more akin to something organic and alive than a machine.

Winthrop's further reflection on the nature of human social order gave primacy to individual persons and then to labor. This secured a point of reference that is profoundly humanistic.

> First for the persons. We are a company professing ourselves fellow members of Christ, in which respect only though we were absent from each other many miles, and had other imployments as far distant, yet we ought to account ourselves knit together by this bond of love, and live in the exercise of it, if we would have comfort of our being in Christ. . . . Secondly for the work we have in hand. It is by a mutual consent, through a special overvaluing providence and a more than an ordinary approbation of the churches of Christ, to seek out a place of cohabitation and consortship under a due form of government both civil and ecclesiastical. In such cases as this, the care of the public must oversway all private respects, by which not only conscience but mere civil policy cloth bind us. (89)

The ends of our work are to function in service of people, not vice versa.

Covenental theology, with its privileging of the public good and regeneration, was further advanced by John Eliot (1604–1690), Richard Baxter (1615–1691), Cotton Mather (1663–1728), and Peter Bulkeley (1717–1800), among many others. It received its poetic idealization in Timothy Dwight's (1752–1817) *Greenfield Hill.* Evidence that this covenental appeal was more than an idle ploy may be seen in the tracts and sermons in which one can observe the struggle to define and rein in settlements and acts that were seen to be unfairly covetous. Parishes reluctantly gave way to expanding settlements to "outlivers" and "outdwellers" (frequently farmers), but this was often done alongside of admonishing others to beware of vice and to recall the greater community of regeneration. Such colossal efforts at direction and education would not have made sense unless there was some successful covenental life.

There is abundant documentation that many of the early colonialists believed they were directed by God to come to the New World by special providence (Miller 1939).[1] There is, of course, the famous depiction of the Americas by Peter Bulkeley in *The Gospel Covenant* (1639–1640):

> We are as a city set upon an hill, in the open view of all the earth; the eyes of the world are upon us because we profess ourselves to be a people in covenant with God, and therefore not only the Lord our God, with whom we have made covenant, but heaven and earth, angels and men, that are witnesses of our profession, will cry shame upon us, if we walk contrary to the covenant

> which we have professed and promised to walk in. If we open the mouths of men against our profession by reason of the scandal of our lives, we (of all men) shall have the greater sin. (In Heimert and Delbanco, 120)

And in addition to such high prose, there were many books of different caliber chronicling God's special acts of providence in guiding the colonialists to establish a society dedicated to God on these shores. Increase Mather's (1639–1723) *An Essay for the Recording of Illustrious Providences* is a classic storehouse of references. The guidance of God is also heralded in Robert Cushman's (1579–1625) *Reasons and Considerations Touching the Lawfulness of Removing out of England into the Parts of America*, in G. Mourt's *Mourt's Relation* (1622; "Mourt" is a pseudonym, and the identity of the author is not known), and in William Bradford's (1590–1657) *Of Plymouth Plantation*. Worries that the God-oriented life intended by God was imperiled may be traced in Peter Bulkeley's *The Gospel Covenant* (1639–1640) and Edward Johnson's (1598–1672) *Wonder-Working Providence of Zion's Savior in New England*. All these serve to document the wide-ranging conviction that the colonies were to be viewed *sub specie aeternitatis*.

Bulkeley used the image of a city on a hill, not a farm on the plain, but what he, Winthrop, and others pointed to was an integrated form of life that was satisfying to individuals, the greater polity, and to God. Thus, the farmer as well as the carpenter and cleric all had their role in what Perry Miller has effectively described as a Great Chain of Being in the early colonies (recalling the medieval notion and the related principle of plenitude which applauds the fecundity of goods; *bonum est multiplex*). Human beings find themselves in a good creation, and are called to assume their responsible role in it among the vast expanse of diverse, created goods.

> Out of the same being [God] have proceeded the stars, animals, men; but in some incarnations the being has taken forms superior to others. "The least spear of grass has the same power to make it that made heaven and angels," says the Puritan, but he does not then chant with the author of *Leaves of Grass* that the least thing in creation is equal to any other. A fly is above the cedar because it has another life which the cedar has not, "so the meanest believer is better than the most glorious hypocrite." In

> the regenerate, the author of all life ordains yet another life superior to the other forms, giving them something which the others have not. In the course of providence God often provides men with powers and gifts, bridles their violence, overcomes their pride, teaches them the truth of Scripture. But these things He does by managing secondary causes, sending men to the right teachers, instructing them through their experience or through science. (Miller 1939, 34)

Science, moral and humane teaching—all had a role in further mobilizing and directing our role in God's creation.

One may object that this view of creation does not sit easily with seventeenth- and eighteenth-century New England teachings about the depravity of human nature. And, indeed, it is not easy to see Puritan joy in the goods of creation just after reading Jonathan Edward's (1703–1758) extraordinary *Sinners in the Hands of an Angry God* or Thomas Hooker's *The Soul's Preparation for Christ.* But the prevalent depiction of human sin that was on display in the Great Awakening (perhaps the high-water mark for American teaching on depravity) was often set up in contrast with the beauty of God and creation. Certainly one would be hard pressed to find a greater exaltation of divine beauty than one articulated in Edward's mature theological writing. Sin was deemed by Edwards et al. as such a grave splinter at the heart of creation because it was an offense to the fecundity of the world, rightly enjoyed in accord with divine ordinances.[2] Despite a robust theology of depravity and early colonial hardships, one need not look far to see praise of God for the abundance of trees, wild food, and good soil. Hunting showed God's grandeur, for colonial hunting was able to generate directly from creation stores of turkeys, geese, ducks, partridges, caribou, moose, bear, deer, lobster, fish, clams, oysters. This bounty may have furthered a sense of shame at falling short of deserving God's gifts, but it also highlighted the generosity of God. God has defused goodness in creation *(Bonum deffusivum sui).* Arguably, the final temperament of the Puritan was a kind of cosmic optimism which regarded nature as fundamentally good and held out the hope that sinners repent, undergo regeneration, and live justly (Miller 1939, chap. 1).

The effort to see America as the new Zion was of course to be challenged, partly by the sheer breadth of European immigration. My sketch of early colonial thought places the emphasis on the Puritan settlements, but there were waves of migration that quickly established significant religious diversity. In addition to Roman Catholic migration, the colonies enjoyed an influx from the whole spread of Protestant faiths: Anglican, Baptist, Congregational, Lutheran, Presbyterian, Quaker. A glorious picture of America as a great city on a hill was compromised by episodes of religious persecution. The religious fervor of some of the early American preaching also generated conflict in the colonies; for example, the strife between New Light and Old Light movements, the controversy over the Halfway Covenant, and debates between predestination Calvinism and Arminianism. Most disturbing of all, of course, is the failure of the early Christian colonialists of all denominations (except some of the Quakers) in the seventeenth and eighteenth centuries to consistently respect Native American entitlements to land and labor, and, tragically, colonial intolerance and the practice of slavery. In 1790, the southern United States had a total population of 1,961,000—including 657,327 slaves (Patterson 1982). In the midst of this splintering and grave injustice, it is difficult to ferret out a constructive, positive vision of the goodness of creation and justice. Still, I believe one can see in the extant colonial books, tracts, sermons, diaries, and covenants some of the great themes and principles of justice and fairness outlined above, themes and principles that would come to be refined and used to condemn the legacy of exploitation in the colonies and the early American Republic.[3]

Minds and Bodies in the Colonies

The last section highlights the overarching theme of a God-given integration of labor and land, governed by conceptions of the common good. It may well seem that the Christian theology in the colonies would have prevented the wayward dualisms such as that between mind and body that were noted at the outset of this chapter, the dualities that have plagued modernism and have been the *bete noire* of pragmatism. In the next section I propose that theism did indeed serve to ameliorate such dualities, and in a final section, "Farming and the Mind of God," I lay out a colonial understanding of farming that integrates land and labor in a theistic framework. But just as the profession of Christian theism did not prevent certain ills, some mild and others so grave that their scars are still very much in evidence, it did not altogether smooth over the philosophical tension between the duality of mind and body. In what follows I note in very broad terms how the colonialists sometimes gave primacy to physical bodies and an inchoate materialism and sometimes took to the opposite task, by giving such prominence to the mind that full-blown philosophical idealism flourished in the major colonial colleges.

The colonialists' preoccupation with meeting the material needs of their settlements went hand in glove with an emphasis in the colonial colleges on the importance of science, both applied and theoretical. Works by Isaac Newton (1642–1727) were widely used in the colleges, and the American contribution to Newton's work, while meager (for example, Newton duly recorded Harvard's Thomas Brattle sighting the Comet of 1680), was seen as a small sign of American promise. Newton's laws were taken to be the clearest, systematic guide to the structure of the cosmos, and held in such esteem that Newton, along with John Locke, were generally singled out as two of the leading lights in early American reflection and use of science. John Locke's work on the nature of perception and matter helped further the Newtonian goal of securing the parameters of the physical science. Locke adopted and refined Galileo's division between primary and secondary qualities, delimiting the physical world in terms of size, shape, and weight, whereas the secondary world of color, smell, sound, and so on are identified as our contribution to the material world as it appears to us. Locke's philosophy of perception thereby enhanced a view of nature that privileged geometry, mathematics, and quantification. While scholasticism would have some role to play in American thought, the American colonies began with modern science without having to throw off an older, Aristotelian science steeped in scholastic categories.

Some of the first American philosophers were attracted to a materialist understanding of the world. In *The Principles of Action in Matter,* Cadwallader Colden (1688–1776) prepared the way for a more positive conception of matter than Locke or Newton allowed. Working from within a Newtonian framework, and in sympathy with Thomas Hobbes (1588–1679), Colden developed a conception of matter sufficiently powerful so that it might, *in theory,* account for much more than Newton and Locke had supposed. Colden construed matter in positive, potent terms; matter was not something that was identified principally in terms of geometry and resistance, but as a causally active source that could provide a rich explanatory framework for describing and explaining the world. While I shall note in the next section that Colden stopped short of a thoroughgoing materialism, he did go quite far in that direction. He conceded the contingency of the present cosmos, but he was less sure about whether there was ever a time when there was no physical world. "There may have been a time when the present solar system did not exist," and yet "No time can be supposed, when no system of matter did exist" (in Anderson and Fisch 1939, 121, 122).

Other materialist elements entered the American scene by way of Joseph Priestly (1733–1804). Much admired by Thomas Jefferson (1743–1826), Priestly earned his scientific and intellectual reputation by discovering oxygen, and sought refuge in America to escape prosecution for his more radical antireligious polemics. Priestly identified the mind and body and denied the existence of an afterlife. Thomas Cooper, Priestly's son-in-law, carried on the case for materialism in his *View of the Metaphysical and Physiological Arguments in Favor of Materialism*. He, too, denied the credibility of believing that there is an afterlife on the grounds that the mind and body are identical. If the mind and body are identical or inseparable, then the destruction of the one seems to lead to the destruction of the other. While this form of materialism may not *entail* the denial of an afterlife (some orthodox Christians at the time believed the afterlife is grounded in a resurrection of the body, and others conceived of an afterlife in terms of a re-creation of embodied persons), it was judged by many contemporaries to threaten belief in an afterlife and thus to challenge the moral and religious groundwork of civil society.

Benjamin Rush (1746–1813) sought out a material explanation of human character and moral decision making. In *Influence of Physical Causes upon the Moral Faculty,* Rush argued for the physical underpinnings of the mind and the importance of empirical inquiry into the root causes of our moral activity. Joseph Buchanan (1785–1829) also sought to develop a scientific, material account of human thought and action in his *Philosophy of Human Nature,* a physical, scientific account of human thought and action. Like Priestly, Cooper, and others drawn to a materialist view of human nature, he denied individual immortality. French materialism had an impact on colonial thought through the writings of Etienne Bonnot de Condillac (1715–1780), Claude-Adrien Helvetius (1715–1771), and Baron d'Holbach (1723–1789).

While matter and materialism had a role in colonial thought, the greater emphasis was on the mind and the development of either a dualist or idealist philosophy. In *Philosophical Ideas in the United States,* Harvey Townsend gives idealism the edge:

> There is one dominant note in American philosophy, i.e. idealism. The word must not be taken, however, either as an epithet of

> praise or as a narrowly technical label. It is intended only to characterize the central tendency in our philosophy to approximate the ancient doctrine that the visible is no whit more real than the invisible, in fact that the invisible kingdoms furnish the foundation for the visible. (Townsend 1934, 4)

Harvey Townsend underscored the colonial readiness to appeal to immaterial realities and principles in both practice and theory: "A kind of naive Platonism was the common possession of colonial settlers" (20). This is rightly not taken to be rigorous philosophical Platonism or idealism (!), still, the general cultural climate allowed for the articulation of philosophical idealism in some of the best of the colonial philosophers.

Samuel Johnson (1696–1772), first president of King's College in New York City (later Columbia University), was one of the premier philosophers in the colonies in his day. Along with Jonathan Edwards (1703–1758) he was also an idealist. Johnson was heavily influenced by the Anglo-Irish Bishop George Berkeley (1685–1753), whom he met during Berkeley's sojourn in Rhode Island. Johnson's two most noted philosophical works are *An Introduction to the Study of Philosophy* and *Elementa Philosophica*.

Johnson's idealism was along strict Berkeleyan lines. In his letter to Berkeley, February 5, 1730, Johnson writes:

> As to space and duration, I do not pretend to have any other notion of their exterior existence than what is necessarily implied in the notion we have of God; I do not suppose they are any thing distinct from, or exterior to, the infinite and external mind; for I conclude with you that there is nothing exterior to my mind but God and other spirits with attributes or properties belonging to them and ideas contained in them. (Anderson and Fisch 1939, 59)

Like Berkeley, Johnson appealed to God's omnipresent, omniperceptive power to explain the existence and stability of the world.

> But it may be asked, How do those things exist, which have an actual existence, but of which no created mind is conscious?— For instance, the Furniture of this room, when we are absent, and the room is shut up, and no created mind perceives it; How

> do these things exist?—I answer . . . in short, That the existence of these things is in God's supposing of them, in order to the rendering complete the series of things . . . according to his own settled order, and that harmony of things, which he has appointed. The supposition of God, which we speak of, is nothing else but God's acting, in the course and series of his exciting ideas, as if they (the things supposed.) were in actual idea. . . . That which truly is the Substance of all Bodies, is the infinitely exact, and precise, and perfectly stable Idea in God's mind, together with his stable Will that the same shall gradually be communicated to us, and to other minds, according to certain fixed and exact established Methods and Laws. (Anderson and Fisch 1939, 75, 76)

Because of this profoundly central role of the mind of God in the construction of the cosmos, Johnson characterized nature as the art of God. By his lights, nature is divine *techne,* and our own activity and art are reflections of and participation in God's power.

Johnson's idealist philosophy offered a formal framework for what Herbert Schneider has characterized as a fundamental Puritan thesis. "The *eupraxia,* or skill, exhibited in the works *(euprassomena)* of God or man was the basic category of philosophic analysis and enabled the Puritans to interpret their arts and crafts, including the most mercantile and menial, in the perspective of God's will" (Schneider 1947, 9). Johnson's philosophy may thereby be seen as a formalized extension of the Puritan emphasis on the primacy of will and activity in the very constitution of the world.

Philosophical idealism also has a rightful place in American philosophy due to the figure who is most often hailed as America's first philosopher of European rank: Jonathan Edwards, a thoroughgoing idealist. Edwards also propounded a philosophy of God that had strong Platonic elements. For Edwards, God's excellence and perfection was supreme, unsurpassable beauty. What we take to be the external material world is itself a reflection of the mind of God.

> As to Bodies . . . they are the communications of the Great Original Spirit; and doubtless, in metaphysical strictness and propriety, He *is,* as there is none else. He is likewise Infinitely Excellent, and all Excellence and Beauty is derived from him, in

> the same manner as all Being. And all other Excellence is, in strictness, only a shadow of His. (In Anderson and Fisch 1939, 74, 75)[4]

In Edwards's philosophical work one may trace a highly exalted picture of the regenerative covenant to which God calls humanity. This is especially apparent in philosophical reflection on the Christian belief in the Trinity, where we find, at the very heart of all that exists, a kind of divine community. The divine community of Father-Son-Holy Spirit calls us to a wider fellowship in this world and the next. From the standpoint of mind-body dualists, early colonial philosophers seemed ready either to supplant mind for body or vice versa.

The Theistic Chord

Christian theism was, I believe, a key factor in offering a stable, integrated picture of mind and body in early colonial thought. Theism acted as what Nicholas Wolterstorff has called "a control belief," a belief that serves as a check or boundary on the direction of one's reflection. To those drawn to idealism, Christian theism served to underscore the goodness of the material world, and to those drawn to materialism it suggested that there was more to the cosmos than that which can be captured in a strictly materialist philosophy of nature.

On the materialist front, few colonialists were prepared to argue that the full complexity, apparent design, and the bare existence of the material world was self-explaining. Even materialists like Thomas Hobbes posited a God, albeit construed along material lines. Cadwallader Colden hedged a strict materialism: "God in the beginning created a certain being, to which he gave the power of motion; and distributed this being, in certain proportions, in the several parts of the universe. The granting of this is no negative to the existence of spirits; they may, and undoubtedly both exist, without including any contradiction" (in Mayer 1951, 69). This philosophical emphasis on the material world was welcomed by Benjamin Franklin (1706–1790), Ethan Allen (1737–1789), Thomas Paine, and Thomas Jefferson, though none of them believed in the sufficiency of a material explanation of the cosmos without some contribution by a divine architect. Once one admits God, whether on deistic or theistic grounds, one opens the door to believing that human nature itself may have features that escape a materialist scheme. To many, Baron d'Holbach's picture of human beings as machines or Hobbes's dismissal of consciousness appeared to ignore the reality of thought and feeling that were commonly assumed to be part of what makes human beings created in God's image (see Henry More's trenchant criticism of Hobbes, for example).

In noting how Christian theism restrained full-scale materialism, it should also be underscored that Christian theism was not generally thought of as an obstacle to science in the colonies. Scientific inquiry in the early colleges was often construed as a natural, good exercise of religious inquiry. By studying nature one thereby studies the mind of God. So, Cotton Mather (1681–1724), in his great *The Christian Philosopher,* extols the order and design of the universe as something to be investigated to the glory of God. "The works of the Glorious GOD in the *Creation* of the *World,* are what I now propose to exhibit" (Mather 1994, 17). The text reads like a mix of early science and ecstatic celebration of stars, climate, water, earth, minerals, etc. "Even the most *noxious* and most *abject* of the Vegetables, how useful they are!" (137).

While idealism achieved celebrity with Johnson, Edwards, and Berkeley, its reign was short. At Princeton, for example, idealism enjoyed authoritative teaching only briefly before it was supplanted by the commonsense philosophers who were trained in the Scottish Enlightenment. In a sense, the Scottish philosophers Thomas Reid (1710–1796), Dugald Stewart (1753–1828), and others were able to appreciate the idealist emphasis on mind and yet also the materialist conception of the physical world. The mind-body dualism in Reid's work is perhaps the most compelling of the period. According to Reid, the nature of the material world as a mind-independent reality is evident to us in perception. Moreover, to explain the material world as entirely constituted by the mental is to fail to recognize the goodness of the material world that God has created.

In the first section of this chapter, I listed some of the ways in which modern philosophy has been afflicted by troublesome dualities, the most prominent being mind-body dualism. But in the colonial, theistic climate some of these worries were addressed. Consider briefly colonial thought on mind-body dualism and the problems of metaphysics, ethics, and epistemology.

In terms of explaining mind-body interaction, many colonial thinkers appealed to the greater power of God in explaining how there could be any causal relations whatsoever. If one begins with a materialist concept of the origins of the cosmos it is difficult to explain the emergence of any immaterial reality. But if one is at least open to the possibility that the very nature of the material world, its powers and liabilities, stem from God, then the existence of mind and the nature of mind-body interaction fits into an integrative scheme. While an idealist, Johnson articulated mind-body interaction along dualist line in *Elementa Philosophica:*

> We are, at present, spirits or minds connected with gross, tangible bodies in such a manner, that as our bodies, can perceive and act nothing but by our minds, so, on the other hand, our minds perceive and act by means of our bodily organs. Such is the present law of our nature, which I conceive to be no other than a mere arbitrary constitution or establishment of Him that hath made us to be what we are. And accordingly I apprehend that the union between our souls and bodies, during our present state, consists in nothing else but this law of our nature, which is the will and perpetual fiat of that infinite Parent Mind, who made us, and holds our souls in life, and in whom we live, and move, and have our being, viz. that our bodies should be thus acted by our minds, and that our minds should thus perceive and act by the organs of our bodies, and under such limitations as in fact we find ourselves to be attended with. (In Anderson and Fisch 1939, 66)

To some materialist critics, the appeal to God in explaining mind-body interaction will appear to be an explanation of the obscure in terms of the more obscure *(obscurum per obscurius)*. But contemporary physicists over two hundred years after Johnson's death have yet to provide consensus on a theory of the ultimate constituents of the physical world, and a theistic account of *both* the physical and mental world would place mind-body interaction within a comprehensive, unified framework.[5] Consider the ethical problem of dualism. Does mind-body dualism denigrate the body? While there have been dualists who treated the body as a prison or bare habitation (Pythagoras, Plato), surely one need not do so. If the mind and body are both deemed to be creations of a good God, and if the incarnation of God has united God with a physical body, then it is hard to see the body as something base and valueless. Consider finally the epistemological worry: If the mind and body are separate, how are we to guarantee our knowledge of other minds? On this front, perhaps one of the more dramatic proposals by colonial Puritans is that we may *not* have such a guarantee but by the grace of God. Puritans promoted a general reservation or skepticism about our access to the interior lives of others and self-knowledge based in introspection.[6] The result of sin and the fall is that we may indeed be able to mask our true intentions even from ourselves. In this respect, mind-body dualism actually fits the moral data, rather than philosophies of mind that deny the existence of an interior subjective world. In the absence of sin and in a state of grace, we may well have lucid, unimpeded access to our own minds and, due to the good disclosure of others, access to them as well. But sin may still haunt these claims.

Alan Heimert and Andrew Delbanco summarize the Puritan outlook: "If the Catholic question had been 'what shall I do to be saved?' and the question of the Reformers became 'how shall I know if I am saved?' perhaps the American Puritans asked, more eventually, 'What am I in the eyes of God?" (Heimert and Delbanco 1985, 15). This God-eye orientation is the key, I believe; it is from the God's eye point of view in colonial thought that we find a concord between the mind and the body and some of the other dualities Dewey would protest in the twentieth century. I suggest that the Deweyan critique of modern dualities had the force it did partly because of the recession of theism in late nineteenth- and early-twentieth-century philosophy. In an overriding theistic context, one may find an integral understanding of the mind and body (Taliaferro 1994).

If the mind and body found some kind of integration, however uneasy, in pre-Revolutionary America, what of agriculture and the greater understanding of the role of society, the relation between the individual and nature, a person's land and labor?

Farming and the Mind of God

Douglas Hurt has recently argued for the prominence of economics in the history of agriculture and has taken aim at the supposition that in the colonial years there was anything like agrarianism, which he defines as follows:

> Agrarianism is the belief that farming is the best way of life and the most important economic endeavor. Agrarianism also implies that farmers willfully sought to avoid commercial agriculture and preferred a "moral economy" in which they produced for subsistence purposes rather than the market and economic gain. The agrarian tradition has long been recognized as central to the American experience. (Hurt 1994, 72)

Hurt is convinced that there has never been such a phase:

> Although the belief in the agrarian tradition remains, it is colored with myth. It paints a mental image of a past that never was while denying the reality of contemporary agricultural life. Agrarianism now, as in the past, remains more myth than reality. (77)

Is this right?

I believe the evidence is mixed.[7] To be sure, economic gain was pivotal in early American life, and this was true in agriculture as much as in everything else. Alan Heimert and Andrew Delbanco put the economic factor dramatically: "To say that the early settlements of New England included or tolerated merchant activity is badly to underestimate the case. It is more accurate to say that the first English settlements *were* merchant activities" (Heimert and Delbanco 1985, 185). True, it is difficult to suppose that the Virginia Company of London was moved by anything other than self-interest. But can we have merchant activities, commercial agriculture, or a family farm without a culture, without an overriding framework of how to understand one's self and the land? There are conceptual reasons for thinking this is absurd.[8] And insofar as we search to sympathetically reconstruct what it was like to farm in the context of the colonial cultures, I believe we are led to conclude that farming was carried out *sub specie aeternitatis*, and assessed *ordine ad universum*. However imperfectly executed, there were strong religious strands that bolstered a stewardship model that secured farming as something ordained by God in a good creation. Early farming was not so much intended as activity within a "moral economy" as opposed to "commercial agriculture"; I believe it was largely intended as both. The immensity of pious literature cited above, the existence of counsels of stewardship in sermons and essays, prevalent biblical teaching, etc., all weigh in to construct a picture of farmers who understood their livelihood (including economic exchange) in the context of a generally theistic cosmos. The appeal to theism at many points may be no more than lip service, but it is difficult to explain it away altogether as a bare accessory to the mercantile lives of the colonists.

From the standpoint of colonial theology, Christian theism provided a guarantor for farming. The art of farming was in keeping with the art of God. In biblical texts, God is a farmer, planting a garden in Eden (Gen.); God blesses harvests and may be called on in prayer for rain (Old Testament; for example, Elijah's prayer for rain); God is a sower, planting the Word of God (Matt.); Jesus is even first thought of as a gardener at the resurrection (John 15): the church is fed by bread and wine, the fruits of the earth and labor. All this teaching, prevalent from pulpits throughout the colonies, served as the guarantor of reason and industry that the farmer expended. Theism, in a sense, provided the underwriting for the enterprise. Reason was the important tool of the farmer. What is the origin of reason itself? How is it that we can have confidence in it? As the Cambridge Platonists argued, reason may be trusted for it is "the Light of the Lord." The reason why the Cambridge Platonists, especially More and Cudworth, were so popular in the colonies is because they (like the Scottish school of common sense) located the reliability of reason in the mind and purposes of a good God. Farmers, as well as merchants, mathematicians and politicians, may trust reason because its origin is in God. The mind of a farmer made in the image of God is trustworthy because of the supremely good, anterior mind of God.[9]

Just as explaining the rise of modern science requires taking seriously the philosophical context of this event (whether it be in scholastic theism or in the Renaissance revival of Greek humanism), one needs to take seriously the context of these exchanges and agricultural practices. Miller rightly points out how sciences in the colonies were bolstered by faith.

> Science was not merely tolerated because faith was believed to be secure whatever physics or astronomy might teach, but it was actually advanced as a part of faith itself, a positive declaration of the will of God, a necessary and indispensable complement to biblical revelation. (Miller 1939, 211)

The same thing is true about agriculture.

Christian theism helped provide an integral framework, however imperfect, by which to conceive of land and labor, mind and body, self and nature. Of course Timothy Dwight's poetry is idealized, Jared Eliot's (1685–1763) *Essays on Field Husbandry* may have fallen on deaf ears,[10] Dickinson's "Farmer's Letter" in 1764 may have had more to do with political theory than agriculture, and Thomas Paine's *Agrarian Justice* can be advanced just as well without any reference to God. But I propose that all the emphasis upon the mind of God in the colonies and its integral role in holding together the cosmos itself was not empty rhetoric, but helped set a standard for colonial agriculture to be carried out responsibly and wisely for the common good.[11]

Part II

PREPRAGMATIST AGRARIANISM IN AMERICA

Franklin Agrarius

JAMES CAMPBELL

BENJAMIN FRANKLIN was born in Boston in 1706 and died in Philadelphia in 1790. During his long life, he developed from simple printer into public citizen, inventor and scientific researcher, political figure, colonial agent, revolutionary, ambassador, and statesman. In each of these roles, Franklin demonstrated a concern with advancing the common good. While there are many aspects to his understanding of the common good, in this essay I wish to concentrate on Franklin as an exponent of agrarian thinking. I intend to examine the relationships among his economic, social, and political ideas that emphasize personal virtues like industry and frugality and then connect them up at the national level with the advocacy of a society that is to remain primarily agricultural. Throughout, this discussion will be essentially ethical, reflecting the moral tone of Franklin's agrarianism.

Franklin's place within the American philosophical tradition has been uncertain. Like many other eighteenth-century figures of potential philosophic interest, he was the author of letters and essays rather than the standard type of systematic treatises. Franklin wrote also about issues of practical human concern. Historians have tended to pass over him in favor of thinkers who wrote in the standard forms and examined more specialized philosophical issues.[1] If we consider his contemporary situation, however, Franklin offered what was clearly a philosophical vision: the broad application of reason through the moral and natural sciences to advance human well-being. Moreover, this vision presents a pragmatic understanding of the meaning and value of life. As he wrote in one of his many relevant formulations: "Opinions should be judg'd of by their Influences and Effects" (2:203).[2] His broad sense of pragmatism envisioned advancing the common good through attempts to improve science, religion, morality, and politics. Franklin himself poses the question this way: "What signifies Philosophy that does not apply to some Use?" (9:251).[3]

Rather than as a philosopher, Franklin is better known in American intellectual history for his numerous calls for industry and frugality as a means to a life of economic security. Readers can easily find instances of relevant aphorisms in the various editions of *Poor Richard's Almanack* and suggestive comments here and there in the *Autobiography.* They can also consider longer examples, like his 1748 essay, "Advice to a Young Tradesman, Written by an Old One." This essay begins with a series of admonitions that would have satisfied Poor Richard at his most pecuniary, like: "Remember that Time is Money . . . Credit is Money . . . Money can beget Money" and concludes with the assurance that if young entrepreneurs lead lives of "Industry and Frugality," if they live by the motto "Waste neither Time nor Money, but make the best Use of both," they will soon be on the road to wealth (3:306–308).

We must be careful, however, not simply to equate these fragments with his complete view.[4] Any attempt to understand the meaning of Franklin's message on the relationship of industry and frugality to personal economic security and well-being requires that we first of all locate him within the life of an agriculturally based, semi-frontier society that of necessity concerned itself with building up its material infrastructure.[5] Here, a level of enthusiasm for matters economic—that later and out of context might appear as unbalanced—seems to make more sense. Franklin was writing for individuals like himself, who were starting from scratch, for individuals without the benefits of social status or inheritance, who were attempting to rise by their own personal diligence. He was at the same time communicating to the residents (and future residents) of a young country the gospel that this is a land where nothing is given to anyone and where individuals' efforts will pay off. In this "Land of Labour" (W8:607), people will succeed for the most part on their own efforts.

Industry and frugality are virtues that support each other in Franklin's way of thinking: no person could hope to attain economic freedom without both of them. He describes the virtue of industry as follows: "Lose no Time. Be always employ'd in something useful. Cut off all unnecessary Actions" (A:149). In the *Almanack* for 1751, he expands on this virtue: "the Industrious know how to employ every Piece of Time to a real Advantage in their different Professions. . . . If we lose our Money, it gives us some Concern. If we are cheated or robb'd of it, we are angry: But Money lost may be found; what we are robb'd of may be restored: The Treasure of Time once lost, can never be recovered" (4:86–87). This admonition to remain busy and focused meant recognizing that in agrarian America the commitment to extraordinary levels of hard work had to become ordinary. Its companion virtue, frugality, meant to Franklin attempting to resist the incursions of luxury that continued industry could make possible and to strive to live as simply as possible.[6] He described frugality as follows: "Make no Expence but to do good to others or yourself: i.e. Waste nothing" (A:149). A social version of this appreciation of frugality is contained in his 1771 admonition from London in support of nonimportation that the colonists recognize "all they save in refusing to purchase foreign Gewgaws, and in making their own Apparel" (18:81).

Behind Franklin's call for industry and frugality to secure economic well-being was not the amassing of a limitless fortune to fund a life of private excess.[7] Rather, Franklin was interested in attaining the level of personal economic security that would make possible a higher level of social service. As he writes of himself, "I would rather have it said, *He lived usefully*, than, *He died Rich*" (3:475). On the one hand, his expectations for pleasure were fairly modest. As he writes: "Happiness consists more in small Conveniencies or Pleasures that occur every day, than in great Pieces of good Fortune that happen but seldom to a Man in the Course of his Life" (15:60–61). On the other hand, he recognized the importance of beneficence. His understanding of human nature is social, and one prominent element in it is a conception of our natural debt to our fellows. Consider this comment from the *Almanack* of 1737: "The noblest question in the world is *What Good may I do in it?*" (2:171). Another instance can be pointed to that comes from nearly a half century later. After giving an unfortunate man a loan in 1784, Franklin instructed that the recipient should not expect to repay *him* personally. Rather, after the borrower is back on his feet, Franklin writes, "when you meet with another honest Man in similar Distress, you must pay me by lending this Sum to him; enjoining him to discharge the Debt by a like operation, when he shall be able, and shall meet with such another opportunity" (W9:197). Through such service, individuals can repay in part some of the kindnesses and mercies that they have received from their fellows and from God. Franklin believed that we will be able to do this if we have been good stewards of what we have received, if we have worked with diligence and used only what we needed.

In these key elements of industry and frugality, in a life of simplicity, was to be found a life that was appropriate to the American situation behind the vast frontier. Franklin was never a backwoodsman himself, of course, and he was often not enamored of the actions and attitudes of the actual backwoodsmen. He writes in 1760, for example: "The people that inhabit the [North American] frontiers, are generally the refuse of both nations [France and Britain], often of the worst morals and the least discretion, remote from the eye, the prudence, and the restraint of government" (9:65). Still, he frequently asserts his belief in the superiority of life in semi-frontier America, where, as he writes in "Information to Those Who Would Remove to America" in 1782, existence was "a general happy Mediocrity" (W8:604).

In particular, unlike European society, agrarian American society was not economically fragmented. He writes of his encounters with poverty during his travels in 1771 to Ireland and Scotland, where, in spite of the opulent life of the few, "[t]he Bulk of the People [are] Tenants, extreamly poor, living in the most sordid Wretchedness in dirty Hovels of Mud and Straw, and cloathed only in Rags" (19:7). In America, on the contrary, "[t]here are few great Proprietors of the Soil, and few Tenants," he writes in 1782. "Most People cultivate their own Lands, or follow some Handicraft or Merchandise; very few rich enough to live idly upon their Rents or Incomes" (W8:604). The Americans, as a consequence, led lives that were thus both simpler and happier than those of Europeans. In "The Internal State of America" (ca. 1787), he writes:

> Whoever has travelled thro' the various Parts of Europe, and observed how small is the Proportion of People in Affluence or easy Circumstances there, compar'd with those in Poverty and Misery; the few rich and haughty Landlords, the multitude of poor, abject, and rack'd Tenants, and the half-paid and half-starv'd ragged Labourers; and views here the happy Mediocrity, that so generally prevails throughout these States, when the Cultivator works for himself, and supports his Family in decent Plenty. (W10:120)[8]

Franklin believed that such individuals, intimately connected with the day-to-day realities of making their own living, would demonstrate priorities proper to a simple life and avoid the temptation to pursue material luxuries.[9]

Moreover, America was a land where common folk—like Franklin himself—could rise. "If they are poor, they begin first as Servants or Journeymen," he writes, "and if they are sober, industrious, and frugal, they soon become Masters, establish themselves in Business, marry, raise Families, and become respectable Citizens" (W8:608), just as he had done. Franklin's explicit focus in this self-improvement process was the ability of individuals to rise above the dependencies that they would have known in Europe. His concern was thus with the absolute advance in economic status that was possible for virtually all in America, not with the possibility of occasional individual advances relative to the more limited advances of other Americans.[10] In such a country, he hoped it would be possible to develop a self-directing democracy free of the restraints of European custom. And, of course, underlying all of this discussion of pursuing the common good was his belief that the physical situation in agrarian America represented the nearest possible approximation to utopia in his contemporary world, offering "a good Climate, fertile Soil, wholesome Air and Water, plenty of Provisions and Fuel, good Pay for Labour, kind Neighbours, good Laws, Liberty, and a hearty Welcome" (W9:21). In America, Franklin saw the possibility of happiness held out to those whose efforts were sufficient and whose goals were simple, to the industrious and frugal.

Franklin writes to a European correspondent in late 1775 that "[t]his is a good country for artificers or farmers, but gentlemen, of mere science in *les belles lettres,* cannot so easily subsist here, there being little demand for their assistance among an industrious people, who, as yet, have not much leisure for studies of that kind" (22:287). A few years later he reinforces this point with the observation that "the natural Geniuses" of America who were possessed of special talents in art and architecture "have uniformly quitted that Country for Europe, where they can be more suitably rewarded" (W8:604).[11] As Franklin understood their situation, the natural condition of almost all Americans, at least for the foreseeable future, was to be agriculture. "For one Artisan, or Merchant," he writes, "we have at least 100 Farmers, by far the greatest part Cultivators of their own fertile Lands" (W10:117–118). Franklin, whether in Boston or Philadelphia or London or Paris, was an urban person himself and thus remained part of the 1 percent throughout his life. As he admits to one of his British correspondents in 1784, "I am not much acquainted with Country Affairs, having been always an Inhabitant of Cities" (W9:160). While it is true that Franklin never was a farmer himself,[12] living most of his life in semi-frontier America, he recognized the importance of farming to humankind in general, and to the colonies and the young Republic in particular. And, being Franklin, he acted to advance it. For example, his proposals for establishing both the American Philosophical Society and the Philadelphia Academy, which developed into the University of Pennsylvania, have agricultural components. With regard to the former, in his original "Proposal for Promoting Useful Knowledge among the British Plantations in America" of 1743, Franklin included within the purview of this planned Philosophical Society: "All new-discovered Plants, Herbs, Trees, Roots, etc. their Virtues, Uses, etc. Methods of Propagating them, and making such as are useful, but particular to some Plantations, more general" (2:381). With regard to the latter, in his "Proposals Relating to the Education of Youth in Pennsylvania" of 1749, Franklin notes that the "Improvement of Agriculture" is "useful to all, and Skill in it no Disparagement to any." He consequently suggested that the students in the Academy learn such valuable agricultural skills as "*Gardening, Planting, Grafting, Inoculating,* etc." (3:417). He was active as well in the distribution, both to and from America, of plants and seeds thought to be of value, and in the dissemination of instructional books about agriculture.[13] Franklin also seemed to have come upon the idea of crop insurance to protect farmers against "storms, blight, insects, etc." (W9:674).

A large part of Franklin's interest in agriculture was attributable to his recognition that deliberate and careful farming is the key to advances in human well-being. By means of farming, humanity had been able to make the tremendous advance from the level of subsistence, that a life of hunting and gathering had made possible, to the level of comfortable agricultural surplus that was now supporting a progressing civilization: "From the Labour arises a *great Increase* of vegetable and animal Food, and the Materials for Clothing, as Flax, Wool, Silk, etc. The Superfluity of these is Wealth. With this Wealth we pay for the Labour employed in building our Houses, Cities, etc. which are therefore only Subsistence thus metamorphosed" (16:108). Moreover, because of the settlers' fortunate situation in a New World, their potential for tapping the riches of America's land and waters, "the true Sources of Wealth and Plenty" (16:209), was limitless. He later describes their success as follows: "The Agriculture and Fisheries of the United States are the great Sources of our encreasing Wealth. He that puts a Seed into the Earth is recompens'd, perhaps, by receiving twenty out of it; and he who draws a Fish out of our Waters, draws up a Piece of Silver" (W10:122). In addition to recognizing its actual centrality at the heart of the growing American economy, Franklin was also committed to the moral primacy of agriculture.[14] As he writes, agriculture is "the only *honest Way;* wherein Man receives a real Increase of the Seed thrown into the Ground, in a kind of continual Miracle wrought by the Hand of God in his Favour, as a Reward for his innocent Life, and virtuous Industry" (16:109). In this mood, he continues elsewhere that agriculture is "the most useful, the most independent, and therefore the noblest of Employments" (W9:491). Franklin puts special emphasis upon this independence: "The farmer has no need of popular favour, nor the favour of the great; the success of his crops depending only on the blessing of God upon his honest industry" (W10:3).[15]

The actual and moral primacy of agriculture was the keystone of Franklin's larger conception of a healthy economy that additionally incorporated modest levels of manufacturing and trade. While he may maintain that "the true Source of Riches is Husbandry" (15:52), he recognized as well that producing agricultural surpluses does not occur in an economic vacuum. Farming is hard work and farmers will only cultivate what they anticipate being able to use. They grow as much as they need for themselves and their families; and, should they think that they can sell any surplus for a gain, they will grow it as well. Depending on their location and transportation realities, however, selling the surplus is often difficult. For some farmers, Franklin writes, "the Expence of Carriage will exceed the Value of the Commodity." For these individuals, selling their surplus crops is not an option; and "if Some other Means of making an Advantage of it are not discovered, the Cultivator will abate of his Labour and raise no more than he can consume in his Family." The advantage that was discovered was that this fragile surplus could be transformed into a manufactured product—like "Linnen and Woollen Cloth" (18:273)—that could be used in trade. Manufacturing produces no new value; it changes the form of the old.

> Agriculture is truly *productive* of *new wealth;* Manufactures only change Forms; and whatever value they give to the Material they work upon, they in the mean time consume an equal value in Provisions, etc. So that Riches are not *increased* by Manufacturing; the only advantage is, that Provisions in the Shape of Manufactures are more easily carried for Sale to Foreign Markets.

Thus, when it becomes economically feasible to transport the reworked agricultural products to a desirable market, it becomes economically worthwhile to produce them. This manufacturing might be conducted on a very small scale, using only family members: "Spinning or Knitting etc. to *gather up the fragments* (of Time) *that nothing may be lost*" (15:52). Or it might be conducted on a far larger scale, using the labor of many individuals whom the entrepreneur puts to work:

> if he can draw around him working People who have no Lands on which to subsist, and who will for the Corn and other Subsistence he can furnish them with, work up his Flax and Wooll into Cloth, then is his Corn also turn'd into Cloth, and with his Flax and Wooll render'd portable, so that it may easily be carry'd to Market, and the Value brought home in Money. This seems the chief Advantage of Manufactures.

Regardless of the scale of production, however, Franklin was careful to emphasize repeatedly that manufacturing produces no new value. "When a Grain of Corn is put into the Ground it may produce ten Grains: After defraying the Expence, here is a real Increase of Wealth"; manufacturers, on the other hand, "make no Addition to [wealth], they only change its Form" (18:273–274).

Franklin's understanding of a well-balanced economy was thus dual. The first aspect of his position was that manufacturing should not be allowed to expand too greatly. He was concerned that the Americans not increase manufacturing "beyond reasonable Bounds" so that it became necessary to import food (15:52). In their selection of areas for manufacturing expansion, he was mindful as well that in some industries America stood at a competitive disadvantage. "The buying up Quantities of Wool and Flax, with the Design to employ Spinners, Weavers, etc., and form great Establishments, producing Quantities of Linen and Woollen Goods for Sale," he writes in 1782, "has been several times attempted in different Provinces; but those Projects have generally failed, goods of equal Value being imported cheaper" (W8:610). Any further expansion into these noncompetitive areas would presumably meet with similar failure. The second aspect of his position was that manufacturing was not likely to expand too greatly as long as the fundamental economy of the country remained in balance. "Manufactures," he writes, "are founded in poverty."

> It is the multitude of poor without land in a country, and who must work for others at low wages or starve, that enables undertakers to carry on a manufacture, and afford it cheap enough to prevent the importation of the same kind from abroad, and to bear the expence of its own exportation. But no man who can have a piece of land of his own, sufficient by his labour to subsist his family in plenty, is poor enough to be a manufacturer and work for a master. Hence while there is land enough in America for our people, there can never be manufactures to any amount or value. (9:73)[16]

In Franklin's mind, it was to the advantage of America and its residents that it remain a primarily agricultural country into the foreseeable future. In such an agricultural country, balanced with manufacturing and trade in modest amounts, the citizens would be more virtuous and happy.

Some commentators on Franklin's economic thought have found his assertion of the primacy of agriculture to be in conflict with his earlier position that suggested a labor theory of value (see Carey 1928, 146–147; Conkin 1976, 105–106). This earlier view, presented in "The Nature and Necessity of a Paper-Currency" (1729), argues that labor—not gold or silver—is the proper measure of value. As he writes there, "Silver it self is of no certain permanent Value, being worth more or less according to its Scarcity or Plenty"; and, if we are to seek a permanent *"Measure of Values,"* we should look to *"Labour."*

> Suppose one Man employed to raise Corn, while another is digging and refining Silver; at the Year's End, or at any other Period of Time, the compleat Produce of Corn, and that of Silver, are the natural Price of each other; and if one be twenty Bushels, and the other twenty Ounces, then an Ounce of that Silver is worth the Labour of raising a Bushel of that Corn.

Thus, if we are attempting to determine the relative worths of goods, Franklin believed that such comparisons should be conducted in terms of the units of labor necessary to produce them. Once equivalencies of worth are determined, exchanges of comparable worth are facilitated by money. "Men have invented Money, properly called a *Medium of Exchange,* because through or by its Means Labour is exchanged for Labour, or one Commodity for another." We must remain careful, however, to remember that monetary values are derivative: "Trade in general being nothing else but the Exchange of Labour for Labour, the Value of all Things is, as I have said before, most justly measured by Labour" (1:148–150).[17]

The key to recognizing that this support for a labor theory of value was not opposed to his later views on the primacy of agriculture is to be found in Franklin's claim that value is "most justly measured" by labor. The two views are thus fully compatible when we distinguish between the *origin* of value—the productivity of the earth—and the *standard* of human values—as represented in units of labor. Franklin suggests this interpretation when he writes that "[t]he first Elements of Wealth are obtained by Labour, from the Earth and Waters" (W9:246). Value, thus, is not created by labor, although it is obtained by and to be measured in labor. Agricultural labor is, moreover, its primary form.

> Food is *always* necessary to *all,* and much the greatest Part of the Labour of Mankind is employ'd in raising Provisions for the Mouth. Is not this kind of Labour therefore the fittest to be the Standard by which to measure the Values of all other Labour, and consequently of all other Things whose Value depends on the Labour of making or procuring them?

What would result from Franklin's view here is the following sort of equivalency: "If the Labour of the Farmer in producing a Bushel of Wheat be equal to the Labour of the Miner in producing an Ounce of Silver, will not the Bushel of Wheat just measure the Value of the Ounce of Silver?" In this way—and enhanced by the fact that "[t]he Miner must eat, the Farmer indeed can live without the Silver" (16:47)—the value of all products is to be pegged on a labor scale. "Necessaries of Life that are not Foods, and all other Conveniencies, have their Values estimated by the Proportion of Food consumed while we are employed in procuring them" (16:107).[18]

The labor theory of value as presented by Franklin emphasized a measurement in units of time. It also recognized, of course, the need for the labor in question to be put to some good use. Franklin's understanding of the importance of labor to advancing human well-being can be seen in his belief that "if every Man and Woman would work for four Hours each Day on something useful, that Labour would produce sufficient to procure all the Necessaries and Comforts of Life, Want and

Misery would be banished out of the World, and the rest of the 24 hours might be Leisure and Pleasure." It was Franklin's belief that the prevalence of poverty in his contemporary situation was the result of the unfortunate fact that the legitimate contributions of those who were doing useful work were being squandered by those who did not work and by those whose work was wasted. As he writes, "the Employment of Men and Women in Works, that produce neither the Necessaries nor Conveniences of Life, who, with those who do nothing, consume the Necessaries raised by the Laborious." For example, in a productive household the individuals busy themselves throughout their days and years with the activities of agriculture broadly conceived. "Spinning . . . hewing Timber and sawing Boards . . . making Bricks, etc." fill their working hours. All of these are useful forms of labor and result in a surplus. Franklin stresses, however, that not all labor is this productive: "if, instead of employing a Man I feed in making Bricks, I employ him in fiddling for me, the Corn he eats is gone, and no Part of his Manufacture remains to augment the Wealth and Convenience of the family; I shall therefore be the poorer for this fiddling Man" (W9:246). While Franklin's stance here sets an impossibly high and narrowly pecuniary standard of usefulness—he was known to "fiddle" a good bit himself[19]—still it suggests his combined moral and socioeconomic point. "Look round the World and see the Millions employ'd in doing nothing," he writes, "or in something that amounts to nothing, when the Necessaries and Conveniences of Life are in question" (W9:247). If they too were to become productive by doing a significant amount of useful labor, the "necessaries and conveniences" of life would become more widespread and human well-being would be advanced.

In his early writings, Franklin was favorably inclined toward commerce, seeing trade as the generally beneficial result of a number of economic factors: locales and peoples are differently productive, and the voluntary exchanges of surpluses are to "the mutual Advantage and Satisfaction of both" and thus commerce is an activity "highly convenient and beneficial to Mankind" (1:148). As the years went on, however, he began to question its value with a special emphasis upon two lines of inquiry. The first dealt with the balance between trade, which he saw as itself unproductive, and any values that trade brought. The most obvious advantage of trade is variety, which he describes as follows: "each Party increases the Number of his Enjoyments, having, instead of Wheat alone or Wine alone, the Use of both Wheat and Wine" (16:108). One potentially negative aspect of this enjoyment was its ability to undermine the life of industry and frugality. "Foreign Luxuries and needless Manufactures imported and used in a Nation," he writes, "diminish the People of the Nation that uses them" (4:231). At times, Franklin saw this aspect to be so strongly negative as to make trade unnecessary. In September 1775, just as the colonial troubles were about to boil over into war, Franklin thought that the Americans were willing, if necessary, to surrender their seacoast and forego external commerce. The "internal Country" would be enough, he writes. "It is a good one and fruitful . . . Agriculture is the great Source of Wealth and Plenty. By cutting off our Trade [the British] have thrown us *to the Earth,* whence like *Antaeus* we shall rise yearly with fresh Strength and Vigour" (22:199). He continues in the same theme in 1780 that America "can very well subsist and flourish without a Commerce with Europe, a Commerce that chiefly imports Superfluities and Luxuries" (W8:107). Further, there was the actual expense of the trading, which became a more relevant factor as the focus of consideration moves from necessities to conveniences. "What is the Bulk of Commerce," he wonders in 1784, "but the Toil of Millions for Superfluities, to the great Hazard and Loss of many Lives by the constant Dangers of the Sea?" Especially with regard to such items as tea, coffee, sugar, and tobacco that made up a large portion of America's imports and exports, Franklin believed that trade was largely wasteful. "These things cannot be called the Necessaries of Life, for our Ancestors lived comfortably without them" (W9:247). Moreover, there were even cases in which the specific "products" involved in the trading make the commerce, however profitable, morally wrong. The primary instance of immoral commerce was the slave trade.[20] The first line of criticism that Franklin made against commerce, then, was that it was itself unproductive and tended to undermine a life of industry and frugality.

Franklin's second line of criticism against commerce was that it tended to be unfair to one of the trading partners. As a result, at the same time that the unneeded goods "diminish" the importing country, they also "increase the People of the Nation that furnishes them" (4:231). This inequality was the result of an imbalance, either in the level of political power between the participants in the exchange or in their relative knowledge of values. While it was possible for any particular instance of trade to be fair, defined as a trade in which "equal Values are exchanged for equal the Expence of Transport included," it was just as possible for it to be unfair. Setting aside mercantilist and other political interference, the bulk of unfair trades resulted when one of the partners in the exchange did not know the value—in terms of the amount of invested labor—of the material to be received in the exchange. While this may occur with regard to agriculture, it was especially likely in the case of manufactures, where

> there are many expediting and facilitating Methods of working, not generally known; and Strangers to the Manufactures, though they know pretty well the Expences of raising Wheat, are unacquainted with those short Methods of working, and thence being apt to suppose more Labour employed in the Manufactures than there really is, are more easily imposed on in their Value, and induced to allow more for them than they are honestly worth.

Formulating this point about unfairness more strongly, Franklin writes "the Advantage of Manufactures is, that under their shape Provisions may be more easily carried to a foreign Market; and by their means our Traders may more easily cheat Strangers" (16:108–109).

These two criticisms of the commercial life bring us back to Franklin's emphasis upon the primacy of agriculture. As he writes, "there seem to be but three Ways for a Nation to acquire Wealth: The first is by *War* as the Romans did in plundering their conquered Neighbours. This is *Robbery.* The second by *Commerce* which is generally *Cheating.* The third by *Agriculture* the only *honest Way*" (16:109). Franklin believed that by means of a balanced economy that places primacy in agriculture, and complements this with a small amount of manufacturing and trade, America would grow into a powerful and progressive country of happy and virtuous citizens.

The other side of Franklin's hopefulness about agrarian America as a potential utopia and the likely success of even average individuals who were willing to work hard was his apparent harshness toward the poor. I say "apparent" harshness not because I have any doubts about either the intended or the likely short-term effects of his program, but because he himself saw the eventual result to be beneficence. "I am for doing good to the poor," he writes, "but I differ in opinion of the means" (13:515). Many individuals, he maintains, fail to see that maintenance programs that provide short-term help also cause long-term damage to those who are helped and to the society in general.[21] As Franklin understood the human situation, our natural inclination to "a life of ease, of freedom from care and labour" (4:481) was held in check under normal circumstances by the opposing realities of physical need and social virtue. Short-term help, however, weakens this check by satisfying the physical need while at the same time undermining shame; and, as a consequence, the inclination to live an easy life stands unopposed. Writing on the laboring poor in 1768, Franklin asserts that such short-term help is harmful:

> I fear the giving mankind a dependance on any thing for support in age or sickness, besides industry and frugality during youth and health, tends to flatter our natural indolence, to encourage idleness and prodigality, and thereby to promote and increase poverty, the very evil it was intended to cure; thus multiplying beggars, instead of diminishing them. (15:104)

In consequence, anything that we might do to diminish individuals' inducements to industry, frugality, and sobriety would ultimately harm them.[22]

Franklin believed in general that we need to be extremely cautious in any attempts to restructure our world. "Whenever we attempt to mend the scheme of Providence and to interfere in the Government of the World," he writes, "we had need be very circumspect lest we do more harm than Good" (4:480). Sometimes, as in the case of lightning rods, our efforts result in improvements in human well-being. In the case of illness, he suggests intervention as well. In 1751, Franklin's support for the Pennsylvania Hospital as an institution of organized charity was based in part on the fact that health and illness result, at least to some extent, from a natural lottery: "We are in this World mutual Hosts to each other; the Circumstances and Fortunes of Men and Families are continually changing . . . how careful should we be not to *harden our Hearts* against the Distresses of our Fellow Creatures" (4:149). With regard to efforts to palliate poverty, however, he anticipated that harm rather than good would ultimately result from our misguided efforts:

> To relieve the misfortunes of our fellow creatures is concurring with the Deity, 'tis Godlike, but if we provide encouragements for Laziness, and supports for Folly, may it not be found fighting against the order of God and Nature, which perhaps has appointed Want and Misery as the proper Punishments for, and

> Cautions against as well as necessary consequences of Idleness and Extravagancy. (4:480)

This was the crux of his criticism of various "poor laws." As he wrote in 1766 with regard to one particular British law, "you offered a premium for the encouragement of idleness, and you should not now wonder that it has had its effect in the increase of poverty" and the creation of a class that is increasingly "idle, dissolute, drunken, and insolent." He continues that "the more public provisions were made for the poor, the less they provided for themselves, and of course became poorer. And, on the contrary, the less was done for them, the more they did for themselves, and became richer." Franklin thus believed that the correct program would not attempt to palliate poverty, but rather would leave the poor exposed to its ravages unmitigated: "the best way of doing good to the poor, is not making them easy *in* poverty, but leading or driving them *out* of it." By holding individuals responsible for their own conditions, and the conditions of their families, they will be made better: "more will be done for their happiness by inuring them to provide for themselves, than could be done by dividing all your estates among them" (13:515–516).

The initial harshness of Franklin's position on poverty should be clear: no palliatives will be offered to assist those who are deficient in the virtues of industry and frugality. As I suggested earlier, however, this harshness may be only apparent when viewed over the long term. One key element in his longer view was that in the potential utopia of the New World, where the Frontier beckons and where individuals' well-being depends primarily on their own efforts, where there are boundless woods that "afford Freedom and Subsistence to any Man who can bait a Hook or pull a Trigger" (13:232), Franklin believed that there were alternatives not present in the more developed economies of Europe. Equally valuable possibilities were present in the semi-frontier areas of agrarian America as well. Using his own experience as an indication of what was possible for an industrious and frugal American, Franklin urged his fellows to work hard on the seemingly minor aspects of their present situation—"a penny sav'd is a penny got" (1:241–242)—with faith in eventual prosperity.[23]

A further mitigation of the apparent harshness of Franklin's discussion of the poor is his larger theme of the public ownership of wealth, a kind of partial and sketchy socialism that sounds little like the individualism of many of Poor Richard's familiar aphorisms. Shortly after his retirement from the printing business to free up more time for his philosophical experiments, Franklin criticized those who pursue wealth without end and who believe there is importance in amassing a great *"dying worth."* He asserts, on the contrary, the superiority of claims of human need to claims of property ownership, writing that "what we have above what we can use, is not properly *ours,* tho' we possess it" (3:479).[24] Late in his life, Franklin develops this suggestion of public ownership of wealth further:

> All the Property that is necessary to a Man, for the Conservation of the Individual and the Propagation of the Species, is his natural Right, which none can justly deprive him of: But all Property superfluous to such purposes is the Property of the Publick, who, by their Laws, have created it, and who may therefore by other Laws dispose of it, whenever the Welfare of the Publick shall demand such Disposition. (W9:138)

Franklin continues that, because "Private Property . . . is a Creature of Society," it is therefore "subject to the Calls of that Society, whenever its Necessities shall require it, even to the last Farthing." Moreover, he writes that these "Contributions" to alleviate public need "are not to be considered as conferring a Benefit on the Publick, entitling the Contributors to the Distinctions of Honour and Power, but as the Return of an Obligation previously received, or the Payment of a just Debt" (W10:59).

Thus it would seem that Franklin's call for the deliberate pursuit of personal improvement, while often individualistic in its aphoristic form, should always be seen as being more socially oriented in its ultimate goal of advancing the common good. In this way, Franklin could assert, for example, the primacy of society on questions of "superfluous" property. "He that does not like civil Society on these Terms, let him retire and live among Savages," he writes. "He can have no right to the benefits of Society, who will not pay his Club towards the Support of it" (W9:138). Moreover, we need to keep ever-conscious the fact that this was not Franklin's analysis of our current situation but of his. His ideal was of a prosperous middle class whose members lived simple lives of democratic equality. Those who met with greater economic success in life were responsible to help those in genuine need; but those who from a lack of virtue failed to pull their own weight should expect no help from society. With simplicity, the citizens of this new agrarian land would direct their individual energies toward endeavors that would result in personal improvement. With democracy, the citizens of this new agrarian land would be free to direct the choices of the society toward the common good. And with equality, the citizens of this new agrarian land would not suffer the deadening restraints of custom that would keep the industrious and frugal from prospering and advancing. As long as America maintained its healthy balance of agriculture, manufacturing, and trade, its citizens would remain prosperous, happy, and free.

Thomas Jefferson and Agrarian Philosophy

PAUL B. THOMPSON

No figure in American intellectual history is both more central and more marginal to the main themes of the present volume than Thomas Jefferson, author of the Declaration of Independence and third president of the United States. Jefferson is central because he is more closely associated with the formulation and defense of agrarian ideals than any other figure in American history. It is impossible to avoid Jefferson in any work that purports to discuss the importance of farming to any larger cultural or political project in an American context. Yet Jefferson is marginal because the agrarianism with which he is associated was neither Jeffersonian to the extent that it is American, nor original to Jefferson to the extent that it reflected his actual opinions. There is little evidence that Jefferson's agrarian ideas had a particularly profound influence on the transcendentalists, Hegelians, and pragmatists that established the philosophical tradition in America during the century following his death. Many of these figures expressed admiration for Jefferson, and they were indeed influenced by Jefferson's naturalism, as well as his commitments to individual liberty, religious tolerance, and general education. But these aspects of Jefferson's thought complement a fairly conventional European agrarianism, typical of eighteenth-century political thought.

Nevertheless, Jefferson has been a magnet for those who advocate a special moral or political status for farmers and for country life. Partly this reflects Jefferson's enormous popularity. "He was America's Everyman," notes Joseph J. Ellis (1997, 7). But more than that, Jefferson has long been identified as the special advocate of the American farmer.

The importance of Jefferson as spokesman for rural America may be more celebrated now than at any time in the past. When 70 percent of Americans were family farmers, Jefferson was read as an advocate of the people against aristocracy, and as a supporter of individual liberty against government power. After FDR these two themes seemed less obviously complementary. Today, there are multiple schisms in Jeffersonia. There is, on the one hand, the historians' Jefferson. This is a contested Jefferson, to be sure, as documented by the contributors to Peter Onuf's *Jeffesonian Legacies* (1993). Yet whether they see him as budding capitalist or radical democrat, historians remain likely to reinforce the idea that Jefferson advocated agrarian settlement as an adjunct to his broader views of democracy (see the discussion of Griswald in "Two on Jefferson's Agrarianism" in this volume). On the other hand, advocates for farming interests of all manner read the passages extolling farming more literally, and reject (indeed, never consider) the possibility that Jefferson was using the farmer as a stand-in for entrepreneurs or for the common man.

Two post–World War II intellectuals who led this rereading of Jefferson are discussed in Gene Wunderlich's contribution to this volume. John Brewster and A. Whitney Griswold were, in a modest sense, products of the classical period in American philosophy, and each chose to use Jefferson as a token for a political philosophy that placed country life at the very heart of democracy. In this they are representatives of a much broader line of thought on farming, rurality, and democracy that may have begun with Liberty Hyde Bailey and that lists Harold Briemyer, Jim Hightower, and Wendell Berry among its leading contemporary figures. In the final chapter of this volume, Jeffrey Burkhardt argues that a new generation of philosophers is developing agrarian thought into a more comprehensive philosophy of public science and agricultural ethics. Though each of the authors contributing to this twentieth-century agrarianism would find much to criticize in the writings of the others, virtually all of them have used Jefferson's name in association with ideas that they did not derive from Jefferson's writings. Contemporary agrarianism probably owes more to Franklin, Emerson, and Thoreau than to Jefferson's political philosophy, and the chapters by Campbell, Corrington, and Anderson lay a surer philosophical footing for neo-agrarianism than the present one.

Jefferson will not go away, however limited his actual relevance to pragmatism, naturalism, or twentieth-century conceptions of democracy in America, and the chapters that follow contain many references to

Jefferson. As such it is important to offer an interpretation of Jeffersonian agrarianism and to examine its philosophical pedigree. I will argue that Jefferson offered several memorable expressions of an agrarianism that was neither original nor even peculiarly American. The gradual decline of agrarian thinking since 1800 makes it seem like an exceptional view, but many (perhaps most) thought that way in Jefferson's day. Jefferson offered a succinct, poignant, and persuasive statement of agrarian thinking that is also ambiguous in a way that has allowed people with very different moral visions and political agendas to appropriate Jefferson's words for their own purposes. If there is a link between Jefferson's agrarianism and later developments in American thought, it is to be found in that ambiguity (or the common fount of European political economy) and not in any uniquely Jeffersonian genius.

Agrarianism involves the claim that farming and/or country life contribute unique attributes to human moral development and community. As Montmarquet shows in *The Idea of Agrarianism,* the specific nature and philosophic import of this claim is subject to many different interpretations. Those who speak of Jeffersonian agrarianism imply that Jefferson formulated or articulated a distinct, original interpretation of the agrarian claim. Lacking that, one would expect that he might have provided a particularly cogent or sophisticated statement of an agrarian philosophy that had been promulgated in less persuasive form by others. Though Jefferson wrote at length on agriculture, the passages that make up his agrarian canon are few in number. It is therefore possible to examine each of them in some detail.

The Agrarian Canon

The textual basis for attributing an original or systematic agrarian philosophy to Jefferson is remarkably thin. The most frequently cited passage comes from his *Notes on the State of Virginia.*

> Those who labour in the earth are the chosen people of God, if ever he had a chosen people, in whose breasts he has made his peculiar deposit of substantial and genuine virtue. It is the focus in which he keeps alive that sacred fire, which otherwise might escape from the face of the earth. Corruption of morals in the mass of cultivators is a phaenomenon of which no age nor nation has furnished an example. (Jefferson 1984, 290)

An even shorter passage from a 1785 letter to John Jay is also frequently quoted. "Cultivators of the earth are the most valuable citizens. They are the most vigorous, the most independent, the most virtuous, & they are tied to their country & wedded its liberty & interests by the most lasting bonds." (Jefferson 1984, 818). Quotation from the *Notes* often simply identifies farmers as "the chosen people of God," omitting not only the qualifying phrase "if ever he had a chosen people," but also the subtler point that Jefferson does not capitalize the pronoun "he" in referring to God. Quotation from the letter to Jay often downplays the fact that Jefferson is attributing value to farmers *as citizens* rather that tout court, and almost never mentions the fact that the letter itself is about sea power (Wunderlich 1984).

The *Notes* and Jay passages can be supplemented with others less frequently quoted. For example, Garry Wills cites Jefferson's letters on the importance of developing American educational institutions in an essay on contemporary Americans' attraction to rustic authenticity. In a letter to Peter Carr of August 10, 1787, Jefferson wrote, "State a moral case to a ploughman & a professor. The farmer will decide it as well, & often better than the latter, because he has not been led astray by artificial rules" (Jefferson 1984, 902). Wills includes this quote with others from 1785 that make no use of examples or metaphors from farming. Wills cites them to argue that for Jefferson the exceptional moral character of Americans, their fitness for democracy, "was to be maintained by closeness to the native soil" (Wills 1997, 32). There is no reason to suggest that Wills is misrepresenting Jefferson, but it is remarkable how Jefferson's prose needs to be filled in with Wills's more explicitly bucolic phrasing. The expression "native soil" is used repeatedly by Wills, but not by Jefferson. As if he were filling in what Jefferson must have meant, Wills links the two Jefferson quotes with his own sentence, "The native wisdom of our fields can only lose by exposure to European sensibilities" (Wills 1997, 32).

As I will argue below, Jefferson believed in a version of agrarianism, but these passages do not support the claim that he thought of himself as expositing or crafting a novel philosophy (as he clearly did with respect to his views on religious tolerance or natural rights). They do not even support the claim that Jefferson endorsed belief in the exceptionalism of the American moral character and tied it to "the native wisdom of our fields." It is, thus, worth taking some pains to consider each passage, beginning with the letters to Jay and Carr. John Jay, of course, was a founding father himself, an author of the Federalist Papers and chief justice of the

Supreme Court. Peter Carr was the son of Jefferson's sister Martha. His father was Dabney Carr, a boyhood friend who passed many hours with Jefferson on the present site of Monticello during their youth. Both swore that should one be the first to die, the other would see that a shaded spot on the mountain would be his final resting place. Jefferson buried the elder Carr at that site in 1773, and Peter grew up in Jefferson's family. Although Jefferson wrote both letters with the full expectation that they would eventually be read by others, it is nevertheless important to bear in mind to whom they were originally addressed.

The Letters to Carr and Jay

Jefferson's letters to Carr are full of moral advice of a peculiarly fatherly kind. Two years before the letter cited by Wills, Jefferson is advising Carr, "Encourage all your virtuous dispositions, and exercise them whenever an opportunity arises; being assured that they will gain strength by exercise, as a limb of the body does, and that exercise will make them habitual" (Jefferson 1984, 815). The passage cited by Wills appears in a letter where Jefferson is offering specific advice on Peter's education. The letter contains discussions of Italian, Spanish, and religious education. The "ploughman" quote comes from a paragraph on the value of moral philosophy, which begins with Jefferson's summary evaluation: "I think it lost time to attend lectures in this branch" (Jefferson 1984, 901). Jefferson is less praising farming than deriding formal education in morals.

Yet it should also be noted that Jefferson had already advised Carr of the necessary reading on morality in the earlier letter—"Epictetus, Xenophontis Memorabilia, Plato's Socratic dialogues, Cicero's philosophies, Antoninus, and Seneca" (Jefferson 1984, 816). The Carr letter of 1787 was thus a singular piece of advice. Jefferson was not praising rusticity at the expense of education. Indeed, in *Notes on the State of Virginia* Jefferson writes:

> There is a certain period of life, say for eight to fifteen or sixteen years of age, when the mind, like body, is not yet firm enough for laborious and close operations. If applied to such, it falls an early victim to premature exertion; exhibiting indeed at first, in these young and tender subjects, the flattering appearance of their being men while they are yet children, but ending in reducing them to be children when they should be men. (Jefferson 1984, 273)

From this Jefferson adduces not that children should turn to husbandry, but that they should occupy themselves by learning Greek and Latin.

The letter to Jay is written from Paris during the same week as Jefferson's earlier letter to Carr. Jefferson is suggesting that the new republic can neglect the formation of a strong navy to the extent that it relies on farming, rather than trade, for its principle sources of wealth. The praise of farming is almost an aside, and is comparative. Farmers are better citizens than manufacturers or traders because the basis of their economic interest (land) cannot be alienated from the polity. Traders and manufacturers can pull up stakes when times get tough. By the time that he was writing Jay, Jefferson had witnessed capital flight to Canada in the years before and during the revolution. He had ample reason to advocate policies that gave precedence to landed wealth. He was using an argument with a European provenance in an American collocation. Placed in conjunction with Europe's established aristocracy, the connection between landed wealth and political responsibility was an argument against democratic reform. In America—even plantation Virginia—it was an argument for democracy.

Not insignificantly for prose written in 1785, Jefferson's advocacy of farming over trade and manufacture favored the primacy of what became the House of Representatives over the lordlike Senate. The view that political power should be apportioned according to population was being advocated by "large states" (for example, Virginia, Pennsylvania) and being opposed by "small states" (for example, Rhode Island, New Jersey) in the ongoing constitutional debate. The bicameral legislature was the eventual compromise. Large states would have dominance in the House, but the Senate would balance that power by apportioning an equal number of seats to each state. Today we draw no necessary connection between farming, geographical size, and population. Farming states such as the Dakotas fare better in the Senate than the House. But for the framers of the constitution, this was an "argument for democracy" that happened to favor the interests of Virginians in a crucial constitutional debate. Jefferson was, of course, a landowner himself, on top of it all. There is no reason to think that Jefferson did not believe what he wrote, but modern readers should not underestimate Jefferson's mastery of what we now call "spin."

The Notes on the State of Virginia

The *Notes* provide the best basis for regarding Jefferson as an agrarian, and this work also provides a sound basis for reading the letter to Jay as expressing authentically agrarian sentiments. The agrarian passages from the *Notes* are embedded in a discussion of manufacturing which begins with the admission that trade and manufacture have never had great importance within the domestic economy of the state of Virginia. Jefferson notes, "The political œconomists of Europe have established it as a principle that every state should endeavour to manufacture for itself: and this principle, like many others, we transfer to America, without calculating the difference of circumstance should often produce a difference of result" (Jefferson 1984, 290). The difference that Jefferson observes consists of the fact that land is not available in Europe, hence European states have no recourse but to employ "the surplus of their people" in manufacturing. Then comes the passage quoted above.

The *Notes* were written in 1781 and published in 1787, and thus contemporaneous with the 1785 letter to Jay. The agrarian passages in the *Notes* can be read as a more extended development of what Jefferson was stating in that letter. America does not suffer from the same compulsion toward manufacture as Europe, hence there is no reason to encourage development of manufacture. Those engaged in manufacturing cannot be expected to exhibit the same degree of political loyalty as those whose livelihood depends on an asset (land) that cannot be alienated from the territory of the nation. Hence it is politically wiser to develop agriculture, at least until such time as it becomes necessary to promote a mercantile or manufacturing economy. In this vein, the *Notes* passages on agrarianism conclude:

> It is better to carry provisions and materials to workmen [in Europe], than bring them to the provisions and materials, and with them their manners and principles. The loss by the transportation of commodities across the Atlantic will be made up in happiness and permanence of government. The mobs of great cities add just so much to the support of pure government, as sores do to the strength of the human body. It is the manners and spirit of a people which preserve a republic in vigour. A degeneracy in these is a canker which soon eats to the heart of its laws and customs. (Jefferson 1984, 291)

This is a rather hard-nosed agrarianism addressed as some rather specific policy questions (for example, the development of urban trading centers and sea power). Jefferson does not attribute special efficacy to the native soil of America or the state of Virginia, nor does he offer his agrarian observations within the context of a treatise on morals or politics. Indeed, the agrarian passages from the *Notes* occupy a few hundred words in a work of many thousands consisting primarily of an inventory on the physical and cultural geography of the state. Nevertheless, the *Notes* leave little doubt that Jefferson believed farming would produce citizens of superior moral character.

There is, however, little consensus on how to read the *Notes* within the broader context of Jefferson's life and thought. On the one hand, recent historical scholarship has fastened on passages in which Jefferson discusses the institution of slavery. Jefferson is forthright against slavery in the *Notes,* but not every contemporary reader is favorably impressed. Paul Finkelman points to a section in which Jefferson discusses a plan to revise the administration of justice in Virginia that had, as one of its elements, emancipation of slaves born after passage of the revisal act (see Jefferson 1984, 263–264). Jefferson revised the *Notes* in 1783 and again in late 1784 or early 1785, after his arrival in France. Yet, Finkelman notes:

> By the time Jefferson left for France it was clear that no one would introduce his proposed emancipation-colonization scheme. Nevertheless, Jefferson did not revise his account of the emancipation amendment. There is no indication why Jefferson persisted in telling his European readers that this law would be introduced when he knew it had not been, and would not be proposed. (Finkelman 1993, 196)

A more charitable reading of the *Notes* is put forward by Joseph Ellis, who calls them "[p]art travel guide, part scientific treatise and part philosophical meditation" (Ellis 1997, 85). But like Finkelman, Ellis judges the passages on slavery to be the main focus of attention in eighteenth-century Paris.

On the other hand, other Jefferson scholars give pride of place to the geographical passages. Merrill Peterson reads the *Notes* as a response to philosophers such as Buffon who believed that America would not support an advanced civilization (Peterson 1970, 66). Garry Wills follows this reading in *Inventing America*. According to Wills, Buffon believed:

> The climate [in America] was too moist and hot to nurture vigorous life; men and animals were small and reproduced only with difficulty. Much of Jefferson's *Notes on the State of Virginia* is devoted

> to a refutation of those charges. America's Indians are said to reproduce themselves well and to be physically and mentally full grown. . . . In another part of the *Notes* Jefferson had made an injudicious reference to the "infecundity" of the Indians' women. He deleted this, lest Buffon have a weapon against America . . . The antiquity and number of flourishing Indian tribes, the ease of Indian self-government, are among Jefferson's favorite themes in the book—along with description of the large and manifold animal life supported by the continent. (Wills 1979, 162)

Jack P. Greene lists the *Notes* as one of several treatises written between 1775 and 1815 that sought to reconstruct the later eighteenth century's understanding of Virginia and Virginians. In general, these works sought to celebrate Virginia's contribution to the Revolution and to establish "unanimity, moderation, generosity patriotism and political expertise" as the character and genius of the Commonwealth (Greene 1993, 242). Greene ranks John Daly Burk's four-volume *History of Virginia* (1804–5 and 1816) and Edmund Randolph's *History of Virginia* along with the *Notes* as three works of special importance in affecting this reconstruction. Greene's reading complements Wills's with an interpretation more focused on the domestic rather than the European audience for the *Notes.*

Joseph Ellis rates the fame of that brief passage from the *Notes* second only to the opening lines of the Declaration of Independence from Jefferson's entire literary estate. Recounting an observation attributed to Douglas L. Wilson, Ellis notes that "American agriculture has never quite recovered from this resounding complement" (Ellis 1997, 137). Yet whether the larger significance of the *Notes* is seen in its condemnation of slavery or in its contribution to the rehabilitation of Virginia's public image, the book was clearly not a tract on agrarian philosophy. It is tempting to conclude that the long paragraph on the virtues of farming is simply an anomaly, relatively unimportant to Jefferson and the larger subject matter of this work. In fact, geography and slavery are both relevant to the agrarian passages from the *Notes,* but the book must be read within the broader context of eighteenth-century agrarianism for that to become clear.

Jefferson's Other Agricultural Writings

Given the above, one might think that Jefferson's agrarian canon has more to do with egalitarian democracy than agriculture as such. Nevertheless, there is one sense in which contemporary advocates of farm interests, largely ignored by the historians who monitor the sacred halls of official Jeffersonia, may be correct to read Jefferson more literally. Jefferson knew quite a bit about farming in the same way that the average professor of agricultural science today knows about farming. This is not to say that either can farm. Jefferson had significant practical acquaintance with farming through his plantations at Monticello and Poplar Forest, but he was notoriously unsuccessful with the business side of these operations. Nevertheless Jefferson was unarguably among the preeminent agricultural scientists of his time. He had few equals on the American continent in applying scientific principles to the selection of crop and plant varieties, and he is credited with a modification of the plow blade, which increased the efficiency with which energy is applied to the task of turning soil (Wills 1978, 143).

Writings on agricultural science and practice that do not introduce agrarian themes counterbalance the agrarian canon. The largest single compilation is Edwin Morris Betts's *Thomas Jefferson's Farm Book* (1953), where documents, records, and correspondence that Jefferson produced in connection with the farming operations at his plantations are collected. The *Farm Book* is full of notes on crop rotations and yields, along with practical correspondence on subjects such as disputes with neighboring farmers. It documents Jefferson's application of scientific principles to agriculture in great detail, but one searches in vain for any sentiments remotely like the famous passage from the *Notes.* Perhaps it is not surprising that writings so focused on practice provide little occasion for musing on agrarian ideals. However, letters to Hector St. John de Crèvecoeur—author of *Letters from an American Farmer* and an authentic agrarian of the first order—would appear to provide the ideal place for Jefferson to explore agrarian themes. Yet the two Crèvecoeur letters collected in Jefferson's *Writings* belie that hope. The letter of 1788 deals with despotism and geopolitics (Jefferson 1984, 927–929). The letter of 1787 takes up farming and Homer, but is focused on the likely source of a particular innovation in wheel making. Jefferson opines that a journalist he has read "supposes the English workman got his idea from Homer. But it is more likely the Jersey farmer got his idea from Homer, because ours are the only farmers who can read Homer" (Jefferson 1984, 878). This is no transcendent agrarian sensibility rising mysteriously from the American soil, but agricultural science applied by educated husbandry men in practical affairs.

An 1803 letter to Sir John Sinclair is of singular interest. Jefferson was president when the letter was written, and roughly a third of it discusses Napoléon Bonaparte as a threat to world peace, offering assurances for U.S. neutrality in any conflict between France and Great Britain. The pretext, however, is agricultural science. Jefferson begins by noting the salutary effects of gypsum on Loudon County, Virginia soils. He concludes by discussing the desirability of a national agricultural society. Significantly, this society is not to promote agrarian ideals or agrarian political philosophy, but to foster and disseminate the application of scientific principles to agriculture (Jefferson 1984, 1132–1133). It is an idea that prefigures the public support of agricultural experiment stations and state extension services, one of the most potent forces for modernization in the history of agriculture.

Though Jefferson advocated agrarian ideals in a rhetorically potent but somewhat offhand fashion, he advocated scientific agriculture unambiguously. He developed a strategy for soil conservation by planting along the contour of sloped land (he called it "horizontal ploughing") (Jefferson 1984, 1405–1406). He argued for professorships in the agricultural sciences at American universities (Jefferson 1984, 838). The tension between agriculture and manufacturing that is described in the *Notes* and the Jay letter of 1785 was somewhat resolved in his mature thought. His 1818 *Report of the Commissioners for the University of Virginia* includes a summary description of the aims of university education: "To harmonize and promote the interests of agriculture, manufactures and commerce, and by well informed views of political economy to give a free scope to the public industry" (Jefferson 1990, 134). If agrarianism is to be set against modernization as the central theme in American agricultural history, as Robert Kirkendall (1984) suggests, the balance of Jefferson's writings place him on the side of the modernizers. Ironically, the writings on science and education foretell Dewey's pragmatism far more than Jefferson's agrarian canon.

Indeed, Jefferson's mature thought concludes with a substantial recantation of the sentiments expressed in the crucial passage of the *Notes.* An 1816 letter to Benjamin Austin discusses the earlier rationale for developing agriculture and importing manufactured goods at some length, and details why circumstances and experience have led Jefferson to change his opinion. "We must now place the manufacturer by the side of the agriculturist," Jefferson writes, and "experience has taught me that manufactures are now as necessary to our independence as to our comfort" (1984, 1371). Yet the Austin letter should not be read as a nullification of the sentiments expressed in the *Notes.* Even at the end of his life, Ellis argues, Jefferson believed in preserving as much of the agrarian character of America as possible. According to Ellis, Jefferson thought that despite the necessity of developing a manufacturing base,

> America should remain a predominantly agricultural economy and society. Domestic manufacturing was permissible, but large factories should be resisted. Most important, the English model of a thoroughly commercial and industrial society in which the economy was dominated by merchants, bankers and industrialists should be avoided at all costs. "We may exclude them from our territory," he warned, "as we do persons afflicted with disease." (Ellis 1997, 258)

Jefferson was a modernizer but also an agrarian. How are these ideas to be reconciled? Clearly they posed no contradiction in Jefferson's mind. In order to see why not, it is crucial to attend the broader context of eighteenth-century agrarian thought.

Eighteenth-Century Agrarianism in Europe

Agriculture does not figure prominently in the thought of Marcuse or Arendt. Farming never comes up in the writings of Rawls or Nozick. Food production was not a major theme in the political writings of Sartre, or even Foucault. None of these late-twentieth-century political theorists believe that the mode of producing sustenance from land or sea plays a prominent role in the formation of social institutions or political character. They are all about as unagrarian as one can imagine. If these philosophers are the benchmark for political theory at the millennium, every major political thinker of the eighteenth century would be classified as agrarian by the standards of the twenty-first century. Eighteenth-century political thought was awash with the idea that land and its cultivation (or the absence thereof) is a central determinant of political culture.

Eighteenth century agrarianism came in several flavors, however. One emphasized soil and climate as a determinant of national character and political morality. Agriculture was important primarily as a medium for translating physical geography into human character. A second form of agrarianism emphasized the role of agricultural production in the formation of national wealth, and in turn tied agricultural production capacity to military might. Yet another form attempted to match social and political institutions to the mode of material production, not necessarily favoring agriculture, but providing a basis for Jefferson's view that farming makes virtuous citizens. Finally, there were budding philosophies that prefigured romantic celebration of sublime nature. We shall consider them in reverse order.

Rousseau's Anticipation of Romantic Nature

The last of these is arguably the least agrarian, and least at home in the eighteenth century. Romantic celebrations of nature's transformative and saving power became influential in the landscapes of German painter Albert Bierstadt and American painters Thomas Cole and Frederick Church, all of whom worked after 1820. John Robert Cozens painted imposing mountain scenes between 1776 and 1800, but the paintings had neither agrarian subject matter nor subtext (Schama 1991, 459–462). Nineteenth-century landscape artists may have been expressing nineteenth-century philosophical themes consistent with Emerson, Thoreau, or Schelling, and we still tend to think of agrarianism in these romantic terms today. These themes were not entirely absent from eighteenth-century political thought. Jean-Jacques Rousseau (1712–1778) was the eighteenth-century precursor of the nineteenth-century transcendentalists and romantics. Rousseau famously argued that noble savages once lived peaceably in the state of nature, and that civilization and cultivated rationality are bought at the price of social inequality, conflict, and moral corruption. Rousseau's noble savage is implicit in the romantic ideal of recovering virtue and innocence through return to nature.

Farming is not prominent in either Rousseau or in later European romanticism, however. Rousseau does take up farming in *Émile* under the heading "Moral Teaching through Action," but his lesson is not the one an Emerson, Thoreau, or Dewey would teach. Rousseau does not celebrate husbandry as a natural avocation. Instead, he assists little Émile in planting some beans.

> We come every day to water our beans and are delighted to see them spring up. I increase his pleasure by telling him that this spot *belongs* to him. In explaining this word I make him feel that he has spent his time, his work, his trouble, in short his whole

> person upon it, that he has in this plot a part of himself, which he may keep against all comers, just as he might wrest his arm out of the hands of anyone who wished to hold it against his will. (Rousseau 1762, 101)

Unfortunately, Émile has chosen a spot where the gardener has already planted exotic melons, and the beans are torn out. Recognizing the gardener's prior claim, Émile and Rousseau pledge not to dig again until they can find a spot where no one has been at work before. "Then you may throw aside your tools, gentlemen;" the gardener replies, "for there is no ground here uncultivated." Protesting that this leaves them with no place to plant, they are met with the gardener's rebuke, "What has that to do with me?" (Rousseau 1762, 102). Agriculture has a role in "moral teaching through action," but for Rousseau it is to create an experience of the injustice owing to private property, and not to recapture authentic moral sensibility through an encounter with nature.

Nineteenth-century European philosophers argued that artistic contemplation of nature could bring a person full circle, as it were. They accepted Rousseau's suggestion that settled civilization corrupts, but saw in art the power to redeem, hence the philosophical importance of the landscape tradition. As Corrington shows in this volume, Emerson formulated an agrarian version of this ideal by arguing, "That uncorrupted behavior which we admire in animals and in young children belongs to [the farmer], to the hunter, the sailor—the man who lives in the presence of nature" (Emerson 1870, 153). However, there is no reason to think that this extension of Rousseau's thought was implicit in Jeffersonian agrarianism. On the contrary, it is unlikely that Jefferson ever gave a second thought to the implications of Rousseau's books for agrarian democracy. If he had, he would have read Rousseau as an adversary, rather than an inspiration.

Adam Smith and the Theory of Political Institutions

The idea that a mismatch between material modes of production and social institutions could spell political disaster was implicit in much of eighteenth-century philosophy, but it received a particularly influential treatment by philosophers of the Scottish Enlightenment. The Scots had an impressive object lesson before them. Scotland exchanged parliamentary independence for free trade with England in the Union of 1707.

Institutional reform by way of religious liberalization, university reorganization, and encouragement of voluntary incorporation facilitated trade and precipitated economic growth. David Hume, Adam Ferguson, and Adam Smith all contributed to the theorization of this transformation in the Scottish economy and in intellectual life (Robertson 1987, 470). Smith and Ferguson both traced the acquisition of customs and institutions through a sequence of stages that corresponded to a particular mode of subsistence: hunting, pastoral, agricultural, and commercial. Some of the passages in which the development of social institutions is traced sound particularly agrarian in tone. “In pastoral countries,” Smith writes in *The Theory of Moral Sentiments,* “all the different branches of the same family commonly chuse to live in the neighbourhood of one another. . . . Their concord strengthens their necessary association; their discord always weakens, and might destroy it” (Smith 1759, 222). This stands in contrast to commercial countries,

> where the authority of law is always perfectly sufficient to protect the meanest man in the state, the descendants of the same family, having no such motive for keeping together, naturally separate and disperse, as interest or inclination may direct. They soon cease to be of importance to one another; and in a few generations, not only lose all care about one another, but all remembrance of their common origin. (Smith 1759, 223)

Smith is describing, not endorsing, in these passages. *The Wealth of Nations* (1776) makes it clear that European nations have no choice but to embrace private property and the rule of law, and with them a system that turns greed to social good. Nevertheless, the contrast between agricultural countries and commercial countries runs throughout Smith's work, and it would be easy to read him as laying the groundwork for agrarianism wherever population growth and commercial development have not already rendered it impossible.

In *Inventing America,* Garry Wills argued that Jefferson was more deeply influenced by the Scottish Enlightenment than by Locke's *2nd Treatise of Government* (Wills 1979). Wills stresses the Scottish emphasis on moral empathy, evident in the above passage from Smith, rather than the link between mode of subsistence and institutional development. Yet the latter theme was deeply influential throughout Europe. Consider Immanuel Kant's 1876 essay, “Speculative Beginning of Human History.”

Echoing (but not citing) Smith, Kant describes the pastoral life as most leisurely and secure but supplanted by the appearance of agriculture and permanent civilization.

> When subsistence depends on the earth's cultivation and planting (especially trees), permanent housing is required, and its defense against all intrusions requires a number of men who will support one another. . . . Culture and the beginning of art, of entertainment, as well as of industriousness (Gen. 4:21–22) must have sprung from this, but above all, some form of civil constitution and of public justice began. (Kant 1983, 56)

In describing why herdsmen do not take to civil society, Kant offers a comment that foreshadows Jefferson's concern about Tory manufacturers deserting the new United States like rats leaving a sinking ship. "Since the herdsman leaves behind nothing of value that he cannot find in other places," Kant warns, "it is a simple matter for him to move his herd to distant parts, once it has done its damage, and in so doing he avoids having to make good on the damage" (Kant 1983, 56).

The Physiocrats

The works of the Scottish Enlightenment and Rousseau's *Discourse on the Origin of Inequality* (1755) each offered a hypothetical anthropology in which language, social solidarity, and private property are analyzed as responses to the need for survival. Among the French, however, it is the Physiocrats who mirror (indeed, anticipate) Smith's interest in the link between social institutions and economic development. Promoting social reforms that would stimulate French agriculture, Francois Quesnay (1694–1774) and the marquis de Mirabeau (1715–1789) developed a prototype of classical political economy that advocated free trade and absolute respect for the sanctity of private property. Unlike Smith, Quesnay and Mirabeau believed that land productivity—understood as soil fertility and climatic suitability measured on a per-acre basis—fixed a nation's capacity for wealth and social development (Fox-Genovese, 1976).

Physiocracy spawned a lot of the agrarian rhetoric that is today dismissed as misguided. Soil takes on a unique and almost mystical capacity to create wealth in the Physiocrat doctrine, but we should not be too hasty. On the one hand, by emphasizing agriculture to the exclusion of all other productive activities, Quesnay and Mirabeau unleashed rhetoric of questionable economic pedigree. It centered on the view that manufacture simply rearranged and reorganized the composition of existing material goods, while tilling the soil was capable of truly producing something new. There is a certain common sense to this view, if one neglects the fact that even agriculture simply "rearranges" water, soil minerals, and organic material. There is also a certain deep wisdom to the view as well, since agriculture captures solar energy in a way that few manufacturing activities can match. Nevertheless, attributing this kind of economic significance to agricultural production, while denying it to manufacturing, is a dubious ground for agrarian philosophy.

On the other hand, Physiocracy was an argument for reforming land rents that penalized more productive farming methods and bringing new lands into production. This part of the argument was probably based on an astute comparison of English and French economic and political development. Social theorist Jack Goldstone argues that Great Britain's civil wars of 1642–49 were caused in part by the social pressure of population growth coupled with stagnation of agricultural productivity. By the close of the seventeenth century, Britain had undertaken land reforms and adopted tax policies that stimulated agricultural production, easing these tensions in the long run. Ironically, the greater power of the French monarchy and French nobility allowed France to forestall such reforms for a hundred years. The population pressures mounted, culminating in bread riots and the French revolution (Goldstone 1989). Quesnay and Mirabeau veiled the comparison to the English system behind rhetoric that praised agriculture as the sole source of national wealth. With respect to the wisdom of land and tax reform in eighteenth-century France, at least, history proved them right.

French Natural Philosophy

Franklin, not Jefferson, was the most prolific advocate of Physiocrat philosophy among America's founding fathers (see Campbell, this volume). But the Physiocrats are probably best read as one strand in a complex web of eighteenth-century French philosophy that did have a significant influence on Jefferson. Montesquieu's *De l'esprit des lois* (1748) put forward the view that human needs for survival and procreation establish the basis for norms. The most general of these are nearly universal, but different historical and natural environments establish a second order of quasi-natural needs. In this way, climate and geography have the capacity to shape national character, providing different peoples with differing solutions to the basic problems of survival (Pangle 1973).

In *Inventing America,* Garry Wills traced environmental determinism through several French natural philosophers. Buffon (discussed above) thought precise calculation of human needs and productive capacities could be inferred from the study of soil and climate. According to Wills, several in Jefferson's circle sought to develop and apply such quasi-statistical approaches to the design of ideal political institutions, including Buffon, Condorcet, and Condillac. Wills argued that the marquis de Chastellux was the member of this group to have the greatest influence on Jefferson. Chastellux wrote a two-volume study on happiness that utilized detailed quantitative methods to compare the relative happiness of peoples under a variety of climatic conditions and political institutions. In the end, only two factors were positively correlated with happiness: agriculture and population. Wills summarized Chastellux's conclusion as follows: "Farming is the ideal state for man. It is the golden mean placed between wandering like nomads and being packed into corrupt cities. In order to advance agriculture, men must tame and enrich the land, use climate to advantage, and perfect wild nature by applied science" (Wills 1979, 160).

If Wills is right, *Notes on the State of Virginia* was intended to be read as work in the broad philosophical tradition of Montesquieu, Buffon, and Condorcet. It was not a work aimed at advancing the social theory of these masters, but at rebutting some of their views on America's potential for civil society. The rebuttal consisted of correcting their mistaken factual assumptions, of providing better data. When Jefferson penned the famous agrarian passages in *Notes,* he was merely pointing out that low population density in the Americas would permit the kind of agrarian society that was becoming increasingly impossible in France and England. In fact, Jefferson's prose would have been equally consistent with the ideas of Smith and Ferguson or Quesnay and Mirabeau. With respect to theory, he was simply saying what virtually everyone already believed about the relationship between farming, citizenship, and the institutions of government.

Jefferson's Legacy

Garry Wills also ties the most unsavory aspect of Jefferson's legacy to his fascination with the various forms of environmental determinism advanced by *les philosophes.* Jefferson was a slave owner and a racist. Though sharply critical of the institution of slavery, Jefferson clearly believed the black slaves of Virginia to be of inferior character and intelligence. As Wills notes, this combination of attitudes was not unusual among Europeans who were persuaded by the calculations of men like Buffon and Chastellux. They had assured themselves that Africa could not produce a civilized society, in large part because the African climate was thought inhospitable to settled agriculture. Without agriculture, Africans would never move beyond hunting and pastoral societies, hence would not develop the norms, institutions, and moral personality required for civility. To this general philosophy, Jefferson (who had direct experience with slavery these European intellectuals lacked) added the observation that the conditions of slavery were even less likely to engender norms and moral character than subtropical environments (Wills 1979).

Jefferson took environmental influence on character quite seriously. It was in no sense a rationalization of simple racist beliefs. He argued that the practice of slavery had a detrimental effect on both master and slave, with the former becoming inured to tyranny and practiced in domination. Edmund S. Morgan sees a silver lining in Virginians' practice of slavery, however: "Virginians may have had a special appreciation of the freedom dear to republicans, because they saw everyday what life without it could be like" (Morgan 1975, 376). Nevertheless, Jefferson's attitudes toward race reflect one of the great flaws in eighteenth-century agrarian thought. A theory of moral and political development based on soil and its cultivation became a vehicle for the myth of blood and soil. There is more than a little wisdom in Montesquieu's or Smith's evolutionary approach to social and political institutions. Their theories are authentic forerunners of Dewey's social thought. Yet evolutionary and developmental social theory too often gave way to prejudice and self-congratulatory nationalism and racism. Nothing in agrarian theory suggests a genetic determinism with respect to moral and political character, but agrarian just-so stories were concocted to bolster hierarchies based on blood lineage.

Jefferson may have been less one-sided in applying the principles of environmental determinism, and he did not attach the sense of finality to blood lineage that is characteristic of the most incorrigible forms of racism and nationalism. Wills argues that Jefferson believed blacks left alone under favorable climatic conditions would develop social institutions and personal virtues of their own. This opinion, however, merely testifies to Jefferson's prejudicial belief in the inherent distinctness of the races. Had he followed out the logical implications of environmental determinism, he would have found no basis for thinking that human populations sharing a common environment could maintain distinct moral and cultural identities forever. Jefferson's view could have been based only on a dogma of blood lineage that remained ever unquestioned.

Jefferson's attitudes toward blacks and slavery are a crucial part of his legacy for those of us who continue to struggle with the issue of race in American society. The fact that racist ideas are bound up in eighteenth-century agrarian political philosophy provides all the more reason to untangle assumptions and implicit beliefs that have agrarian roots. These roots are not uniquely American. American racism did not spring forth solely from American soil. The philosophical substance of agrarian philosophy (including its racist elements) as exposited and advocated by Thomas Jefferson was a commonplace of eighteenth-century thought, and its most prolific advocates were European. Though these ideas are arguably a part of Jefferson's legacy, Jefferson did not invent them. Jefferson's agrarian legacy consists primarily in the fact that he was one of many who pointed out their implications for a relatively underpopulated and industrially underdeveloped America. He was also in a unique position to popularize and act on the implications of agrarian thought.

Although Jeffersonian agrarianism was not particularly Jeffersonian, neither was it idle philosophy in the American political context of the late eighteenth century. It mattered in political choices that have set the course of American history in innumerable ways. Many of these ways are bound up in the ideological rift that created the two-party system in American politics. Alexander Hamilton was the central figure on the other side of this rift, and he held views contrary to Jefferson's agrarianism in several crucial respects. Hamilton argued for policies that would hasten industrial development, and he believed that civic virtue would attach more readily to men of wealth and attainment than to rude farmers, who must of necessity attend to their own interests. This latter belief animated political conflicts between Jefferson and Hamilton until the latter's untimely death in 1804.

The ideological cleft between Hamilton and Jefferson underscored American political conflicts at least until the Civil War. Hamilton's elitism provided a rationale for nepotism in filling of offices in the executive branch during the terms of Washington and Adams. The favored families were acquiring a taste for civic duty. Jefferson the democrat threw them out on assuming the presidency. In debates over the budget, Jefferson stood against Hamilton's plan for an industrial center in New Jersey. The contest between their ideals was energized by the old-fashioned political clash of interests that began at the second constitutional convention. "Small" states with fixed borders such as New Jersey and Connecticut vied for power with "large" southern states with an indefinite capacity to expand their farming population. In a nation with upward of 90 percent employed in agriculture, "large" and "small" were defined primarily in numbers of farmers. Jeffersonian democrats were expected to win majorities in the house, while the Senate would shelter elites. There the limitless population of farming states could not overrun small industrial states. It is ironic that farm states such as the Dakotas enjoy far more relative political power in the Senate today.

Of course, if democratic majorities were to rule, and if farmers were to participate, there are necessary constraints on the size and scope of the federal government. By the time of Jefferson's death in 1826 the House/Senate, large states/small states cleft was becoming a states' rights/federal powers debate. Agrarian philosophy's capacity to rationalize black slavery connected with its argument for a direct democracy of farmers. It is possible to block this connection by attending closely to the difference between blood heredity, on which agrarianism is silent, and soil heredity, which is exclusively social and historical, affording no basis for racial divisions. Agrarian political philosophy became embroiled with the South's defense of slavery through a historical and geographical accident. There was no logical or philosophical basis for such a connection, but there can be little doubt that agrarian arguments gave many southerners a sense of justification and even moral righteousness. By the Civil War, eighteenth-century agrarianism had become associated with the themes that continue to tarnish it to this day.

In the debates where he participated, Jefferson was not quoting eighteenth-century agrarian themes idly, nor were they a metaphor for some deeper vision of everyman democratic populism. He was advocating an agrarian future for America. By this he meant an America where the vast majority of people would be farmers and the greatest measure of economic wealth would derive from agriculture. Practically, this meant policies that would open new lands to farming, such as the Louisiana Purchase and the Lewis and Clark Expedition. Such policies were pursued at the expense of those that would subsidize industrial development. In all these respects, the contemporary agrarian activists who read Jefferson as having a very literal interest in agriculture are right. But in Jefferson's time adopting such a stance implied no halt to scientific and technological advance. Jefferson was a modernizer *and* an agrarian. It is simply impossible to extend Jeffersonian agrarianism to the early-twentieth-century context of Liberty Hyde Bailey, the great advocate of "country life," much less to our own times when Wendell Berry and Jim Hightower campaign against modernization and on behalf of family farms. All of these men owe a debt to the larger tradition of eighteenth-century agrarian thought, but it is a tradition that has been dramatically reshaped since Jefferson's death.

It is equally impossible to trace any clear path from Jefferson's agrarianism to the ideas of Emerson, Thoreau, and finally Dewey. Obviously, the same eighteenth-century thinkers who influenced Jefferson influenced these American philosophers. Any discernable intellectual lineage almost certainly stems from Smith, Montesquieu, and the Physiocrats, rather than Jefferson himself. However, in choosing democracy and the common farmer over industry and wealthy elites, Jefferson took a course not unlike the one that Dewey would take in formulating pragmatist social philosophy a century after Jefferson's death. To the extent that Jeffersonian agrarianism celebrated the primacy of nature, it connects with Emerson and Thoreau. To the extent that it was democratic, it was of a piece with Dewey's political thought. These thin threads may be the sole connections between Jeffersonian agrarianism and the American philosophical tradition.

Yet we should not forget those happier aspects of Jefferson's life and thought that foretold an intellectual tradition to blossom long after his death. As Morgan (1975) writes, "Jefferson himself, whatever his shortcomings, was the greatest champion of liberty this country has ever had" (376). Jefferson established a legacy for American philosophy in the individualism implicit in his conception of inalienable rights from the Declaration of Independence. His authorship of the Virginia Statute on Religious Toleration rendered religious and civic virtue separate, anticipating the ethical naturalism of the American tradition. In founding the University of Virginia, Jefferson struck out for public education in a manner that likewise anticipates many of Dewey's mature themes. It was the declaration, the statute, and the university that Jefferson chose to commemorate on his tombstone, and not agrarianism. Perhaps he was the best judge of his legacy, after all.

Emerson and the Agricultural Midworld

ROBERT S. CORRINGTON

EMERSON'S UNRELENTING inquiry into the structures and potencies of nature place him at the forefront of those philosophers who have advanced our understanding of nature's complex forms of interaction. His sensitivity for the metaphors buried within ordinary affairs enabled him to exhibit the intimate correlation between the human process and the unlimited scope of nature. Each local scene carried within itself a lesson or fable that had archetypal power reaching down into the heart of the world. In translating these native metaphors into poetic utterance, Emerson allowed nature to exhibit its own inner rhythms and dynamisms. More importantly, this compelled the finite human process to open itself to that which forms its own measure and origin.

Parallel to his search for the elusive potencies of nature is his exploration of human paradigms and their correlation to the deeper currents of the world. These "representative men" serve to guide us toward the genial forms of empowerment that make the human process unique within the innumerable orders of the world. At turns Emerson celebrates the poet, the artist, the statesman, and the farmer as living symbols of nature. Each type represents a clearing onto the spiritual forces that unify and quicken the natural orders that we encounter. In emulating these paradigmatic individuals we learn those secrets of the spiritual life that would otherwise remain beyond our ken.

In Emerson's early writings he celebrates the poet as the "new-born bard of the holy Ghost" (1836, 254) who will replace the minister or priest as the ideal model for self-transformation. The poet participates in the powers of nature and compresses them into the measured cadences of poetry. The minister lives out of the dead letter and kills the spirit, thus alienating the congregation from the eternal dynamism of nature. In his later reflections, Emerson allows that the farmer, never exhibiting his reflections in verse, serves as a paradigm of how nature interacts with the human process:

> The glory of the farmer is that, in the division of labors, it is his part to create. All trade rests at last on his primitive activity. He stands close to nature; he obtains from the earth the bread and the meat. The food which was not, he causes to be. The first farmer was the first man, and all historic nobility rests on possession and use of land. (1870a, 133)

Like Adam, the farmer helps to give shape to the world and to bring forth all that sustains the other orders of life. God's primal creation is recapitulated by the farmer who stands rooted in nature. Emerson shifts the emphasis away from the revelatory power of language toward the practical and efficacious work of the farmer. As an amateur farmer himself, he goes so far as to speak of "we farmers" when describing the activity of creation and preservation. The act of farming, in its quiet but sure reconstruction of local habitation, is itself a kind of inquiry into the ultimate structures of the world. In tilling a field or in planting a fence around fruit trees the farmer is learning how the eternal rhythms of the world can enhance the needs of the human process. Poetic inspiration gives way to active transformations that unveil nature and show its way to the self.

Hence both the poem and the small farm represent clearings within which eternal truths appear. The metaphor for these microcosmic structures is that of the "midworld." Unlike the unlimited 'realm' of nature *per se,* the midworld is circumscribable by the human process. It can be the subject of study and reflection and thereby point toward the macrocosmic world within which it appears. By the same token, the midworld points to the much smaller realm of the self and illuminates not only its most pervasive features but its correlation to nature as a whole. For Emerson, "The midworld is best" (1844, 337) because it provides the bond between the self and its world. Without this bond the human process would become alienated from those very forces that provide wisdom and insight.

Agriculture is an activity that reveals the basic contour of the midworld and, in turn, of nature. Emerson eulogized such activity in language that mirrored his earlier writings on the nature of poetic creation. In either case, what was sought was the method of nature as it permeates the cosmos and thereby enters into the core of the self. Nature's internal laws become manifest in all orders of human activity but emerge in greater purity in those actions that directly alter and compress the forces of the environment. In order to show how Emerson's thought evolved toward his deepened sense of the centrality of agriculture it is necessary to detail his understanding of the method of nature itself.

Four essays stand out as conveying Emerson's sense of nature's internal logic and consequent outward expression. Of initial importance is his 1836 work, *Nature,* where he shows how nature and spirit enter into the ecstasies of the human process. His 1841 address "The Method of Nature," delivered in Waterville, Maine, advances beyond his 1836 formulations by showing in more detail how the particular participates in a universal end. His 1844 essay "Experience" betrays a growing reticence about the ability of the self to enter fully into nature's mysteries. Finally, his 1860 essay "Fate" paints a much starker and more brooding picture of those forces of nature that can and will diminish the human process. In briefly tracing through these essays, we will gain access to the changing roles of nature and the human process as they converge on the midworld that sustains their relation. Once this has been understood it will be possible to examine his understanding of how the seemingly uninspired and antipoetic realm of agriculture preserves our contact with nature.

In *Nature* Emerson struggles to articulate the most pervasive features of the world and to show how the spirit moves among the orders of nature as their animating principle. He distinguishes between two possible definitions of nature itself, one derived from the German post-Kantian Fichte, the other derived from common sense. The former definition sees nature as the not-me—that is, as all that obtains outside of the internal acts of the self. The latter definition sees nature as being constituted by essences unchanged by the self. Emerson wishes to broaden his understanding of nature in such a way as to do justice to both interpretations while finding a conception that is even more encompassing. Emersonian naturalism insists that the spirit is pervasive within nature and enters fully into the inner dynamism of the human process. Art is envisioned as the "place" where nature and its truth becomes most fully manifest.

As is often noted, Emerson uses the image of the circle or the horizon to define the measure of nature as it enters into human awareness. Metaphorically, nature can be seen as the location for the endless intersection of circles of varying size. Any given circle will be tempted to equate itself with nature per se, and this temptation must be resisted by the poet and seer. Emerson shows the import of the concept of horizon as it relates to finite personal and economic interests:

> The charming landscape which I saw this morning is indubitably made up of some twenty or thirty farms. Miller owns this field, Locke that, and Manning the woodland beyond. But none of them own the landscape. There is a property in the horizon which no man has but he whose eye can integrate all the parts, that is, the poet. This is the best part of these men's farms, yet to this their warranty-deeds give no title. (1836, 188)

The poet moves past all finite boundaries toward the currents of being that animate the world. If the farmer is tempted to claim ownership of the landscape, the poet knows that all such claims are quietly overturned by the eternal powers of nature that can never be owned. The poet is the primary agent of nature's revelation because of the gift of language that is itself rooted in the forces of nature. All poetic utterance is symbolic of nature, which is itself symbolic of the spirit. For the young Emerson, language is the most forceful source of revelation.

The poet is not, however, a passive recipient of the truths of nature but has the godlike ability to transform the world through poetic speech. Emerson's early idealism insists that the human mind constitutes the very shapes of reality and can reconstruct the meaning and texture of the world at will. The poet makes the world conform to thoughts:

> He [the poet] unfixes the land and the sea, makes them revolve around the axis of his primary thought, and disposes them anew. Possessed himself by a heroic passion, he uses matter as symbols of it. The sensual man conforms thought to things; the poet conforms things to his thoughts. The one esteems nature as rooted and fast; the other as fluid, and impresses his being thereon. To him, the refractory world is ductile and flexible; he invests dust and stones with humanity, and makes them the words of the Reason. The Imagination may be defined to be, the use which the Reason makes of the material world. (1836, 210)

The mere sensual man remains in a passive state and allows brute matter to impress itself on his awareness. The poet, on the other hand, uses the Imagination, itself a mighty power, to shape a world in his image. The imposition of form actually unhinges the world and allows its inner dynamism to emerge more clearly. In the attempt to own the land or the outer landscape, the sensual man betrays the deeper currents of being that themselves shape and reshape the orders of the world. The poet imitates this protean activity by imposing new and varied forms on experience. In this sense, the poet is an agent of nature's own drive toward eternal self-transformation. By using the human Imagination, the poet participates more directly in the movement of the world as it allows old forms to die and compels new forms to take their place.

The various positive historical religions derive their power and validation from their imitation of the forces of nature. Jesus becomes a kind of popular poet of the spirit who, because of his lack of proper poetic speech, points toward nature in a less compelling way than will the new poet. All things preach to us, and there is no longer any need to derive revelation from archaic and shopworn religions. The Hebrew and Greek texts of the Bible are fragmentary and incomplete compared to the kind of direct revelation available to the common person who allows the spirit into his or her life. The human process can be made divine when it enters into the realms of spirit:

> Who can set bounds to the possibilities of man? Once inhale the upper air, being admitted to behold the absolute natures of justice and truth, and we learn that man has access to the entire mind of the Creator, is himself the creator in the finite. (1836, 217)

If we can recapitulate creation it follows that we do not need historical or textual forms of mediation to link us to the inner life of God. Such forms of mediation still have some educational value but only insofar as they give way to a direct encounter with the spirit. Nature is the source and the validation for all human religions and should be approached directly without the mediating structures that all too frequently become ends in themselves.

In 1836 Emerson privileges the poet as the new seer who will transform communal and personal life. Language does mediate between an alienated self and nature, but this unique form of mediation has a special kind of transparency. No other human activity can convey the inner logic of nature like poetic utterance. Practical activities, the provenance of the sensuous man, remain too bound to specific and finite horizons and circles. Nature speaks through the poet and not through the work of the crafts or of industry. Of course, Emerson learned to broaden his conception of natural revelation as his own reflections on nature went past this early idealism with its excessive emphasis on Imagination and the world constituting power of the poet.

By 1841 Emerson had deepened his conception of nature so as to show the ways in which finite orders relate more directly to the infinite potency of nature. In his Waterville oration, "The Method of Nature," he develops a Neoplatonic theory of emanation that shows how all particulars struggle toward the universal. Nature is unrelenting in its sweep:

> Every natural fact is an emanation, and that from which it emanates is an emanation also, and from every emanation is a new emanation. If anything could stand still, it would be crushed and dissipated by the torrent it resisted, and if it were a mind, would be crazed; as insane persons are those who hold fast to one thought and do not flow with the course of nature. Not the cause, but an ever novel effect, nature descends always from above. (1841, 190–191)

While his aggressive idealism remains intact, if slightly muted, he begins to modify his perspective to show how nature enters into and shapes the orders of the world. The dynamism of nature is more clearly drawn than it was in 1836, and Emerson acknowledges how all particulars are caught in the downward movements of nature. Nature itself is in a constant process of giving birth and is in rapid metamorphosis. All finite ends are momentary and merely use particulars for their fulfillment. Nature is not so much the screen on which the poet may project any picture at will as it is the eternal seedbed of all possibilities.

Nature uses the human process for its own universal end, and there is no longer any room for the proud boasting of the world constituting poet. This shift in Emerson's perspective is profound and enabled him to develop a less eulogistic and narrow conception of nature and of the type of self that correlates to nature's laws. Instead of the sovereign power of the Imagination we have a more humble awe before the infinite:

> Is a man boastful and knowing, and his own master?—we turn from him without hope: but let him be filled with awe and dread before the Vast and the Divine, which uses him glad to be used, and our eye is riveted to the chain of events. (1841, 209)

All selves are caught in the vast emanating powers of nature and must obey the momentum of nature's end or they will be crushed. Nature is indifferent to finite human ends unless they in turn serve the larger end of nature. This striking sense of human vulnerability to the infinite sweep of nature foreshadows the pessimistic naturalism of Santayana who himself came to see the utter indifference of nature to the needs and aspirations of the human process.

By 1844 Emerson's idealism had cooled even more and made it possible for him to probe into the demonic and destructive forces of the world. Many critics trace this inner transformation to the death of his son Waldo in 1842, but the inner logic of his expanding conception of nature would have moved him in this direction regardless of specific finite losses. His own intellectual honesty compelled him to deepen his understanding of the method of nature and to distance himself from his earlier emphasis on the sheer luminosity of human speech. In his 1844 essay "Experience" he presents a stark and lonely image of the human process and its relation to nature. Instead of the "transparent eyeball" of *Nature* we see an alienated self caught on an infinite stairway in which neither the origin nor goal of the stairs are visible. This metaphor compresses in one image his growing sense that it is impossible to penetrate into the mysteries of nature and thus impossible to find a secure or determinate location for the self.

The once luminous orders of the world are now little more than surfaces that deflect our blows and leave us without a sense of the inner being of the world. "Life is a bubble and a scepticism, and a sleep within a sleep" (1844, 337). The clear and distinct world of the Imagination darkens and becomes bereft of natural illumination. The self is a stranger in an indifferent world that seems uncongenial to the aspirations of the human process. Further, nature is no longer the benevolent source of truth but represents a battle field of contending and incompatible forces:

> Nature, as we know her, is no saint. The lights of the church, the ascetics, Gentoos [Hindus] and corn-eaters, she does not distinguish by any favor. She comes eating and drinking and sinning. Her darlings, the great, the strong, the beautiful, are not children of our law, do not come out of the Sunday school, nor weigh their food nor punctually keep the commandments. (1844, 337)

Nature is beyond good and evil and pursues elusive ends that pulse through the self but refuse to reveal their whence or wherefore. Our ignorance of the vast sweep of the world is one of the most striking results of cumulative experience and serves to alienate us from the heart of nature. The rhythms of nature are fragmented into pulses that move the self in a direction that cannot be fathomed. The ethical concepts of human community are little more than pale denials of a premoral universe.

The fitful starting and stopping of experience rides on the back of an even more fitful nature. The law of compensation seems more muted as the origin and goal of all action becomes shrouded in mystery. How can a fitful and seemingly indifferent nature reward or punish action if it is impossible to gauge the ultimate upshot of such acts? For good or ill, we must adjust to a world that is less rational and measured than we thought:

> Nature hates calculators; her methods are saltatory and impulsive. Man lives by pulses; our organic movements are such; and the chemical and ethereal agents are undulatory and alternate; and the mind goes antagonizing on, and never prospers but by fits. (1844, 339)

The mind no longer partakes of the currents of the universal being but stands amid an impulsive world. The self is antagonistic and alienated from nature and can only move through the world in a fragmented way. Unity seems just beyond the reach of the human process and the laws of nature seem to recede further and further from view.

By 1844 Emerson had developed an increased skepticism, combined with a more fundamental world weariness that seemed to cut off the self from those springs of power that were so obvious to the younger Emerson. The power of spirit is likened to that of a mighty river that has been dammed upstream somewhere. The loss of innocence and power compels the self to ride on the chaotic pulses of a less than compassionate and less than rational nature. The very concepts of origin and goal are effaced so that the location of the self remains veiled in mystery. It is as if nature begins to mock the imperial self that once seemed so sure of its place in the world.

The mocking quality of nature is conveyed in Emerson's 1847 poem "The World-Soul," where he depicts the utter distance between the human soul and its world. The following stanza gives a succinct expression of this sense of distance:

> Alas! the Sprite that haunts us
> Deceives our rash desire;

> It whispers of the glorious gods,
> And leaves us in the mire.
> We cannot learn the cipher
> That's writ upon our cell;
> Stars taunt us by a mystery
> Which we could never spell.
> (1847, 24–25)

The Sprite, a personification of nature's outer face, infects the self with a restlessness that cannot be stilled. Not only is nature receding from the self, it also taunts the self in its very act of withdrawal. Our habitat is no longer the luminous horizon of *Nature* but a "cell" from which we cannot escape. Nature does write its own inner history and logic in a special cipher script but fails to provide the hermeneutic key that will make such a script come alive. All attempts to decode the mystery fail and frustrate the self even further. For Emerson, the act of writing a poem is no longer the act of creating a midworld between the self and the world but an act of resignation that acknowledges that the midworld remains elusive.

By 1860 Emerson had moved further toward a sense of nature that was not only beyond good and evil but in some deeper sense even hostile to the needs of the self. Behind the congenial façade of American life lies a blood-soaked panorama that threatens to expose itself:

> The way of Providence is a little rude. The habit of snake and spider, the snap of the tiger, and other leapers and bloody jumpers, the crackle of the bones of his prey in the coil of the anaconda,—these are in the system and our habits are like theirs. You have just dined, and however scrupulously the slaughterhouse is concealed in the graceful distance of miles, there is complicity,—expensive races,—race living at the expense of race. (1860, 381)

Like Thoreau, Emerson came to acknowledge the utter waste and cruelty of nature and its lack of compassion for its endless stream of victims. Buried within the heart of human nature is the same tendency toward conquest and destruction as is manifest in the snake and tiger. Nature feeds off of itself as it moves toward an elusive end that may not in itself redeem the entire process of mutual devouring.

Nature's "tyrannous circumstance" binds the self to a small round of dreary existence. Positive power may struggle against circumstance but its deliverances are feeble and often deflected from their proper orbit. Finitude seems to absorb the impulses toward transcendence and thereby deaden the power of spirit. The self is caught in an endless round of disease, famine, suicide, and loss of power. Fate overwhelms the purposes of the individual and shrinks all horizons. Temperament freezes the personality into a cycle of mere repetition, where novelty and growth are foreclosed. Emerson paints a dark picture of human prospects amid a hostile natural environment.

Yet this stark vision is not without some melioration. Fate can be challenged by power and intellect, and the destructive forces of the world can be harnessed to serve human interests. Human praxis creates its own kind of midworld in which vast torrents become mild and regular irrigation ditches; where disease and its spread can be conquered by engineering and good sanitation. The midworld of action is not that of the world-constituting poet but one that recognizes the absolute power and supremacy of the not-me. Societies absorb and redirect the energies of nature and turn them toward their own use. The midworld becomes the locus for instrumental control and for the taming of an irrational and indifferent nature. Human ends are preserved only insofar as the potencies of nature are refocused around personal and communal needs.

The evolution in Emerson's thought was profound in the period from 1838 to 1860. His conception of nature evolved to reflect his more successful probes into the innumerable orders of the world. If he overstressed the luminous power of horizons and language in his early work, he more than made up for it by coming to stress the dark and fitful events of a nature that was always just beyond human apprehension. The eulogistic tone of 1836 gave way to a world weariness and growing reticence about human powers. Nature was no longer the nurturing mother but the indifferent fate and power that surrounded and mocked the self.

The conception of the midworld followed this progression closely. The midworld of the poem gave way to a more practical and less luminous midworld of human activity. Instead of bringing nature to luminous self-transparency within language, the practical midworld dams rivers and rebuilds local landscapes to meliorate weather conditions. The practical understanding of the midworld helped Emerson find a less inflated and less honorific conception of nature per se.

His concern with the dignity of farming can be traced to this transition from the centrality of language toward the centrality of human productivity. Of all forms of human making, farming is the original. In carving out a field, or in draining off swampland, the farmer is harnessing the infinite potencies of nature for ends that might not otherwise be realized. The farmer is as sensitive to the intrinsic shapes of the world as to the possibilities for change and transformation:

> He bends to the order of the seasons, the weather, the soils and crops, as the sails of a ship bend to the wind. He represents continuous hard labor, year in, year out, and small gains. He is a slow person, timed to nature, and not to city watches. (1870a, 134)

The farmer sets the measure for the rest of society by slowing his pace to that of nature. Insofar as his actions are parallel to those of nature, the farmer becomes a "representative man" for the entire social order. Emerson eulogized rural over city values, yet did so in order to preserve a practical midworld against constant corrosion and self-effacement.

Interestingly, the farmer and the poet are seen to stem from the same "old Nature" and serve in different ways to open out the truths of nature. Even though Emerson grew skeptical about our ability to learn the deepest secrets of the world, he still retained a muted faith in those individuals who sustain some microcosm of intelligibility. The farmer is a hero of the stature of Achilles:

> But he stands well on the world,—as Adam did, as an Indian does, as Homer's heroes, Agamemnon or Achilles, do. He is a person whom a poet of any clime—Milton, Firdusi, or Cervantes—would appreciate as being really a piece of the old Nature, comparable to sun and moon, rainbow and flood; because he is, as all natural persons are, representative of Nature as much as these. (1870, 148)

The selection and enrichment of a plot of land is an act of creation of the same ontological worth as the sacking of Troy or the life of a wandering hero. Yet the farmer works quietly and at a much slower pace than the ancient heroes. His relation to the world is one of acceptance rather than defiance, and his own spirit is without the kind of manic inflation that characterizes the tragic hero. As a representative of nature, the farmer is his own midworld and keeps the truths of nature from fading into oblivion. Consider how the farmer improves on nature by using nature's own laws to transform a hostile environment into one that is friendly:

> Plant fruit-trees by the roadside, and their fruit will never be allowed to ripen. Draw a pine fence about them, and for fifty years they mature for their owner their delicate fruit. There is a great deal of enchantment in a chestnut rail or picketed pine boards. (1870, 142)

Protection from traffic and wind allows the fruit tree to fulfill its own *entelechy* and become useful to the farmer. The pine fence replaces or augments the poem as a fitting symbol of the midworld. Less dramatic than the self-luminous horizon of *Nature,* the simple fence points toward the potencies of nature that impinge on human commerce within the world. Agriculture is not only useful but represents a form of exploration of the orders of the world.

The exploratory power of agriculture is manifest in the simple act of laying down tiles for drainage. As the land is drained and converted to agricultural use, a rich subworld is discovered that remained hidden until human industry brought it into the open:

> By drainage we went down to a subsoil we did not know, and have found there is a Concord under old Concord, which we are now getting the best crops from; a Middlesex under Middlesex; and in fine, that Massachusetts has a basement story more valuable and that promises to pay a better rent than all the superstructure. (1870, 145)

The midworld of farming reveals features of nature that would otherwise have remained beyond human ken. Note that none of this activity requires poetic speech or the solemn cadencies of poetic language. Emerson came more and more to understand that the simplest activities could reveal as much as the most exalted. Farming, unlike modern industry, still conforms to the deeper rhythms of nature and mirrors nature's laws.

Agriculture thus guides both the poet and the ordinary person toward the deeply embedded truths of the world. If the midworld of farming is less intense and dramatic than that of the poem, it compensates by being more secure and efficacious within the community. The farmer, like Adam, builds and maintains the microcosm of truth that can still guide and transform society. Were a society to ignore its farmers, the true poets of the landscape, it would risk breaking all contact with nature. From this would follow an increased alienation from the potencies that still infuse the world. If the farmer settles for less wisdom and less clarity than the poet, it is because human praxis must always start, *in medias res* and work more quietly toward origins and goals. We may indeed be on an endless stairway that hides its beginning and end but our practical actions enable us to transverse this infinity with measured step and sure gaze. The farmer says little compared to the poet, yet lays down deep traces of nature through the midworld of agriculture. In learning to read the more modest script of agriculture we gain that wisdom that knows the limits of the human while remaining open to transcendence.

Wild Farming

Thoreau and Agrarian Life

DOUGLAS R. ANDERSON

HENRY DAVID THOREAU is, in the popular domain, often cast among the defenders of a simple agrarian lifestyle. Attentive readers of his various works, however, have noted ambivalence in his assessment of farm life. Thoreau clearly does not work under the influence of a Jeffersonian romance with agrarianism. As James Montmarquet notes concerning Thoreau's view: "Virtue is certainly not the inevitable outcome of the efforts of farming; it requires special effort and attention in its own right" (see "American Agrarianism," this volume). The dividing line in this ambivalence marks a distinction between that which is wild and that which is tame or civilized. In *A Week on the Concord and Merrimack Rivers* Thoreau honors the farmer as a fixture of the wild dimension of human existence. In his tale of the wilderness farmer Rice he says:

> He was indeed a coarse and sensual man and, as I have said, uncivil, but he had his just quarrel with nature and mankind, I have no doubt, only he had no artificial covering to his ill-humors. He was earthy enough, but yet there was good soil in him, and even a long-suffering Saxon probity at bottom. (Thoreau 1961, 218)

In *Walden,* on the other hand, the farmer is occasionally chastised for being trapped by the conventions of the civilized world:

> The farmer is endeavoring to solve the problem of a livelihood by a formula more complicated than the problem itself. To get his shoestrings he speculates in herds of cattle. (1966, 22)

Thoreau's ambivalence seems to turn on an ambiguity he found in the practice of agrarian life as he knew it. That is, he saw an overcivilized dimension in farming that made it routine and enslaving *and* he saw in it the possibility of a wilder dimension through which human freedom might be lived. The farm thus constituted, for Thoreau, a borderland or midworld between the wild and the tamed. Both actually and figuratively, the farm stands between the city and the wilderness. As such, it marks a place through which one must travel when moving in either direction. Thoreau used this ambiguity, or duality, in agrarian life for two purposes—as a metaphor to awaken us to the possibilities of growth for selves and cultures and to indicate the need for a preservation of the wild side of farming as an actual condition for self- and cultural growth. To effect both purposes, Thoreau found himself engaged in describing and defending what might be called "wild farming."

I take my bearing from my belief that Thoreau's central project was to explore the conditions of the growth of persons in a finite setting, and consequently to rethink the conditions for the amelioration of human cultures. Agrarian life suited Thoreau's purpose here insofar as it provided, in its status as a borderland, metaphorical exemplification of both our enslavement and our freedom: the hindering and enabling of our personal creativity and growth. Indeed, *Walden* is easily read as a contemporary fertility myth. As Stanley Hyman suggests, "the whole two years and two months he compresses into the cycle of a year, to frame the book on the basic rebirth pattern of the death and renewal of vegetation, ending it with the magical appearance of spring" (Hyman 1954, 177). Through this myth, and in his other writings, Thoreau aims both to exercise the metaphorical power of and to identify a need for the instantiation of his own conception of wild farming.

Wild Farming as Metaphor for Self-Cultivation

In a journal entry Thoreau wrote: "As for farming, I am convinced it is too tame for me—and my genius dates from an older era than the agricultural" (1984, 2:22). On other occasions, however, he found in farming a lived proximity to nature's wildness not available in the cities whose "seeming advantages" he was able to forego "without misgiving" (1984, 3:359). As James McIntosh points out concerning *A Week*, "Thoreau would think not only of himself and his brother but also of the haymakers as involved in nature, blended with the grass and in touch with its secret life" (McIntosh

1974, 144). This dual possibility of peril and promise in agrarian life provided Thoreau a figure through which to diagnose and treat the ill health of particular lives within the structure of American culture.

In many of his later writings Thoreau began with a direct assessment of current conditions. In *Walden,* "Walking," and "Wild Apples" we find Thoreau announcing and repeating a particular concern:

> Most men, even in this comparatively free country, through mere ignorance and mistake, are so occupied with the factitious cares and superfluously coarse labors of life that its finer fruits cannot be plucked by them. (1966, 3)

> But possibly the day will come when it [the landscape] will be partitioned off into so-called pleasure-grounds, in which a few will take a narrow and exclusive pleasure only,—when fences shall be multiplied, and man-traps and other engines invented to confine men to the *public* road, and walking over the surface of God's earth shall be construed to mean trespassing on some gentleman's grounds. (1884, 175)

> The era of the Wild Apple will soon be past. (1884, 304)

The illnesses Thoreau found in American culture were for the most part attributable to the deadening effects of being overcivilized, of becoming a culture of inheritance only. In this, of course, Thoreau was not unique; rather, he was exemplary of the transcendentalist movement. His response to this cultural phenomenon, nevertheless, had its own flavor.

Thoreau's diagnosis was straightforward. The overcivilization he incessantly pointed to and challenged had the effect of taming individuals and the culture. This taming, in turn, led to a soporific state in which individuals found themselves adopting current practices as if somehow they were eternal human necessities—souls became owned by their possessions and blinded by their desires. Conformity and mediocrity were the fruit of success. The overall effect of this "Americanization" of things was a regressive culture constituted of individuals who were neither free nor creative. Such individuals, as Thoreau saw it, had little potential for growth; they were stillborn and, in becoming removed from their own lives, they became what Ortega later described as "dehumanized." It was no surprise to Thoreau, then, that a culture so constituted should have little promise left in it: "While civilization has been improving our houses, it has not equally improved the men who are to inhabit them" (1966, 22–23).

As an antidote to our condition, Thoreau tries to take us in the direction of the wild. As John Burroughs points out, "[h]is whole life was a search for the wild, not only in nature but in literature, in life, in morals" (in Harding 1954, 87). Pursuit of the wild makes at least a cameo appearance in almost all of Thoreau's texts and is a central feature of *Walden,* "Walking," "Life without Principle," and "Wild Apples." Characteristically, Thoreau does not attempt a straightforward delineation of the wild or of wildness. His project is to awaken others to the possibility of their own searches for what is wild, and it is here that the midworld of agrarian life takes a central position in Thoreau's work.

As we have already seen, when the city encroaches on the farm, as it had already done in Thoreau's Concord, the effect is a degeneration of agrarian life; farming becomes too tame. Yet this fact must be taken in conjunction with an assessment of Thoreau's own relation to wilderness. If a bit extreme, I do not believe Fannie Hardy Eckstrom's suggestion that Thoreau was not a wilderness person is far out of line:

> Thoreau never got to feel at home in the Maine wilderness. He was a good "pasture man," but here was something too large for him. He appreciated all the more its wildness and strangeness; and was the more unready to be venturesome. (Eckstrom 1954, 113)

It was precisely in his being a "pasture man" that Thoreau was able to see and appreciate the wild in nature. His life at Walden and his treks beyond local fields confronted him persistently with the possibilities available to human existence. Yet it was also as a "pasture man" that he maintained his links with civilization; he understood that he lived what he called a "border life," that he was "semi-civilized," and that he had "strayed into the woods from the cultivated stock" (Thoreau 1884, 303, 280). As many readers have noted, the persona of *Walden* was not a recluse—his insistence on reading and, especially, on the intellectual subtleties of Thoreauvian reading, demands a cultural and intellectual inheritance. In short, if he was a defender of a certain kind of agrarian life, Thoreau was neither a "back-to-nature" advocate nor a primitivist.

Thoreau's heading toward the wild, then, when taken from the midworld of agrarian life, maintains a continuity with the career of a particular culture and civilization. Thoreau does not defend a mindless wildness. Lewis Mumford gets at this in articulating a distinction between Thoreau's and the pioneer's search for and engagement with the wild:

> He understood the precise thing that the pioneer lacked. The pioneer had exhausted himself in a senseless external activity, which answered no inner demands except those for oblivion. (Mumford 1962, 19)

Thus, the do-your-own-thing attitude Thoreau occasionally projects, like the same attitude found in Emerson's "Self-Reliance," while trying to outflank the constraints of mere conformity, must not ignore whatever constraints are necessary for a recovery of the self and a recivilizing of culture. The metaphorical wild farming toward which Thoreau heads maintains the wild in a conjunctive tension with a discipline of self-cultivation. As Leo Stoller maintains, the "keystone of the doctrine of simplicity is the principle of self-culture" (1962, 49; see also Mumford 1962, 16). The upshot is that Thoreau's borderland existence, his defense of a kind of self-sustaining agrarianism in *Walden* and elsewhere, is not meant to run us all to the woods to become, as Emerson says, "woodchucks." Rather, he wants us to remain attentive to the wild side of things without denying our acculturation.

It is important, for example, to see this in "Wild Apples." It is not a simple nostalgia for the wildness of wild apples that is at stake. Thoreau, following a distinction he finds in Pliny, begins by making it clear that apples themselves *are* civilized:

> Pliny, adopting the distinction of Theophrastus, says, "of trees there are some which are altogether wild (*sylvestres*), some more civilized (*urbaniones*)." Theophrastus includes the apple among the last; and, indeed, it is in this sense the most civilized of all trees. (1884, 269)

What is at stake, then, is an ongoing attentiveness to the wildness that is available to apple trees. We must, Thoreau implores, maintain a place for wild apples. We must, as it were, become wild farmers so that we do not both deaden our own existence and close all avenues of cultural growth. There must be a vigilance for the wild in order that there can be a maintenance of "civiliz*ing*"—"*our* wild apple," says Thoreau, "is wild only like myself, perchance, who belong not to the aboriginal race here, but have strayed into the woods from the cultivated stock" (280). As wild apple farmers, we can keep the very possibility of our culture alive:

> Every wild-apple shrub excites our expectation thus, somewhat as every wild child . . . . What a lesson to man! So are human beings, referred to the highest standard, the celestial fruit which they suggest and aspire to bear, browsed on by fate; and only the most persistent and strongest genius defends itself and prevails, sends a tender scion upward at last, and drops its perfect fruit on the ungrateful earth. Poets and philosophers and statesmen thus spring up in the country pastures, and outlast the hosts of unoriginal men. (287)

Thoreau's insistence on this wild farming is not principally a demand for the maintenance of any actual agrarian life. It serves initially as a figure or metaphor for the conduct of any life; indeed, Thoreau's texts seem to aim not at farmers, but at those who are the keepers of a culture's (in his case, America's) intellectual and spiritual traditions. I do not mean to suggest that Thoreau was merely a professional philosopher of the sort he detested, but that his aim was to reach an audience well schooled in American and European intellectual practices. The figure of wild farming suggests that any life can benefit from the recovery of a Socratic mood that recognizes and attends to its own internal borderland: to the continuity of its own tameness and wildness. In this activity alone does one awaken to one's own possibilities; in this way alone can one achieve the freedom that Thoreau takes to be constitutive of a truly human life.

The metaphor of wild farming seems well suited to accomplish Thoreau's first purpose and thus exemplifies Hyman's insight that, for Thoreau, "the richest function of nature is to symbolize human life, to become fable and myth for man's inward experience" (1954, 174). Indeed, Thoreau points to his own "farming" when he says in *Walden* that "some must work in fields if only for the sake of tropes and expression, to serve a parable-maker one day" (1966, 108). However, there seems to be an inadequacy, if not a degeneracy, in a reading of Thoreau that sees only this purpose in Thoreau's use of agrarian life. The independent, "popular," and unschooled reader is too forcefully struck with the sense that Thoreau is directing one not only toward a reconstructed inner life but toward a consideration of an actual revision of lifestyle. A stubbornness abides in these texts that resists the reduction of Thoreau's project to a merely literary one.

Nature's wildness seems too powerful in effecting Thoreau's own awakening to handle it only as a rhetorical figure. As we noted earlier, Thoreau does not, as does Jefferson, see a necessary virtue in living an agrarian life, but he nevertheless seems to see in farming of a certain sort the ground of a second purpose: to clear the ground for the actual wild farming that stands as a condition of a living culture. "I would not have," he says in "Walking," "every man nor every part of a man cultivated, any more than I would have every acre of earth cultivated" (1884, 202). There must be something (and perhaps someone) to remind us of the wild, creative dimension of human existence. This second purpose is thus ancillary to, and perhaps instrumental for, the initial purpose of awakening us to our own projects of self-cultivation. If the first purpose involves returning us to a Socratic mood, this second purpose seems to indicate the need for a new agrarianism that preserves this mood as a real possibility. As Stoller suggests in reference to some later entries in Thoreau's journal: "If it was essential for the education of man to be in continuous association with nature, then it was essential to find some way to preserve nature itself" (Stoller 1962, 47).

If Thoreau does not make a Jeffersonian demand for a fully agrarian democracy, he also places little faith in an urbanized culture that increases its harvest simply by way of industrialization and technological innovation. Thoreau was not antitechnological, but he did not see wildness as a particularly prominent feature of technology. Nor is wildness to be found in corporate farming; such farming seems merely a limit in the direction of the "civilized" farming Thoreau castigated. A remark from *Walden* brings this much into focus:

> We have no festival, nor procession, nor ceremony, not excepting our Cattle-shows and so called Thanksgivings, by which the farmer expresses a sense of the sacredness of his calling, or is reminded of its sacred origin. It is the premium and the feast which tempt him. He sacrifices not to Ceres and the Terrestrial Jove, but to the infernal Plutus rather. By avarice and selfishness, and a grovelling habit, from which none of us is free, of regarding the soil as property, or the means of acquiring property chiefly, the landscape is deformed, husbandry is degraded with us, and the farmer leads the meanest of lives. (111)

I suspect the meanness of *this* agrarian life is to be found in the irony of its lost opportunity. The wild dimension is lost: forgotten are Dionysus and the maenads, lost to our concern for deeds and legal ownership is the perception of wildness. Corporate and technologically driven farming give birth not to wild farmers but to persons who are so intent on marking off their boundaries that they become like the "worldly miser" in "Walking," who could not see that "heaven had taken place around him, and did not see the angels going to and fro, but was looking for an old post-hole in the midst of paradise" (1884, 170).

I see Thoreau making two suggestions concerning the necessity of wild farming: (1) that we ought to preserve a sort of agrarian life that keeps open an avenue to nature's wildness, and (2) that we should engage in the actual preservation of wilderness. These suggestions are not offered as mutually exclusive; indeed, if Thoreau is right about the borderland effect of farming, they may be seen as integral to each other. No avenue is needed if no wilderness survives and, if there is a wilderness, it will have no efficacy if it has no bridge to civilization and the human will that civilizes.

I remain wary of giving full articulation to that which an author has not himself given full articulation. At the same time, I believe Thoreau often provides more content to his thought than is attended to by literary-oriented scholars. On behalf of the first version of wild farming that I believe Thoreau pushes for, I would adduce as evidence much of what has already been said regarding the figure of the borderland; only now I would read it in a more literal mood. In his own experience, Walden was not a wilderness but a locus of mediation. Of his bean-field, Thoreau says:

> Mine was, as it were, the connecting link between wild and cultivated fields; as some states are civilized, and others half-civilized, and others savage and barbarous, so my field was, though not in a bad sense, a half-cultivated field. (1966, 106)

Thoreau sees virtue in this geographically mediating presence of the wild farm; it is a place where one can live with a reasonably clear perception of the virtues of both the civilized and the wild. It is a place through which one may travel to the wild and to civilization.

Thoreau also sees some importance in the kind of life such a place might engender, in the actual living of the wild farmer. This living is an engagement, a confrontation, with that which is beyond our control—nature's own possibilities. Living in this way illustrates for a culture the importance of risk. Whereas neopragmatists such as Richard Rorty seek to glorify the reduction of risk in cultures, Thoreau worries over its loss. In a culture such as ours, which Rorty has identified as "postmodern bourgeois liberal"—a culture whose mainstream is so unattuned to risk that it has used the AIDS epidemic as a sort of smelling salts—Thoreau would see wild farming's direct engagement with nature as an important and ongoing reminder. In a passage in "Walking," sounding much like Emerson, Thoreau defends agrarian work: "The callous palms of the laborer are conversant with finer tissues of self-respect and heroism, whose touch thrills the heart, than the languid fingers of idleness" (1884, 168). Later in the same essay, in a less gentle and less Emersonian vein, he addresses the confrontational nature of wild farming:

> The weapons with which we have gained our most important victories, which should be handed down as heirlooms from father to son are not the sword and the lance, but the bush whack, the turf-cutter, the spade, and the bog-hoe, rusted with the blood of many a meadow, and begrimed with the dust of many a hard-fought field. (192–193)

For Thoreau, a living culture will always make room for farmers and lives of this sort as actual reminders of the rudiments of human existence. The risk, confrontation, and engagement with nature these lives involve cannot be maintained at second hand or by memory. A culture that forecloses on wild farming will, by that act, also foreclose elements of its own freedom and power.

The second version of actual wild farming Thoreau intimates is the stewardship and preservation of wilderness. In a sense, Thoreau asks us to cultivate, but not tame, nature's own wild features. The trick to this sort of cultivation can be seen through the distinction between *agape* and *eros*. A cultivator working from agape will attend to that which is cultivated but allow it to take its own course. One who works from eros will dominate the object of attention. It is the agape version that Thoreau seems to have in mind. This project is most clearly evident in "Walking," where Thoreau's appeal for wilderness is integrated with his concern for the recivilizing of selves: "To preserve wild animals implies generally the creation of a forest for them to dwell in or resort to. So it is with man" (1884, 191). Despite the metaphorical edge of "wilderness" and "wildness," Thoreau dwells so heavily on nature's presence in "Walking" that it is difficult to marginalize his concern with the actual environment in which we find ourselves and with the ways in which we attempt to develop this environment. He is overtly worried by the conventional conception of progress, for example, because it fails to meet some basic aesthetic needs:

> Nowadays almost all man's improvements, so called, as the building of houses, and the cutting down of the forest and of all large trees, simply deform the landscape, and make it more and more tame and cheap. (1884, 169)

In light of such direct remarks, it is not surprising that those who have taken up wilderness preservation as a cause hail Thoreau as a progenitor. What such appropriation occasionally fails to remember is that this actual preservation is, for Thoreau, instrumental to his central project of reawakening selves and civilizations.

The romantic flavor of Thoreau's call to preserve wilderness should not obscure the practical side of his character. As one who knew how to make and fix things, he no doubt understood the costs involved in wilderness preservation—both real and opportunity costs. Yet his own economy of the soul and human community demanded that this cost be borne. As Philip and Kathryn Whitford point out, Thoreau reveals in his journal a working knowledge of how woods and woodlots might be better managed (Whitford 1954, 203ff.). More generally, I think Mumford is right in maintaining that "Thoreau saw what was needed to preserve the valuable heritage of the American wilderness" (1962, 18). Thus, for Thoreau, not only must we preserve an agrarian life open to wildness, but we must create an agriculture of the wilderness itself—we must design a wild farming whose harvest *is* wilderness. Again, Thoreau does not denigrate civilization wholesale; rather, he calls for a recivilizing community that outgrows its own blindnesses and destructive tendencies. Only a civilized and awakened culture could evidence its freedom by choosing to keep these kinds of wild farming as integral elements of its wholeness.

Conclusion

That Thoreau had two purposes in his discussions of wild farming is perhaps best illustrated by his own life and his wish to "combine the hardiness of . . . savages with the intellectualness of the civilized man" (1966, 8). If writing was his occupation, it cannot be denied that he remained, in his own way, conversant with nature's wildness even if he was not a bona fide "mountain man." Thoreau by and large enacted what he demanded. Such an attempt to live one's own philosophy is rare enough among intellectuals; to bring it off with the sort of success Thoreau achieved is even more rare. As a wild farmer of sorts, Thoreau was able to say:

> For my part, I feel that with regard to Nature I live a sort of border life, on the confines of a world into which I make occasional and transitional and transient forays only, and my patriotism and allegiance to the State into whose territories I seem to retreat are those of a moss-trooper. (1884, 207)

In his border life, Thoreau undertook to cultivate his own particular field of endeavor—an idiosyncratic occupation:

> Surveyor, if not of higher ways, then of forest paths and all across-lot routes, keeping many open ravines bridged and passable at all seasons, where the public heel had testified to the importance of the same, all not only without charge, but even at considerable risk and inconvenience. (1966, 253)

No doubt those who pursue Thoreau's insights concerning wild farming will bear similar inconvenience and risk; but in Thoreau's own peculiar economy of the self, it may not be a bad start toward preserving one's own wildness and opening oneself to one's own freedom.

Part III

PRAGMATISM AND AGRARIANISM

Dewey and Royce

Provincialism, Displacement, and Royce's Idea of Community

THOMAS C. HILDE

I want to know the way that leads our human practical life homewards, even if that way proves to be infinitely long.

—Josiah Royce, *The Philosophy of Loyalty*

To be is to be in place, that is, in some place rather than another place. What is less obvious, however, is that every place seized is a place denied.

—John J. McDermott, *Streams of Experience*

In a world of wandering and of private disasters and unsettlement, the loyal indeed are always at home.

—Josiah Royce, *The Philosophy of Loyalty*

A RECURRENT FEATURE of agrarianism in its various manifestations, especially contemporary and popular, is the notion that we can recapture a healthier form of community if we are somehow able to return to healthy—because smaller, nurturing, and placed—farming practices and rural relationships. Agrarianism, quite possibly more than any other life philosophy, relies on the notions of community and place in ways central to its own articulation, despite agrarianism's multiple and often contradictory forms (see "Agrarianism as Philosophy," this volume).[1] Wendell Berry and others have argued that agriculture is not merely a vocation, but is or could be again a culture in itself in which practices, skills, experiences, and local knowledge are shared, inherited, and communally reconstructed in accordance with "nature as measure." If agrarianism at its heart involves a judgment of value, it is a value both applied to and derived from a certain type of avocation: farming. As culture, farming is grounded in an informal history of a community or set of communities, the health of which is thus vital to the health of agriculture and the converse. Farmers, for writers such as Berry, are the artist-guardians of the rootedness of an otherwise overly economically minded and mobile modern America. This is the heart of the agrarian ideal: that the special relationship farmers have to the land and to each other is one of possibly greater value to individuals and to society than others, and that this relationship is the foundation of what is best in American culture.

Often, however, the idea of community is not explicit in articulations of agrarianism. This is unfortunate because it is the form and definitions of community which are going to be determinative of a healthy set of agricultural relations of agriculture-as-culture. Wendell Berry, Wes Jackson, and other prominent agrarians maintaining a contemporary dialogue between the farm and contemporary urban/suburban life suggest the importance of these relations (see, for example, Berry 1977 and Jackson 1985). Much of this work is insightful and eloquent, especially the ongoing general contention that "a people cannot live long at each other's expense or at the expense of their cultural birthright—just as an agriculture cannot live long at the expense of its soil or its workforce, and just as in a natural system the competitions among species must be limited if all are to survive" (Berry 1977, 47).

It is clear that agrarians such as Berry are as concerned with how we live as are pragmatists, if not more so. Yet, as James Campbell has noted, Berry, at least in his earlier works, makes the problematic claim that through their relations to and placement on the land farmers develop a healthy temperament and sense of personal fulfillment which is apparently inaccessible to others (Campbell 1990, 42). The key to the good life lies in the rhythms and hearty labors of rural life. According to Berry, good farmers are better able to grasp the interrelatedness of natural and human things. They maintain an "attitude of humility and reverence" toward nature (Campbell 1990, 42). But, as Campbell also asks, why should the practitioner of agriculture be in a position that is more advantaged, more culturally significant, more fulfilling than, say, a teacher, urban social worker, pilot, or geophysicist? I might also add, why not novel combinations of urban and rural occupations? The claim on Berry's part paradoxically renders the community of agrarians or farmers a community exclusive of those who do not understand the inclusiveness of things because they are not agrarians. In later writings, Berry maintains that community involves, among other things, a "respect for all stations and occupations" (Berry 1993, 119). He continues the theme that community must be "placed," however: "if the word community is to mean or amount to anything, it must refer to a place (in its natural integrity) and its people. It must refer to a placed people" (168). According to Berry, community must also involve the development of virtues of respect, loyalty, and "goodwill," to the maintenance of this community and its locality (119–121). But, whatever the explicit concrete causes, agrarian community has suffered its dissolution through "internal disaffection and external exploitation," and provincial life must now be restored for the health of individuals, of places, and thus the health of the larger society (125). Yet, as Campbell notes in his earlier essay, "in our increasingly industrialized and urbanized world, we are clearly in serious trouble if selfhood and personal fulfillment are possible only on the land" (Campbell 1990, 41).

While Berry's critique of polluting, upscaling, and displacing modern farming practices retains a potent critical force in a massive consumer society, we must nevertheless ask what sort of placement we may actually expect of our mobile world of displacements. In Berry's terms, it is clear that the most wholesome of placed communities is the agrarian community. But can community—whatever it may be—truly take place only on "the land" as things are now? To what extent can we still view community in agrarian terms, or is it even possible that the terms through which we conceptualize the idea of community are as archaic as the horse-drawn plough?

Certainly, the search for a home place, for community, for identity, or for redemption is as significant and problematic in American thought, action, and literary imagination as are the historical actualities and interpretations of the grandiosely individualistic frontier quests that still provide the founding mythology for American industrial enterprise. But while this individualist myth is present throughout American culture, the idea of community is often rendered explicit by theorizing across that which divides human selves from human community, or, in other words, through its affective absence. This is central to lingering romantic strains in an agrarian idea of community: the sense of loss, the sense that things are not as they were and, moreover, that they are worse off for it.

If we know ourselves as selves primarily through others, as pragmatists and other social philosophers, including Royce, have it, then community is a question of viewing one's practices, language games, hopes and histories in and through the others who constitute a given association. It is also a matter of distinguishing one's associations through the observation of what they are not. This latter oppositional aspect of community is, in fact, complicated through historical distances and the warp of our own perspectives external to communities to which we do not or no longer belong. It is complicated by the perhaps rarer identifications with particular communities which cannot or will not have us. This may be especially the case in an era of world travel, displacement and modern nomadism, television, the disneyfication of some of our warmest sentiment, and the Internet: we come to consciousness often through viewing our ostensive linguistic or moral communities from afar. No matter to what extent community may, at least partially, comprise actual shared memories, these distances may be manifested as a romantic nostalgia for things lost or past. And contemporary life is filled with displacements of all sorts. Perhaps a longing for communal identification implies the perspective of lack or loss of community or perhaps a deracination, or at least movement of place or culture in the first place. But if Royce and others are right, it is through active participation in community that the lives of human individuals are rendered meaningful. If it is truly the case that contemporary communities are increasingly mere virtual communities, then both agrarians such as Berry and philosophers such as Royce are, in effect, pointing to and critiquing the problem of meaninglessness in contemporary life. This criticism is central to the role of at least agrarian communitarianism in contemporary culture and society.

Royce was not an agrarian, however, and it is perhaps only his ideas of community and provincialism that place him on territory familiar to agrarianism. Royce's primary intellectual devotion was to the abstractions of an absolutistic, idealistic metaphysics and its logic.[2] Although he did briefly mention agriculture in his 1908 cultural-geographical study, "The Pacific Coast" (Royce 1969, 1:181–204)[3], he did not write about the practice, nobility, politics, goodness, or spirituality of farming in particular. The experience of brute nature is not as important to his philosophical and literary work as it is to that of Thoreau or Emerson. Royce's expressed religion ("a literal evangelical piety," as Ralph Barton Perry put it; 1938, 14) stems originally from his mother's Bible and German idealism rather than from naturalistic accounts of experience. And though generally conservative, he did not call for a political "return to . . ." of any specifically agrarian stripe. Salvation was never a matter of salutary trips into the New

England woods for Royce, but a matter of community regardless of where it was to be found. When he did require mental rejuvenation after a near collapse, he traveled across the Pacific Ocean to Australia (see Oppenheim 1980). In regard to concrete human practices and the human relationship with nature, even Royce's writings on the frontier are devoted primarily to early California mining rather than agriculture, a productive practice rather than the reproductive one generally thought to distinguish agriculture from other human practices (see Berry 1977, 217–218).

Yet, unlike his peers among the philosophical elite in New England, Royce's own life experience was truly unique, for he was indeed a child of the California frontier, and his philosophy began not toward an unquestioned experience of wilderness and rural culture, but within it. Being in a place that occupied the fantasies of many American minds was simply a matter of fact. The realization of what this meant in terms of his philosophical life was reflexive for Royce. California represented the ostensive communal place he came to live from beyond its actually lived, and actually placed, experience. It is in New England that he came to consciousness of his Californian identity, although what that meant exactly remained to be discovered.[4] This awareness of provincial community external to its actual lived experience is a central theme to which I will return in the following section, and one I believe places a pragmatic Royce in the context of the present volume as a corrective to agrarian conceptions of community.

Furthermore, despite the systematic effort toward an idealistic metaphysics, Royce found it necessary—unlike some other pragmatists and their own imperatives—to actually delve into the concrete conditions of lived experience (*California* 1886, *Studies in Good and Evil* 1898, *Pacific Coast* 1900, *Provincialism* 1902, etc.). He explored climate, mining technologies, historical events, and even tried his hand at a novel with California as its setting. However much Royce's absolutist language was underscored by the biblical themes of sin, atonement, and salvation, he found it necessary to examine the concrete conditions and real problems of community on its own terms as well as its concrete moral bonds. This concern was not only philosophical but also deeply personal in the face of loss of placed community, and this is an aspect of Royce's writings I wish to educe here. However much of an overstatement it may be to try to find a specifically agrarian sentiment or influence in Royce, his social, historical, and moral writings reflect a deeply personal concern regarding a particular historical moment in which agrarianism and the frontier of the United States were giving way to full-fledged industrialism and its displacements. In this respect, his mistrust of industrial modernity was closer to that of Thoreau and Emerson than it was to Dewey's qualified techno-optimism. Royce's response to this is twofold: it concerns the variety of creative individual insights and convictions as well as the well-being of the communal "province." That is, given "a world of wandering and of private disasters and unsettlement," Royce may be interpreted as philosophically responding, however obliquely, to the meaning of the loss of agrarian-based society and to what this entails for human community (1969, 2:890).

My point in this essay is rather simple: Royce wrote on California and on concrete community as someone alienated from that community. As such, paradoxically, this made him wise to the ontology and moral force of community. Perhaps though, alienation is too strong and misleading a word here. Rather, Royce's idea of concrete community was constructed out of its personal existential loss, but he forged this very loss into a central force in the moral ideal of community. A stronger claim, which I can only suggest here, is that loss lies at the very heart of at least the modern idea of community, and this loss is reflected in Royce's philosophy. We know now that the reaction to this loss, when it takes the form of a return to heroic purity or to mere decentralization or, conversely, to a Rousseauvian synthesis of selves, is at least as potentially dangerous as any loose collection of self-serving or detached individuals. It is the loss in the idea of community, nevertheless, that is the theme in Royce which, I believe, enables him to speak to the idea of agrarian community as both a sympathizer to its problems and a critic of its latent desires. In order to elicit this aspect of Royce, it is necessary to draw on works in his oeuvre that are rarely discussed by philosophers, particularly his writings on California and his 1902 essay "Provincialism."

Royce's Provincial California and Displacement

During the gold rush of 1849, six years before Josiah Royce's birth, his mother, Sarah Royce, migrated to California across the North American plains, mountains, and desert. She, her husband, and their infant daughter arrived in a California mining camp in October of that year, having successfully endured the same infamous hardships as most other forty-niner pioneers. Also like many others, Sarah Royce attributed the success of the family's voyage to providence, having found that the hand of God existed in the brush and vistas, the wagon tracks of those who had earlier passed through, and the loneliest of moments, guiding the family toward an unfamiliar land. Her account of the voyage was written in the 1880s to assist her son in his history of California and was published in 1932 as *A Frontier Lady: Recollections of the Gold Rush and Early California.* She writes, "only a woman who has been alone upon a desert with her helpless child can have any adequate idea of my experience . . . but that consciousness of an unseen Presence still sustained me" (S. Royce 1932, 45). But Sarah Royce also attributed the family's success to a previously unsuspected personal courage, and a longing for a "home-nest." For the Royces, California represented a loyalty to a home as yet unseen, blended into the classic frontier combination of secular hope and religious faith. It became home especially to the extent that the religious morality and Christian faith the family had brought from the East found sustenance in communities which they built from the crudeness of mining camps, and in "institutions which lie at the foundation of morality and civilization" (S. Royce 1932, 114). This same concern with moral community runs throughout Josiah Royce's philosophical and public policy work, and it finds its roots in his earliest writings.

Royce was reared in the mining camps of northern California and educated mainly by his mother during his early years. He was somewhat of an outsider later in school and was bullied by schoolmates, who, in his sardonic words, taught him "the majesty of the community" (1969, 1:33). In his adult life, he studied briefly in Germany then at Johns Hopkins University, where he received his doctorate and, more importantly, met William James. Royce returned to California and eventually landed an English teaching position at the new University of California in Berkeley.

If we examine Royce's early letters from his native California to potential colleagues in the academies of the eastern United States, we find a Royce longing to escape his "metaphysically bereft" and "execrable" provincial surroundings. In an 1879 letter to William James, Royce writes bluntly, "there is no philosophy in California . . . there could not be found brains enough [to] accomplish the formation of a single respectable idea that was not a manifest plagiarism. Hence the atmosphere for the study of metaphysics is bad. And I wish I were out of it" (Royce 1970, 66). Royce writes that he had "given up trying to follow [California's] madness or to predict its behavior" (1970, 93). In his later speeches and letters, he notes that despite the sublime landscapes and seascapes of California, it took him a long time to learn anything from the place. In fact, it appears that it required the move to Massachusetts, which he made in 1882 through James's assistance in securing him a position in philosophy at Harvard University.

After the passage of only one year in the northeast and with the aid of his mother's recollections, Royce embarked on a study of Californian history during its main gold-boom years. This book, *California from the Conquest in 1846 to the Second Vigilance Committee in San Francisco: A Study of American Character* (1948), is an account that runs from the disordered lives and "seemingly accidental doings of detached but in the sequel vastly influential individuals" through these detached individuals' recognition of "the fearful effects of their own irresponsible freedom" to the first orderings of community (Royce 1948, xxxi, 295). In his writings on California, Royce is largely concerned with discerning the material conditions for social order and moral growth, and this practical concern leads him into geographical studies, accounts of historical events, technological developments, race relations, and the politics of early California. Since, for Royce, "moral growth is everywhere impossible without favorable physical conditions," (1948, 223), he sets out to discover how the various conditions of place may be related to the growth of moral communities.

Underlying Royce's account is an assertion regarding the importance of social order or community, and that the character traits of Californians during the period were instantiations of a broader national character. The "struggle for order," in practical-material terms, required, for example, the technological shift from the solitude of pan mining to what Royce reads as the cooperative process of sluice mining. In other writings on California, Royce takes up the romantic theme of environmental determinism: "it is our American character and civilization which have been already moulded in new ways by these novel aspects of the far western regions" (1969, 1:181). These "novel aspects" included a climate that enabled agriculturalists and even businessmen to work during the entire year, and more generally a natural environment that "suggests in a very dignified way a regularity of existence, a definite reward for a definitely planned deed" (1969, 1:198). The intimacy with relatively unsettled nature created, according to Royce, "a sense of power from these wide views, a habit of personal independence from the contemplation of a world that the eye seems to own" (1969, 1:195).

There is a definite early tendency in Royce toward romanticizing the frontier of late-nineteenth-century America and the character of its inhabitants: the experience of such a place could only lead to the development of particular sorts of individuals. In more concrete terms, according to Royce, the very remoteness of California led to a burgeoning sense of communal identity once mere association had formed, as distinct from the rest of the country. Yet, beyond this simple environmental determinism lies a social imperative that echoes Sarah Royce's account of the necessity of "civilizing institutions." Individuals need community, Royce insisted. They need social order through which to realize themselves. Otherwise, individuals are a "chaos of desires," left to the whims of arbitrary authority and impersonal forces, what Royce will later refer to as "the detached individual . . . an essentially lost being" (1969, 2:1154). The rugged frontier individual is a wanderer, "probably a born wanderer, who will feel as restless in his farm life, or in his own town, as his father felt in his" (1948, 393). Furthermore, as the final lines of *California* summarize,

> we are all but dust, save as this social order gives us life. When we think it our instrument, our plaything, and make our private fortunes the one object, then this social order rapidly becomes vile to us; we call it sordid, degraded, corrupt, unspiritual, and ask how we may escape from it forever. But if we turn again and serve the social order, and not merely ourselves, we soon find that what we are serving is simply our own highest spiritual destiny in bodily form. It is never truly sordid or corrupt or unspiritual; it is only we that are so when we neglect our duty. (394)

Royce's chief concern here, one expressed more clearly and fully in later works, is the danger of this detached individual. The detached individual is one who "naturally hates restraint," who cannot view himself as a member of or devote himself to a larger community. Conversely, one cannot be loyal to that which repeatedly denies one's very capacity to be loyal: any society that abnegates the capacity for loyalty is one that does not merit the distinction of community. For Royce, human salvation comes through human community, and Royce's *California* represented a historical moment in which frontier individualistic egoism in the grand project of conquest was at the crucial juncture of either developing into or resisting community. *California* was a case study in this practical tension.

The *California* project engulfed Royce for three years as a "labor of love" (1948, xxxiv). This great preoccupation would seem to run counter to his philosophical interests as well as his sentiments for the state based on his earlier letters. Royce had found in Cambridge, however, that he wore the indelible mark of the provincial, and through the *California* project discovered the fact that he was indeed "a native Californian . . . who neither can nor would outgrow his healthy local traits" (1948, 393). In later works and letters there is a clear tone of pride in the assertions of his own provincialism and his fluency in the idioms and culture of the frontier. Royce's writings from this point on became a broad conversation between his metaphysics and logic on the one hand, and his social philosophy, geographical writings, and ethics on the other (see McDermott, 1985). This dialectic, as it were, reflects his own lived dichotomy of being a Californian, a rough-edged provincial, while at the same time living in the erudition of the northeastern United States, where he had been "unable to become and to remain . . . provincial" (1969, 2:1071). It also reflects his desire "to know the way that leads our human practical life homewards, even if that way proves to be infinitely long." (1969, 2:858). For the later Royce, this home is no longer merely his distant California, nor placed community more generally, but the idea of community itself and its ethic of loyalty.

Provincialism, Loyalty, and Community

The significance of the idea of community to Royce's work is expressed most famously towards the end of his life in his 1916 talk delivered at the Walton Hotel in Philadelphia—John Dewey was present, among other noted intellectuals—in which he reflected on his own life experience and work. It was here that he famously avowed that his philosophical oeuvre had ultimately pointed toward the idea "that we are saved through the community." This is what he had already sensed as a youth: "I strongly feel that my deepest motives and problems have centered about the Idea of the Community, although this idea has only come gradually to my clear consciousness. This was what I was intensely feeling, in the days when my sisters and I looked across the Sacramento Valley, and wondered about the great world beyond our mountains" (1969, 1:34). John Clendenning points out in his biography of Royce that at the end of Royce's life the philosopher realized something he had known inchoately all along—that, in Clendenning's words, "it was the idea of the community, not the absolute, which formed the center of his thought" (Clendenning 1985, 393).

Royce's writing on community took on a concrete form in the short essay "Provincialism," published in 1908 as part of the book *Race Questions,*

Provincialism, and Other American Problems (see 2:1067– 1088). Here Royce argues that a province is not merely a particularly well-defined geographical place nor defined by a particular practice or set of practices in a particular place. The province, rather, is "any one part of a national domain, which is, geographically and socially, sufficiently unified to have a true consciousness of its own unity, to feel a pride in its own ideals and customs, and to possess a sense of distinction from other parts of the country" (1969, 2:1069). Royce finds loyalty to the province to be crucial to the identity of individuals as well as national identity in the face of "evils" of mobile, industrial modernity. The province is not merely agrarian and opposed to the metropolis in Royce's account since, of course, the largest of cities (for example, New York or Tokyo) may also contain local communities, whether clearly delineated or in a process of displacement. In its moral sense, "wise provincialism" serves, in a rather broad way, as a local consciousness in contradistinction to both homogenizing and fragmenting or displacing forces of the larger industrial society. A specifically placed agrarian province *might* serve in this capacity, but it does not do so exclusively or necessarily.

The problem—an old one—remains how to retain what is valuable in those local and communal ways of life that may be subsumed by industrial capitalism or operational thinking or the disaffection and exploitation endemic to agrarian community, and not lapse into reactionism or into dangerous and excessive forms of communal exclusivity. Royce described tendencies antagonistic to wise provincialism as nonacceptance of strangers into the province, the problems of homogenization of media and communication, and what Royce calls the "mob spirit," the tendency of the ideal causes of nations or communities to become hypostasized into universal and exclusive truths acted upon emotionally. But, for Royce, this signals the question of the general moral and psychological health of individuals. The danger for individuals is detachment, the sort of detachment or displaced loyalty that Royce found in early California, an individualism run wild without directing its loyalty toward anything other than personal gain and, most importantly for Royce, without directing its loyalty to the cause of the community. The detached individual is unable to understand the "social laws" that shape him/her and therefore finds any social ordering, such as it may be, a restraint upon natural desires. For Royce, this is tantamount to finding oneself adrift, without possibilities for discovering one's own causes in life, and thus, at the whims of the destructive tendencies set forth through the mob. This will bring us to the ethic of loyalty. As John J. McDermott writes, "that person who lacks loyalty and concern is fair game for seduction by those nefarious movements which seek to wreck the community on behalf of some political, social or religious ideology, all of them self-aggrandizing." (McDermott 1985, 164).

Royce's provincialism is not a reactionary sectionalism. He tempers the romantic environmental determinism of the California writings with a provincialism that seeks beyond its boundaries. Indeed, ever vigilant in regard to tendencies toward the narrowness of exclusivity, Royce considers it crucial that the province promote intellectual and geographical travel away from the province in order to return with fresh cultural resources rather than have the province suffer the cultural atrophy of "leveling tendencies" or of mob homogenization. As with Royce and his California, understanding of the communities to which one belongs requires leaving them, whether temporarily or even losing them altogether. The traveler here can be viewed perhaps in terms of a navigator of a Thoreauvian "midworld" of sorts (regarding Thoreau's view, see Anderson, this volume). In Royce's philosophy of community, this reflexivity is itself paradoxically constitutive of the idea of community in the form of wise provincialism: that community requires the very tension (displacement) that would seem to undermine it. But this reflexivity is two sided—in terms of the view from the external position (for example, Royce writing from Cambridge about his now-dear California) and in terms of the view toward that which is external in a physical, moral, or intellectual sense (for example, the impetus to travel or wondering "about the great world beyond our mountains"). It points toward what Royce viewed as the necessity of a double placement of the province: its local place and its placement within a larger ongoing, developing ideal community. This double placement is often absent in agrarian ideas of community. Royce's philosophy of community provides a corrective here.

What then is the Roycean notion of community? Its importance is certainly clear for Royce: "man, the social being, naturally, and in one sense helplessly, depends on his communities. Sundered from them, he has neither worth nor wit, but wanders in waste places, and, when he returns, finds the lonely house of his individual life empty, swept, and garnished" (Royce 1968, 131). Although much of Royce's late concern with what he called "the community of interpretation" (and much of his brilliance) was epistemological and metaphysical (derived from his studies of Peirce's logic or semiotic), we are here concerned with community as a moral ideal. Perhaps the most important way to get at the question of Royce's idea of community is by examining his elucidation of three basic conditions for the existence of community. First of these conditions is "the power of an individual self to extend his life, in ideal fashion, so as to regard it as including past and future events which lie far away in time, and which he does not now personally remember" (253). The capacity for awareness that we are historically situated, to view ourselves as living within a context of histories undergone and of hope, is a central aspect of community. The second condition is "that there are in the social world a number of distinct selves capable of social communication, and, in general, engaged in communication [that is, interpretation]" (255). And the third condition is that "the ideally extended past and future selves of the members include at least some events which are, for all these selves, identical" (256). In other words, community is a matter of ideally and imaginatively extending oneself conjointly with others through time, through a "community of memory" and a "community of hope." The former comprises shared memories and experiences undergone; the latter shared ideal goals, or "causes," anticipations, and hopes. Community involves an understanding of those events of the past and future as ones that make up the identity of others as well in an interpretative project of unfolding community; that is, a present that interprets the past to the future. For Royce, placement of community is situatedness in a temporal social and moral sense. In effect, a community is not merely a cooperative human association, but one "conscious of its own life" through time, and which reflects this consciousness in its concrete actions (263). Yet, Royce admonishes,

> the love of a community . . . is . . . discontent with all the present sundering of selves. . . . Such love . . . restless with the narrowness of our momentary view of our common life, desires this common life to be an immediate presence for all of us. Such an immediate presence of all the community to the members would be indeed, if it could wholly and simply take place, a mere blending of the selves,—an interpenetration in which the individuals vanished, and in which, for that very reason, the real community would be lost. (1968, 267–268)

Community requires individuals capable of seeking their own causes even if community is the central cause toward which one directs one's loyalty. As Royce suggests,

> the ideal extension of the self gets a full and concrete meaning only by being actively expressed in the new deeds of each individual life. Unless each man knows how distinct he is from the whole community and from every member of it, he cannot render to the community what love demands,—namely, the devoted work. (268)

Discerning solely the relations between individuals and community was not enough in itself for Royce, however, since there is nothing said thus far that precludes destructive, harmfully exclusive, or oppressive communities. Due to their very nature, specific communities can often be destructive, and the quest for some ideal vision of community can blind the vision in other respects. The ethical imperative underlying community is loyalty to loyalty, a notion Royce had developed earlier in his *The Philosophy of Loyalty* and proceeded to refine in the context of the late post-Peircean logic of what is perhaps his most extensive philosophical work, *The Problem of Christianity.* Loyalty to particular causes, while important, is not enough in itself as an ethical principle. Given that one can be loyal to execrable causes which may be destructive of the possibility of loyalty on the part of others, loyalty to loyalty serves as the ethical principle by which individuals may seek the very cause—ultimately, for Royce, the community—that provides individual lives with the fulfillment of their own individuality while nondestructive of that of others.

This is the moral imperative underlying Royce's version of the argument that individuals are such only through the community, an argument he shares with other social pragmatists. It is also central to his more concrete "provincialism" in that, as Royce thought, social disintegration in a vast modern nation could be countered only by the nurturing of provincial loyalties. These loyalties would be formed partially through practical means respectful of the principle of loyalty to loyalty. Thus derives part of the necessity of travel beyond one's provincial communities: the understanding of oneself in regard to that to which one projects one's loyalties, and the conflicts that such loyalties may imply. Just as one comes to individuality through the social, wise communities or provinces are formed through a move—a reflexive or interpretative move—beyond the bare outline of simple association or of shared sentiment. This can be read in light of Royce's studies into the triadic structure of interpretation in *The Problem of Christianity.*[5] But here I wish merely to note that this imperative that underlies community and that ultimately provides the grounds for the idea of community, obtains from an understanding of that which is external to it.

The external may be manifested in terms of the reflexive move which Royce points toward in the idea of traveling away from the province in order to return to it funded with, at the least, new intellectual and moral resources. Moreover, Royce the idealist certainly thought this external could be manifested in terms of an ideal. In "Provincialism" he tries to remind us that "everything valuable is, in our present human life, known to us as an ideal before it becomes an attainment, and in view of our human imperfections, remains to the end of our short lives much more a hope and an inspiration than it becomes a present achievement" (1969, 2:1085). In terms of the province this means, Royce says, "the better aspect of our provincial consciousness is always its longing for the improvement of the community" (1969, 2:1085). In terms of human community, it means an ongoing, "infinite" interpretative project of moral agents.

Ultimately, given Royce's long-term taste for the absolute, loyalty points toward the universal human community in which human antagonisms are dissolved in part through embracing the categorical imperative of loyalty to loyalty. This is the crowning development of Royce's "evangelical piety" since it is the larger human community that Royce interprets as the Church, a sort of Augustinian heavenly city. For the Royce of *The Problem of Christianity*, God is the ideal community, although through the lens of Peirce's logic the content of this claim might be possible to render in more secular language. The fact remains for Royce, nevertheless, that one must reach toward a larger community rather than solely toward placed local communities. His religion—for all of its biblical allusions and identification of the community as the Church—is the one that more closely approximates what Dewey referred to as "religion as a sense of the whole" which is nonetheless "the most individualized of all things, the most spontaneous, undefinable and varied" (MW14:226). Returning to the above-mentioned external referent in the idea of community, might it even be manifested in terms of displacement or loss?

Loyalty, Loss, and Agrarian Community

Not till we are lost . . . not till we have lost the world, do we begin to find ourselves, and realize where we are and the infinite extent of our relations. In *The Philosophy of Loyalty*, Royce writes that "the memory of those who sorrow over loss is . . . fond of precious myths, and views these myths as a form in which the truth appears" (1969, 2:967). I think we can view the agrarian idea of community in similar terms: the very loss of community elevates it to truth and a contemporary society of "sundered selves" to falsity. Curiously, the shared memory that comprises a condition of community is in fact one now often manifested as grief at the loss of community, whatever one's take on the causes (late capitalism, political ineptitude, moral irresponsibility, fast food, television, etc.; Berry's version is summarized as "internal disaffection and external exploitation" [1993, 125]). The truth of human being appears as community, and the question by the variously disaffected and exploited becomes one of how to arrive at that truth. For agrarians such as Wendell Berry, that truth is attained through placement and a loyal, nurturing responsibility for the place. These two aspects converge in the activity of farming. For others, such as Royce, the truth of the myth has a critical power and that critical power itself comprises the central aspect of community in a broader society that is simply no longer agrarian. Royce writes,

> such loyalty to a lost cause may long survive, not merely in the more or less unreal form of memories and sentiments, but in a generally practical way. And such loyalty to a lost cause may be something that far transcends the power of any mere habit. . . . [It is] a loyalty to a cause whose worldly fortunes seem lost, but whose vitality may outlast centuries, and may involve much novel growth of opinion, of custom, and of ideals. (1969, 2:965)

Indeed, he thinks, "loyalty to lost causes is . . . not only a possible thing, but one of the most potent influences of human history . . . the cause comes to be idealized through its very failure to win temporary and visible success" (1969, 2:966). For Royce, the prime example is the history of Christianity, which he views as an idealized cause given the earthly triumph of "the enemies." But the lost cause is also exemplified by the lost nationalities of early-twentieth-century Irish and Polish. Although only implicit in Royce's writings, California represented a lost cause in regard to his own provincialism, once he had left its formerly "execrable" surroundings. For our purposes, we could say that it is further exemplified in the idea of agrarian community, especially in its romantic strain.

James Montmarquet writes of agrarianism's romantic strain that, "agrarianism today is a philosophy seeking its 'niche' in a largely nonagrarian world. The power of the romantic critique of this civilization retains a kind of force, even as the possibility of a truly agrarian alternative belongs to a long buried past" (Montmarquet 1989, 216). But agrarian romanticism also commits the *philosophic fallacy* that Dewey so detested in that it imagines life to be in the plain dirt of a place, and it imagines that the life of a particular community grew from this idealized soil. In a nonagrarian society, this is bound to impart either a sense of fatalism or dangerously regressive tendencies to agrarian notions of community. Royce almost committed the same fallacy in his conceptualization of California, the province, and the community. But he stopped short because he realized that—given the isolation and the displacement native to any discussion of modern community—the province or the community must be an ongoing activity as well as an ideal. Royce understood that the impetus toward the activity of a developing community is the acknowledgment that the force of community shows its face most clearly in its absence. That is, as an ideal—not a static ideal, not just a memory, but as an ideal activity or striving. It is in these terms that Royce's idea of community retains its own critical force.

For some modern agrarians, loss in regard to community and its necessary placement (in Berry's formulation) is romantic in that, while eloquent in its portrayal of definite forms of culture or community as absent or dying, it does not provide transformative or reproductive possibilities confronted with and in lived experience. This idea of community—whose history shows it to be an ambiguous term—is pervaded by the same sort of sentiment as that directed toward a lost love, a love separated by global temporal and spatial distances believed to be impossible to traverse often due to the most banal of reasons. People nevertheless adhere to these reasons as if their well-being and growth could only be defined through a submission to the supposed contradictions of a global life. Herein lies the passionate heart but also the futility and even the danger of contemporary romanticism, and it is a strain endemic to agrarianism. This strain in agrarian community indicates part of a larger polarization in contemporary life that we are better off wending our sentimental ways past if the purported necessity of community's connection to place cannot be rearticulated in terms of mere sufficiency. Royce understood this, yet occasionally relapsed into static idealizations of his own absolutist brand.

For Royce, the lost cause signifies a progressive discovery that experience is permeated by losses and transformations. Indeed, the loss at the heart of community is not a stolen ideal, but represents a reinterpretation of a horizon toward which human beings might grow, or perhaps may never attain. But this is the nature of Royce's community—through loss and through loyalty we learn to express human commitments to our human selves. Community does not require any specific placement, despite the fact that we are indeed always somewhere, whether communally situated or not. Royce's life example suggests that the idea of community itself may just require the displacement that enables us to imagine it in all its shining absence.

"Every place seized is a place denied," but both are integral to the idea of community; those gained, those lost, and those in which we imagine the finest of human developments. On the Roycean view, it is neither the occupation nor the place that matters as much as it is our capacity for human loyalty. Royce's own words are worth repeating, "in a world of wandering and of private disasters and unsettlement, the loyal indeed are always at home" (1969, 2:890). It is not that if we have specifically placed types of community, virtues such as loyalty follow. Grown out of the commitment to an ethic of loyalty, community itself represents an ethical and social imperative that does not rely on any particular place, real or imagined. It is rather that—whatever or wherever our places, agrarian or otherwise—if we maintain an ethic such as that of loyalty to loyalty, we may potentially engage in whatever was best in what we might have once meant by community. Loss is constitutive of the very idea of community.

Does Metaphysics Rest on an Agrarian Foundation?

A Deweyan Answer and Critique

ARMEN MARSOOBIAN

CONTEMPORARY PHILOSOPHERS would answer the title question of this essay in such vehemently negative terms as to suggest that the query it poses is ludicrous, naive, absurd, or perhaps all three.[1] This essay is aimed less at establishing a positive or negative answer than at demonstrating why the question makes sense. The argument will follow a path laid out in the writings of John Dewey. In establishing the meaningfulness of such a question for Dewey's philosophical thought, two other tasks will also be accomplished. First, the argument demonstrates the link between a given culture's way of situating itself with respect to nature and the intellectual life or philosophy of that culture. As agriculture emerged as a dominant characteristic of Western culture's situation, it was bound to have profound effects upon patterns of Western thought. Dewey was sensitive to these effects, and his ideas about them became a cornerstone of his metaphysics. Second, Dewey's development of this agrarian theme was part of an argument that was intended to lay out the grounds for criticism of the very patterns of thought that would lead contemporary philosophers to slight the title question in the first place. Although it will not be possible to complete the basis for such a critical argument in this essay, it will be useful to summarize how Dewey thought that metaphysics had become wrongheaded in its tendency to ignore the material conditions of life and culture.

None of this is to suggest that Dewey would have answered the title question in the affirmative. His answer would have been hedged and qualified, as the following discussion will show. Dewey was not an agrarian in the sense sometimes attributed to Jefferson or Emerson. He did not think that farm life was an essential foundation for the formation of good moral character (or, at least, there is no evidence of such an opinion in his writings). Dewey was concerned with ethics, but did not see any particular lifestyle as the root of moral value. Dewey was committed to the idea that intelligence and philosophy have a vital role in the formation and application of action guiding beliefs; ethics could not be merely a matter of uneducated intuition, no matter how firmly that intuition may be based upon a lifetime or agrarian (or any other) values.

Dewey maintains that if philosophy is to have a meaningful normative role in contemporary culture, it must be founded upon a thorough genetic analysis of its dominant intellectual categories. The sources of our own approach to philosophy must be made transparent if we are to be confident of their application. Dewey attempts to demonstrate that the values philosophical inquiry has promoted in the past are not the values of today. Bringing philosophy to a present awareness of its normative role is to be achieved by uncovering the values inherent in its past. Dewey thus maintains that the philosophical attempt to provide a "value-free" analysis and description of "reality"—that is, metaphysics as commonly understood—is a heavily value-laden affair. In the pages that follow, our concern will be to demonstrate the importance of Dewey's insights into the relation between value and metaphysics.

Philosophy, Ethics, and History

Dewey maintains that philosophy has never been, despite some explicit claims to the contrary, a pristine description of existence: "It is a generic definition of philosophy to say that it is concerned with problems of being and occurrence from the standpoint of value, rather than from that of mere existence," (LW8:26). This concern takes the form of a generalized criticism. In *Experience and Nature,* Dewey reflects this concern:

> [P]hilosophy is inherently criticism, having its distinctive position among various modes of criticism in its generality; a criticism of criticisms, as it were. Criticism is discriminating judgment, careful appraisal, and judgment is appropriately termed criticism wherever the subject-matter of discrimination concerns goods or values. (LW1:298)

It is not just axiology (ethics and aesthetics) that concerns itself with value, all branches of philosophy, including metaphysics, do.

Dewey contends that a certain blindness has developed in the current interpretation of our philosophical ancestors. We are blind to the true subject matter of philosophy. Dewey identifies its logic: if "goods or values" are not an inherent part of all contemporary "problems" of philosophy, then their presence in past philosophy is only a sign of a certain lack of sophistication in our predecessors' philosophic method.

On the contrary, Dewey contends that the superficiality of such "problems" grows in direct proportion to their remoteness from the cultural context of human values, beliefs, and institutions:

> The [cultural] beliefs and their associated practices . . . constitute . . . the immediate primary material of philosophical reflection. The aim of the latter is to criticize this material, to clarify it, to organize it, to test its internal coherence, and to make explicit its consequences. At the time of origin of every significant philosophy, this cultural context of beliefs and allied institutions is irretrievably there; reference to it is taken for granted and not made explicit. (LW6:18)

Unfortunately most philosophical discourse dissembles. This cultural context and its inherent values are not made explicit precisely because of the mistaken belief, so Dewey claims, that philosophical description and generalization must be value-free, that is, objective. A stronger claim can be made: cultural reference is not merely taken for granted, but intentionally concealed. Consequently metaphysics, as Dewey claims, purports itself to be the description of "the generic traits manifested by existences of all kinds," but in reality emphasizes certain traits over others for prescriptive ends: "while purporting to say that such and such is and always *has* been the purport of the record of nature, in effect they proclaim that such and such *should* be the significant value to which mankind should loyally attach itself" (LW3:7). It is not merely the fact that metaphysical and normative discourse have been confused that troubles Dewey. Metaphysics, through its supposed neutrality, has served to guarantee certain values as real and thus hinders genuine philosophical inquiry. To be genuine, such inquiry must make explicit its choice of values.

The Origins of Philosophical Inquiry

We may clarify this thesis by turning directly to Dewey's conception of the origins of philosophical inquiry in the West. Unlike the Aristotelian claim that philosophy emerged in times and "places where men had leisure," Dewey claims that it was born in conflict. A conflict, he claims, between "deep-sunk customs and unconscious dispositions . . . and . . . newly emerging directions of activity," (LW3:6).[2] Dewey's account of the origins of philosophy is fundamentally an analysis of this conflict. Traditionally, philosophic study of history has examined the intellectual life and products of an age (that is, its most distinctive philosophies) in a social vacuum. Dewey contends that many philosophers "regard it as an adulteration of the purity of philosophy to admit that there is an intimate connection between Greek philosophic literature and anthropological material" (LW5:165).

By contrast, Dewey advocates that the intellectual life of an age be understood in the light of the "primary or gross experience" out of which it emerges. "Gross experience" is not strictly a stage in intellectual development. "Gross experience" characterizes aspects of the experience of modern humankind as well. Yet Dewey's most extensive discussions of gross experience are made in conjunction with his analyses of the intellectual life and values of early humans. The traits or general character of this experience are crucial for understanding Dewey's conception of the role of value in philosophical inquiry. In Dewey's words: "the philosophical tradition regarding knowledge and practice, the immaterial or spiritual and the material, was not original and primitive. It had for its background the state of culture" (LW4:11). Dewey identified a host of traits that characterized the experience of early humans. Customarily he divides these traits into two general groups. The terms *grace* and *deserts* serve to characterize these major groupings. *Grace* characterizes any experience that manifests favor by a superior—that is, something granted not through direct human effort. Though distinct from any religious doctrine of grace, Dewey recognizes that this experience is integral to the atmosphere in which "primitive religion was born and fostered," and is the "religious disposition" lying at the source of any formal religion (LW4:9). *Deserts,* on the other hand, characterizes those experiences in which goods are received as a more or less direct result of our practical or instrumental behavior. Deserts are the natural and proper result of human endeavor.

Grace and deserts are associated with the two main dimensions of experience: the instrumental (doing) and the consummatory (undergoing). The contrast is between experience as seeking means (instrumental experience) and experience as achieving ends (consummatory experience). These are both *structures of experience*—that is, structures of the relation between the human organism and its environment. The association with grace and deserts is a normative one. A philosophy in which experience is of "the given" (that is, of grace), as opposed to one in which experience is made (that is, by deserts), is a philosophy which ultimately denies the centrality of freedom in human experience. Dewey is doing something significantly different than mapping the terrain for philosophy (that is, giving a generic account of the fundamental traits of experience); he is laying the basis for his rejection of the philosophy of the given, which lies at the heart of the Western metaphysical tradition.

The Transformation of Experience and the Rise of Agriculture

Though frequently in evidence in Dewey's major writings, the distinction captured by the terms *grace* and *deserts* can be traced to an early essay entitled "Interpretation of the Savage Mind" (1902). Dewey's concern here is with the genesis of the values pervasive in the experience of early humans. Anthropological discoveries into the dominant daily activities of Australian aborigines serve as the basis for Dewey's analyses. They engender the pervasive values inherent in the emotional, intellectual, and moral life of the group:

> [Occupations] furnish the working classifications and definitions of value; . . . they decide the sets of objects and relations that are important, and thereby provide the content or material of attention, and the qualities that are interestingly significant. The directions given to mental life thereby extend to emotional and intellectual characteristics. (MW2:41–42)

Hunting is dominant in the life of the aborigine and has consequences far beyond the activity itself. "[T]he hunting life differs from, say, the agricultural, in the sort of satisfactions and ends it furnishes, in the objects to which it requires attention, in the problems it sets for reflection and deliberation" (MW2:42). It is precisely these differences in "satisfactions," "ends," and "objects of attention" that occupy Dewey's attention. The accommodation and subsequent conversion of the values inherent in the hunting life into those appropriate for a primarily agricultural society became the prototype of the intellectual conversion he will later call "*the* philosophic fallacy" (LW1:34). The transition must be made clear.

The anthropological thesis that hunting communities preceded agricultural communities is the basis for Dewey's claim that the life of early humans was primarily dramatic, as opposed to speculative, and that speculation when it does take place is shaped by the requirements of the drama. The hunting life was the archetype for the "conflict or problematic situation." The temporally protracted struggles of agricultural life, which were to come later, did not have the same enduring impact upon the mental pattern of early humans. The intensity of the hunt, the immediacy of satisfaction, were all ingredients in the mix that cemented the pattern of the hunt on the symbolic life of humans. Dewey isolates the unique factors that set apart this hunting pattern or schema:

> [I]n all post-hunting situations, the end is mentally apprehended and appreciated not as food satisfaction, but as a continuously ordered series of activities. . . . And hence the direct and personal display of energy, personal putting forth of effort, personal acquisition and use of skill are not conceived or felt as immediate parts of the food process. But the exact contrary is the case in hunting. There are no immediate appliances, no adjustment of means to remote ends, no postponements of satisfaction, no transfer of interest and attention over to a complex system of acts and objects. Want, effort, skill and satisfaction stand in the closest relations to one another. (MW2:43–44)

Early humans, like the Australian aborigines, were not constantly engaged in the hunt. These short periods of intense activity and excitement were contrasted with long periods of leisure. Yet the mental pattern that was brought to these periods of leisure was firmly established in the drama of the hunt:

> The interest of the game, the alternate suspense and movement, the strained and alert attention to stimuli always changing, always demanding graceful, prompt, strategic and forceful response; the play of emotions along the scale of want, effort, success or failure:

> this is the very type, psychically speaking, of the drama. The breathless interest with which we hang upon the movement of play or novel are reflexes of the mental attitudes evolved in the hunting vocation. (MW2:45–46)

A prime example of this dramatic transference is found in the religious and ceremonial life of the aborigine. The *corroborees*—that is, the sacred assembly of the aborigine—is an affair characterized by the traits of grace and not of deserts. Gods are not invoked to provide aid in the hunt, nor are practical hunting skills taught. The ceremony serves "to reinstate the emotional excitations of the food conflict-situations" (MW2:49). The reenactment emphasizes immediacy and intensity. To summarize, the hunt epitomizes the "doubtful, precarious situation." The end, the satisfaction, is contingent upon the often unpredictable appearance of the prey. The response is spontaneous, intense yet frequently short lived. The "goods" of the hunting life are as much favors of the prey as they are results of human effort or deserts. The pantheon of early humans, as well as early cave drawings, was dominated by animal gods. Goods are enjoyed by the grace of these gods. As Dewey remarks in *Experience and Nature:* "Goods are by grace not of ourselves" (LW1:44).[3]

This original ground pattern of the hunt will be transformed with the rise of agriculture. To summarize Dewey's thesis: if value was bestowed upon the immediate dramatic activity of the hunt over and above the worth of its ends, then it is likely that the introduction of the improved instrumentalities of the agricultural life was less auspicious than one would have imagined. The long-term rewards of farming were emotionally and symbolically uncongenial. Value was placed in activities of immediate and intense satisfaction. Rudimentary farming did not fit the bill. The objects of grace were favored over those of deserts. Whether it be the goods of the harvest or those of the developing crafts, the results were the same: such mundane objects and activities were of a diminished imaginative worth in comparison to the symbolic recreations of the hunt. Yet, as Dewey claims, once practice and planned instrumentality (that is, true science) attain an undeniable degree of acceptance in everyday life, a conversion takes place. The desired goal of instrumental behavior—that is, control and stability of one's environment—is converted into a model symbolically and emotionally dominated by the values identified under grace. The sense of control of the precarious situation through its re-creation in symbolic ceremony and ritual is the ground pattern for the imaginative life of early humans.

Evidence for this ground pattern is found in most of Dewey's major works. The opening pages of *Reconstruction in Philosophy* concentrate upon the dramatic nature of early humans' experience and its conversion into symbolic forms (1920). These symbolic forms typify what Dewey calls the inner attitude of grace, which soon came into conflict and eventual accommodation with the outer attitude of deserts. The overtly nonpractical yet dramatic nature of this inner attitude is fundamental to the psychological makeup of early humans. It is through this inner attitude that precariousness is celebrated and thereby controlled through memorial re-creations. Dewey writes:

> Savage man recalled yesterday's struggle with an animal not in order to study in a scientific way the qualities of the animal or for the sake of calculating how better to fight tomorrow, but to escape from the tedium of today by regaining the thrill of yesterday. . . . Memory is vicarious experience in which there is all the emotional values of actual experience without its strains, vicissitudes and troubles. (MW12:80–81)

The precarious is symbolically unified and thus controlled. A unity is imposed or achieved through the dramatic re-creation of the original precarious event:

> The triumph of battle is even more poignant in the memorial war dance than at the moment of victory; the conscious and truly human experience of the chase comes when it is talked over and re-enacted by the camp fire . . . the details *compose into a story* and *fuse into a whole* of meaning. As he resurveys all the moments in thought, a drama emerges with a beginning, a middle and a movement toward the climax of achievement or defeat. (MW12:81; emphasis added)

Early human experience is not, on the face of it, dramatically coherent. There are loose ends, irrelevancies, and trivialities. The drama emerges through a selective reenacting or retelling. Events become meaningful as they fall into place in a "story" with a beginning, middle, and end. Control and security are achieved in the story told "by the camp fire." The contingent and the indeterminate are deemphasized. It is to these aspects of early human experience that Dewey will trace the source of philosophy's later disparagement of practice and action.

Dewey's primary purpose is to highlight the naïveté of any history of philosophy that tries to fit early belief and speculation into a model of dispassionate inquiry and explanation. A case in point is the view of early Greek nature philosophers as protoscientists. The dramatic mental pattern precludes such a possibility.[4] Scientific inquiry, when it did emerge with the rise of agriculture, had to contend with an already deeply embedded mental pattern. The instrumental activities of science, which displayed the outer attitude of deserts, had to contend with the growing institutionalization of the inner attitude of grace into religion, moral codes, and philosophy. The dominant pattern of much of Western intellectual culture had been set. Western philosophy, including its metaphysics, as it developed in Greece of the fifth and fourth century B.C.E. was a reflection of this struggle between the attitudes of grace and deserts. Again and again in his analyses of the history of philosophy Dewey will return to the above patterns of thought and imagination. Within the scope of this essay we cannot hope to present these historical analyses, but I would like to conclude by briefly outlining Dewey's argument as it pertains to the "birthplace" of Western philosophy.

The Dominant Pattern of Western Metaphysics

As we remarked earlier, for Dewey philosophy was not born in moments of leisure. Classic Greek philosophy is a case in point. Dewey traces its development to the conflict between grace as represented in religion and social custom, and deserts as represented in the Sophistic movement. The Sophists epitomized the outer way of deserts. They were the first to apply experimental art *(techne)* in the resolution of human needs. The pattern of instrumental behavior that first arose in the early struggles to make the land sustain life was now directed toward the betterment of that life within the polis. Whether it was the art of speech making or music making, success *(arete)* was by way of human deserts, not by the grace of the gods. Yet the challenge that the Sophists posed for the status quo was not long tolerated. A reaction set in—a reaction which ultimately became known as classic Greek philosophy (whether we call it Platonism, Aristotelianism, or Stoicism).

Dewey contends that the great metaphysical systems that developed out of the teachings of Plato and Aristotle reflect the conversion of the pattern of grace into highly sophisticated schemas of interpreting the world. Their emphases on unity and ends *(telos)* over processes and means reflect a genuine discomfort with the role of instrumentalities for change. In an unusual twist on the typical theme of agrarian conservatism, Dewey argues that classic Greek philosophy and most subsequent philosophies are essentially conservative in nature, not because they reflected agrarian values but because they disparage practice and exalt a highly aestheticized model of knowing. Philosophy, while acknowledging material and efficient causes, often exalts the formal and final cause. Such philosophy is more often concerned with forestalling subsequent progress than with promoting an understanding of the world. In this sense, philosophy is intrinsically prescriptive of certain ends or values. For Dewey these values must be openly acknowledged, freely discussed, and intelligently accepted or rejected. Philosophy understood as the criticism of values should do no less.

Dewey's recurrent call for a reconstruction in philosophy grows out of his insight into the essentially conservative mental pattern he has found lying at the heart of Western metaphysics. His sensitivity to alternative models was in part heightened by his reflections upon agrarian values and their implications. It is this often neglected insight that stands out in Dewey's own practice and extension of philosophy.

The Edible Schoolyard

Agrarian Ideals and Our Industrial Milieu

LARRY A. HICKMAN

ALICE WATERS, the California restaurateur, cookbook author, and proponent of the use of organically grown foods, was recently the subject of a profile on public radio. During her interview she told of driving past inner-city Oakland's King Junior High and noticing that the school was in a state of massive disrepair. Windows were broken, graffiti covered the walls, and trash was strewn throughout the grounds. She decided that she wanted to visit the school, so she called the principal to arrange a tour.

The inside of the building turned out to be as badly maintained as its exterior. What had at one time been the school's kitchen and lunchroom had been converted into storage for damaged tables, chairs, and desks. Since there was no lunchroom, the children usually ate candy or other junk food instead of a nutritious lunch.

But Waters had an idea. After some discussion with the principal, she convinced him to allow her to perform an experiment. The school's kitchen and lunchroom were cleared out and cleaned up. Asphalt in the schoolyard was torn up and hauled away. Soil was brought in and a garden was built. The students learned how to grow herbs and vegetables, which they then prepared in the school's kitchen and served and ate in its lunchroom. Together, she and the children created an "edible schoolyard." The children and their parents were justifiably proud of their edible schoolyard, and their pride began to radiate out to other aspects of the school as well. Parents and children worked together to repair and clean their school.

One of the things that struck me about Waters's story was how closely her experiment resembled another one, undertaken a century earlier at the University of Chicago Primary School by John Dewey.

During the late 1890s Dewey was the head of the department of pedagogy (as well as the head of the departments of philosophy and psychology) at the University of Chicago. He had already become a major figure in the development of the school of American philosophy known as "pragmatism." His version of pragmatism, which he called "instrumentalism," exercised a profound influence on his work in psychology and philosophy. But he often said that its most important implications lay in the field of education, which he characterized as "the development of intelligence."

Dewey's understanding of intelligence grew out of his commitment to Darwinian evolutionary theory. He characterized intelligence as the primary means by which organisms adapt to changing conditions within their environments (MW4:182–183).[1] He thought that the *exercise* of intelligence, or what he sometimes called "mind," is what allows all of us, children and adults alike, to establish a firm footing between the inertia of past traditions and habits, on the one side, and the attractions of a future that is open to enriched possibilities, on the other. "Mind," he wrote, is "a device for keeping track of the increased differentiation and multiplication of conditions, and planning for, arranging for in advance, ends and means of activity which will keep these various factors in proper adjustment to one another" (MW4:184).

Dewey urged educators not to neglect the two sides of education that he termed the "psychological" and the "social" (EW5:85). On the psychological side, the unique interests and powers of children must be assessed, understood, and engaged. Failure to do this, he argued, tends to result in irrelevant curricula, bored students, and classroom experiences that amount to little more than time-serving. On the social side, children's powers can only be developed as they are related to the children's social and cultural histories, and as they are also related to the children's prospects for establishing a place in the world in which they will eventually live.

Dewey rejected the idea, put forward by some educators then as now, that these two aspects of education are in competition with one another. He thought that educational practices that emphasize one of them over the other—the child at the expense of the curriculum or the curriculum at the expense of the child—inevitably prove to be defective. In his view, education that is effective exhibits an organic relation between its psychological and social dimensions.

Dewey summed up his views about education in a short essay entitled "My Pedagogic Creed," which he published in 1897. "I believe," he wrote,

> that the only true education comes through the stimulation of the child's powers by the demands of the social situations in which he finds himself. Through these demands he is stimulated to act as a member of a unity, to emerge from his original narrowness of action and feeling and to conceive of himself from the standpoint of the welfare of the group to which he belongs. Through the responses which others make to his own activities he comes to know what these mean in social terms. The value which they have is reflected back into them. (MW5:84)

In Dewey's view, then, education is clearly much more than simply "a preparation for future living." It is in the truest sense also "a process of living" (EW5:87).

One of the ways in which Dewey put these ideas into practice was by working with his students to construct an "edible schoolyard." The children in his school grew gardens, processed foods and fibers, and learned to cook. In each of these activities, the operative concept was experimentation. The children were encouraged to conduct experiments with the materials in their learning environments and thereby learn firsthand about their properties and uses. Dewey recognized that children are natural experimenters, and that they remain so unless their natural curiosity is thwarted by authoritarian educational practice.

Dewey was thus already doing in the 1890s something similar to what Waters would do a century later. Both Dewey and Waters were attempting to engage their students in terms of the students' own needs and interests. Both of them thought it necessary to begin where their students were, and not where they were not. In order to do this, both Dewey and Waters began with one of the most concrete and time-honored of human activities: the cultivation of plants and the preparation of foods. But they both also realized that even such concrete and quotidian activities can have far-reaching consequences for learning. In a proper educational setting, gardening and cooking can open the doors to broader cultural horizons and therefore to the possibilities of enhanced growth and development on the part of the learner.

Waters's edible schoolyard was designed to solve two sorts of problems. Because her primary interest was oriented toward foods, she articulated the immediate problem in terms of how to improve the children's diet—that is, how to introduce the children to food that would be both appetizing and nutritious and that would therefore enhance their ability to learn. But she also recognized that there was a long-term problem. How could the children, their parents, and their community be helped toward a sense of pride in their school, and how could that pride be utilized as a springboard to effect improvements in the children's educational environment?

Dewey's edible schoolyard was designed to solve a different but related set of problems. Since his primary interest was oriented toward the philosophy of education, his immediate problem was how his edible schoolyard could be used to open the door to wider educational concerns. Since the students would want to know how to recognize edible plants and be able to relate one type of plant to another, for example, they would be led to study botany. His edible schoolyard was also used to open the door to the study of history. The students would learn about the history of the domestication of plants, food preparation, medicinal herbs, and the production and use of fibers.

Economics also came into play: the students would learn about the economy of food production and distribution and the economic advantages and disadvantages of various types of fibers. (Dewey's students carded wool and seeded cotton, and they made cloth of the respective fibers. Then, after they had experienced the difficulty of removing the seeds from cotton, and after they had compared the length of the respective fibers, they were asked why they thought that their grandparents had worn wool rather than cotton garments.)

Dewey's edible schoolyard was thus a tool he used to open the doors to a whole range of related subjects that involved increasing levels of abstraction. His students studied the chemistry of the kitchen and the laboratory; they studied weights and measures; and they studied how people in places different from their own grow and use food and fiber as a part of their unique cultural practices.

Dewey's immediate goal was thus to lead his students to learn more about themselves and their world. But his edible schoolyard also served his long-term goals. It was a tool that he used to engage the native interests and powers of his students and then to encourage them to develop the type of experimental frame of mind that is essential to the growth of intelligence. He recognized that the experience of children is most immediately involved with what is concrete and existential. Their ability to engage their experience in ways that are symbolic and abstract must grow out of their most immediate interests and concerns and then to return to them for what he called its "checks and cues."

Dewey's use of agriculture as a part of his program for primary education is closely connected with other agricultural themes within his work. In the remainder of this essay I will discuss two of those themes and show how they are also related to his philosophy of education. The first is his discussion of agriculture as a feature of the historical development of technology. The second is his analysis of Thomas Jefferson's vision of democracy and agrarian ideals.

Agriculture as a Feature of the Historical Development of Technology

Dewey did not grow up on a farm, but he did grow up in a country in which agricultural tools and methods were still dominant aspects of technological production. He was well aware of the difficult and tedious labor, and the frequent disappointments, that were associated with agricultural production. He was eager to see the burden of agricultural workers made lighter, and he followed with great interest the steady stream of improved agricultural tools and appliances that were becoming available at the end of the nineteenth century.

It was perhaps because he lived in an environment that was permeated by agricultural production, and in which mechanical inventions were rapidly improving the lot of agricultural workers, that Dewey did not romanticize agriculture, as some recent philosophers have had the leisure to do. Nor did he award agriculture pride of place with respect to machine-based industry. He viewed improvements in agricultural production as but one element within a long sequence of advances in human foresight and planning—in brief, as but one important element within the evolution of intelligence.

His historical-anthropological accounts of the growth of technology began with hunter-gatherers, who he described as living in a world of immediate use and enjoyment, largely uninvolved with foresight, long-term planning, or the use of tools that might serve to render their enjoyments more secure. His treatments of these matters tend to occur in the context of discussions about the rise of what he calls specialized "occupations" and the growth of the capacity to make complex value judgments.

Anticipating the work of Lévi-Strauss and others, Dewey was careful not to deny active intelligence to the hunter. But he did think that the type of intelligence exhibited by the hunting mode of life does not require the foresight and care that is characteristic of modes of life that are pastoral, agricultural, commercial, or industrial.

> The earlier forms of occupation, hunting and fishing, call for active intelligence, although the activity is sustained to a great degree by the immediate interest or thrill of excitement, which makes them a recreation to the civilized man. Quickness of perception, alertness of mind and body, and in some cases, physical daring, are the qualities most needed. But in the pastoral life, and still more with the beginning of agriculture and commerce, the man who succeeds must have foresight and continuity of purpose. He must control impulse by reason. He must organize those habits which are the basis of character, instead of yielding to the attractions of various pleasures which might lead him from the main purpose. (MW5:44)

It is a matter of some importance that Dewey thought that the rise of agricultural technology was correlated with improvements in the capacity for value judgments. This point is nowhere clearer than in his discussion of the nature of the type of judgments he termed "practical."

A central feature of Dewey's pragmatism is his view that practical judgments—judgments about things *to be done* as opposed to judgments that are merely *descriptive*—involve two elements that also come into play as a part of historical developments within agricultural modes of life. The first is an enhanced appreciation of the interaction of means and ends. The second is a pragmatic notion of truth that is much richer than the one that involves a mere correspondence between so-called ideas and so-called facts.

The hunting mode of life is characterized by spontaneity. Ends and means are not separate, and neither are they recognized as separable. In agricultural forms of life, however, means are treated as separable from ends: as tools, means become objects of interest in their own right, and that is one of the reasons that they become subject to improvement.

As Dewey described the hunting life, "[t]here are no intermediate appliances, no adjustment of means to remote ends, no postponements of satisfaction, no transfer of interest and attention over to a complex system of acts and objects. Want, effort, skill and satisfaction stand in the closest relations to one another. The ultimate aim and the urgent concern of the moment are identical; memory of the past and hope for the future meet and are lost in the stress of the present problem; tools, implements, weapons are not mechanical and objective means, but are part of the present activity, organic parts of personal skill and effort" (MW2:43–44).

As agricultural forms of life emerge, however, ends and means begin to be viewed as separable, and as instruments of mutual influence.

> [A]ll sorts of intermediate terms come in between the stimulus and the overt act, and between the overt act and the final satisfaction. . . . Even in the crudest agriculture, means are developed to the point where they demand attention on their own account, and control the formation and use of habits to such an extent that they are the central interests, while the food process and enjoyment as such is incidental and occasional. (MW2:43)

In agricultural life, purely hypothetical (descriptive/mythical) forms of judgment begin to be supplemented, and eventually supplanted, by judgments of practice. Successful agricultural production demands that hypotheses be tried out, and it reveals that their truth does not lie in their correspondence with some purportedly preexisting state of affairs so much as in the ways in which some facts have been selected as relevant to the case at hand, other facts have been discarded as irrelevant, and eventual outcomes have proven successful with respect to the bringing about of some desired end.

It is in this account of judgments of practice that we find one of Dewey's clearest statements of his instrumental pragmatism. In his view, "all judgments of fact have reference to a determination of courses of action to be tried and to the discovery of means for their realization. . . . [A]ll propositions which state discoveries or ascertainments, all categorical propositions, would be hypothetical, and their truth would coincide with their tested consequences effected by intelligent action. This theory may be called pragmatism" (MW8:22).

Dewey thought that the most important class of judgments of practice were judgments about value. Whereas hunting modes of life are concerned primarily with the *experience* of value, it is with the rise of agricultural modes of life that *judgments* of value begin to take center stage. In the hunting mode of life certain experiences are *prized*. In agricultural modes of life, because foresight is called for, and because tools and implements are reflectively developed and used, experiences are *appraised* (MW8:26).

Dewey thought that the experiences of individual children tend to recapitulate to a certain extent the history of Homo sapiens, just as their development from embryo to birth tends to recapitulate the history of animal life. His view was not based on a priori considerations, however, but on actual observations of young children. When this idea is coupled with his view that the powers and interests of the child are the legitimate starting places of education, and that children are natural experimenters, then we get his theory of primary education in a nutshell. Under the proper conditions, a child's interest in bows and arrows can provide a springboard for experiments with woods and metals. A child's interest in growing things can offer a springboard for experiments with foods and fibers.

Dewey thought the educational alternative a clear one. On the one hand, children can be asked to memorize and rehearse information *about* these matters, in which case learning will be short term and isolated from environing social conditions. On the other, their native interests can be engaged in ways that promote the development of an experimental frame of mind that will carry over into all areas of their lives. In the former case, their teachers will emphasize isolated judgments of description. (Ironically, this might even include the judgments *describing* certain prized "virtues" that are found in a current best-selling children's book about "the virtues." Of course this approach assumes that the virtues have already been identified and cataloged, and they therefore need not be experimentally appraised by the child.) In the latter, their teachers will emphasize the judgments of *practice* that Dewey calls judgments of *value*—judgments that are involved with things to be done, and with things that the children themselves can do.

Thomas Jefferson's Vision of Democracy and Agrarian Ideals

Another of the agricultural themes in Dewey's work involves Thomas Jefferson's well-known attempts to link the values of agrarianism and democracy, and his fear that the rise of manufacture and trade would result in antidemocratic modes of life. It may seem contradictory that Dewey would praise the growth of industrial technology at the same time that he extolled Jefferson's democratic vision. Dewey resolved this apparent contradiction by probing beneath the surface of Jefferson's agrarianism. He argued that it was not agrarianism per se that made Jefferson's vision of democracy so attractive, but the ideas on which both that vision *and* his agrarianism were based. To honor agrarian over industrial life per se, Dewey thought, was therefore to miss the main point of Jefferson's social philosophy, and what was even worse, to fail to see its continuing relevance.

It was in fact the conditions of liberty and equality that came together in a situation in which land was virtually free and a frontier was open to discovery and settlement that were of interest to Jefferson. It is of course now obvious that land is no longer free and that the frontier no longer exists; the agrarianism of the late eighteenth and early nineteenth centuries is consequently no longer possible. But this recognition does not require us to accept the view that the ideals of liberty and equality are themselves no longer relevant—that is, that they are no longer worth working for.

Dewey thought that Jefferson has been badly misunderstood on this point. The point of honoring Jeffersonian agrarianism is neither to engage in a vain attempt to recover a lost mode of life nor to deplore its disappearance. The point (which is invariably missed by those who romanticize agrarian modes of life) is instead to find ways of honoring the underlying ideals of Jefferson's thought within the context of modes of life that are increasingly industrial (LW11:370).

Dewey suggested that the rise of industry has more or less reversed the original positions of Hamilton and Jefferson: not, however, with respect to basic ideals, but with respect to methods to be employed toward their attainment. Hamilton, of course, had argued for the federal control of common resources, and especially for public control (through a national bank) of public finance. Jefferson had argued for a more nearly classical liberal position according to which government would leave individuals at liberty to cultivate free associations and to organize themselves locally, as equals, with a view to effecting the common good. But Dewey argued that under the present industrial conditions the pursuit of the underlying ideals of liberty and equality that Jefferson had advanced (and that he had linked to agrarianism) is best undertaken with methods that are to some extent Hamiltonian.

In short, within an industrial society the conception of life that underlies Jeffersonian agrarianism can be honored only by also honoring the Hamiltonian notion that common effort is required on a much larger scale than Jefferson had imagined. If the liberty and equality that Jefferson thought were essential aspects of individualism have been undermined by industrial capitalism, which creates and depends on national markets, then they must be reconstructed by systematic efforts on the national level as well as within local communities.

It is unfortunate, Dewey argued, that individualism within our industrial capitalist milieu has come to mean liberty and equality only for individuals with economic power. As he put it, "our existing materialism, with the blight to which it subjects the cultural development of individuals, is the inevitable product of exaggeration of the economic liberty of the few at the expense of the all-around liberty of the many" (LW11:372).

But Dewey clearly rejected the idea that this blighted situation is the fault of industrialism per se. Just as the cultivation of freedom and equality are not tied to agrarianism per se, neither is the capture of public resources by private interests a necessary component, or even a necessary consequence, of industrialization per se. It is the excesses of industrial *capitalism,* or what Dewey called "finance capitalism" that are at fault, and so that is precisely the point at which efforts to enhance liberty and equality must be focused.

As Alan Ryan has suggested in his recent book *John Dewey and the High Tide of Liberalism* (1995), Dewey thought that Jeffersonian democracy, reconstructed for an industrial milieu, translated into a type of guild socialism such as that advocated in Britain by G. D. H. Cole. This would be a socialism in which workers had a greater share in the profits that resulted from their work, a greater say about the conditions under which they worked, and enhanced opportunities for lifelong learning. Because of the different political traditions of the two countries, however, the type of socialism that Dewey thought a proper reconstruction of Jefferson's agrarian democracy would be less centralized in the United States than in its British version. He thought, as Ryan reminds us, that forms of association based upon geographic contiguity would probably have to give way to ones that are based on unity of purpose and function. And this is precisely where the schools would play a central role. It would be the schools that would be among the most potent catalysts of continuing social reform.

Those who search Dewey's work for a roadmap to a revitalized Jeffersonian democracy, or for a grand strategy for its attainment, will of course be disappointed. The type of democracy that Dewey thought desirable cannot by its very nature be specified in advance, since it is more an ongoing method than a finished state of affairs, and since it requires step-by-step, participatory, intelligent action on the part of educated and informed publics. As environing conditions change, so will the particular goals that are deemed worthy of being worked toward. But if Dewey leaves the details of the reconstructed Jeffersonian democracy vague, he is clear enough about the infrastructure that it will require. A principal component of that infrastructure will be an educational system in which children learn the skills of a type of intelligence that is both experimental and social.

These considerations bring us full circle, back to the edible schoolyard. The cultivation, processing, and preparation of food and fiber in an educational setting function in Dewey's work as a kind of metaphor for the cultivation of intelligence. An anti-authoritarianism that recognizes and takes into account the interests and powers of children *as* individuals; expression of those interests and powers in terms of their wider historical and cultural contexts; and the development of an experimental frame of mind: for Dewey these are the goals of education, and they are the goals of democratic life. It is no longer possible to return to an agrarian-based democracy. But it is possible to recognize and honor the ideals on which that form of life was based, and it is possible to attempt to reconstruct those ideals in ways that are appropriate to our current industrial milieu.

Part IV

TWENTIETH-CENTURY AGRARIAN THOUGHT IN THE PRAGMATIST TRADITION

The Relevance of the Jeffersonian Dream Today

JOHN M. BREWSTER

It is timely to take a current look at Griswold's "Farming and Democracy." In this way we may gain improved understanding and appreciation of longstanding and deeply motivating beliefs in our rural traditions and their role as policy directives from colonial times up to now.

As Griswold observed, Jefferson's devotion to the family farm stemmed from his belief in a "causal relationship between [family] farming and the political system of democracy" (Griswold 1948, 19, 47).[1] A prime concern of Griswold was with the validity of this belief. He devoted two chapters to disproving it. We shall not belabor this issue further; a more fruitful question is why such a belief became so much a part of our folklore in the first place.

As we interpret him, Griswold's concern with the supposed causal relationship between farming and democracy grew out of his observation that the New Deal "revived the Jeffersonian Ideal and made the family farm an explicit goal of policy" (1948, 163; see also 4–5, 14–16). He might have added that this revival, like revivals in general, did not include a definition of what was being revived; in this way it generated one of the greatest definitional blizzards of all time. Why this was unavoidable will become apparent as the discussion proceeds.

I take the Jeffersonian Dream to mean Jefferson's affection for and desire to establish and preserve an agriculture of freeholders—full-owner operators, debt free, unrestricted by any contractual obligations to anyone—all in all, pretty much the monarchs of all they survey.

The freehold concept of the family farm was an unusually effective policy directive throughout most of the settlement era. Then the worm turned. An appreciable transformation of freeholders into debtors and tenants was noticeable by 1880. Since that date the relevance of the Jeffersonian Ideal to present problems has continued to decline in the sense that it has failed to generate the kinds of policies and programs required to maintain a high approximation of a freeholder agriculture.

But we know that a predominantly family farm agriculture has not disappeared; that the fostering of this institution remains an important objective of farm policy. In my judgment this will be the case for many years because the family farm is not on the way out, although some dangers confront it now as in the past. Naturally in taking this position I have in mind a definition—a concept—of the family farm which is my gauge for what facts are relevant and what facts are not relevant to the issue.

As used here, a family farm is an agricultural business in which the operating family does most of the work and is a *manager* of ongoing operations of the business as well as a *risk taker* in the outcome (financial returns) of the business venture. In this definition, *operatorship is equated with varying degrees of managerial power and with risk taking involving management and production inputs, including labor.* As used here, managerial power is equated with the operator's prerogatives to negotiate contracts and to make decisions concerning the combination of resources.

This definition applies to the Jeffersonian freeholder and the modern family farmer alike. In both cases, most of the farmwork is done by the operating family who is a risk-taking manager in the outcome of the business. However in the usual case they differ substantially in their degree of risk-taking managerial power. The Jeffersonian freehold was a very high approximation of the self-sufficient firm. By this we mean a firm from which the will and interest of all conceivable participants in the business, except those of the operator, are totally excluded. In the self-sufficient firm, the operator alone is the sole risk-taker in the outcome of production activities he is guiding and coordinating. Thus the managerial power of the Jeffersonian freeholder was absolute, as he was totally independent of all commitments to outside parties concerning the way he used the resources of the farm and the kind of products it produced. In his managerial decisions, he never had to take account of the will or interest of another living soul. The ages had always equated managers of self-sufficient firms as lords—the monarchs of all they survey.

But in industry, this self-sufficient firm disappeared as a representative institution with the close of the handicraft era, and agriculture has been making departures from it since about 1870. And each step in its progressive extinction has aroused new anxieties concerning the future of the family farm.

Family farms are those on which operating families are risk-taking managers and do most of the work. I maintain that such farms are not losing their relative position in American agriculture. I regard these as family farms. But usually the managerial power of their operators is not absolute; it is limited in varying degrees by contractual commitments of the operators with outside participants in the farm business, such as the landlord, the banker, the contractor seeking products of specified qualities, and even the government. Committing certain of their services to the operators, these outsiders seek legally binding commitments of the operator to follow certain lines of behavior as means of protecting them against loss of their stakes in the operators' business. The operator enters into these commitments because he believes the services of the outsiders will enable him to achieve a more profitable business than otherwise. Whether or not this proves to be the case turns on his abilities to guide and coordinate the operations of his business to a successful issue. No one would assume that a modern road builder is not an independent operator simply because he agrees to produce a product that meets the specifications of his customers who send out inspectors to see if he is complying with the specifications he agreed to meet. We do not make this assumption because his contract specifications do not destroy the fact that he remains in some degree a risk-taking manager in the outcome of a business. The same principle applies to farming.

It may be that much of the confusion over what is meant by the family farm stems from the lack of a clear image of the operator's degree of risk-taking managerial power under the contractual foundations of modern farming. We know it lies somewhere between two extremes. It lies to the left of that absolute lordship, which belonged to the self-sufficient firm that excludes all participants in the business except the operator, as did the Jeffersonian freeholder. It also lies to the right of the "directed worker," with whom the farm operator is sometimes equated in literature on "vertical integration." In line with this fact, I equate farm operators with managers and risk takers in the outcome of their business undertakings, and then equate a family farm with a business in which the operating family does most of the farmwork. By a larger than family farm, I mean any agricultural business whose total labor requirement is such that the usual farm operating family cannot supply most of it.

One further definitional matter—in this essay we commonly use the term *proficient family farm*. By this we mean a family business in agriculture with sufficient resources and productivity to yield enough income to meet expenses for (a) family living; (b) farm expenses, including depreciation, maintenance of the livestock herd, equipment, land and buildings, and interest on borrowed capital; (c) enough capital growth for new farm investments required to keep in step with technological advance and rising levels of living. Farms without this level of resources and productivity are inadequate farms, and they are either disappearing or being supplemented by income from nonfarm sources, usually off-farm employment.

Proficient family farms may come into being through the reorganization of inadequate units or because hitherto larger than family farms fall into the category of family farms as a result of substituting capital for labor to the point where the operating family is able to do most of the farmwork.

In using these concepts, I do not mean that they are *the* true ones—their status in this respect will turn on how useful they may prove to be as guides in measuring and interpreting the kinds and directions of changes now occurring among the business units of American agriculture. I expect to modify the concepts I am here using whenever further research investigations warrant it.

In these terms, we will have a family farm agriculture as long as most of our farm production is done by business units in which operating families are risk-taking managers who do most of the work. In keeping with his times, Jefferson thought of the family farm in terms of the restricted freeholder meaning of the term. However, I believe it is in keeping with his larger spirit to expand his dream to include contractual as well as highly self-sufficient firms like the freehold. In Jefferson's time, there was no conflict between an agriculture of proficient family farms and the need of opportunities for farm people. There is a severe conflict today, as only around one million such farms are all that would be needed to supply all the foods and fibers which society wants at reasonable prices.

With these preliminary remarks in mind, this essay centers on five major themes.

First, since colonial times two distinct personal and policy-guiding beliefs have been indigenous to the farm and nonfarm sectors of our society. First is a deeply moving commitment to proficient work as the hallmark of praiseworthy character; the second is an equally firm belief in the natural or moral right of men to acquire all the property they can from the earnings of their work. The first of these commitments stemmed from the revolutionary interpretation of the ethical significance of proficient work by the religious reformers of the late sixteenth and seventeenth centuries, and was a complete reversal of ancient and medieval attitudes toward economic work. In line with this new sense of obligation, people soon found themselves producing beyond the limit required to support their customary needs. But in doing so, they ran into head-on conflict with the age-old belief that the natural or moral right of men to acquire property, as a fair reward for work, is limited to the amount required to produce their customary needs. John Locke, the greatest of the natural rights philosophers, resolved this conflict by demonstrating that the older belief was true only in very primitive, savage societies in which a money economy was totally absent.

Second, from early colonial times American settlers carried the radical belief in proficient work as the badge of superior merit alongside the equally radical Lockean belief in the natural or moral right of individuals to acquire as much property as they can from the earnings of their work. These directives have been stable motivations throughout both the settlement and postsettlement phases of our history.

Throughout the era of cheap land and relatively inexpensive farm technologies, these directives were a powerful generator of the freeholder ideal to which the name of Jefferson is attached. For, if land is free or very cheap, obviously farm families can achieve a greater reward for their proficient work as debt-free, full-owner operators than as renters or owners with varying degrees of credit obligations. This is not necessarily the case, however, if land is increasingly scarce and capital requirements of proficient farms are growing larger. As these conditions came to pass with the close of the settlement era, the same directives called for departures from the freeholds to contractual relationships with creditors, landlords, and others as means of achieving proficient farms. This separation between the farmer's actual status of limited managerial power and his hitherto absolute power was made sufferable by the new idea of an "agricultural ladder" that enabled farm people to envision their departures from the freeholder ideal as actual stepping stones to its fulfillment.

Third, working hand in hand with the foregoing directives to a freeholder agriculture in the settlement era was the belief inherited from feudal times that tenantry and wage status are a badge of inferior character. More specifically, farmers demanded freeholds as the best means of enabling them to earn as much as they could as a fair reward for working as proficiently as they could, and they also demanded freeholds as the best possible means of escaping the ancient onus of tenantry and wage status as a badge of inferior character. But this "marriage of convenience" began cracking up with the disappearance of cheap land and the increasing capital requirements of proficient farms. For these conditions generated increasing conflict between the ancient devotion to the freeholder ideal with the more modern devotions to proficient work as a badge of personal excellence and to acquisition of as much earnings and property as one can get as fair reward for his proficiency. In this conflict, the age-old devotion to a freehold agriculture as a citadel of superior virtue proved an increasingly weak competitor.

Fourth, in developing the foregoing themes, emphasis is placed on the economic or materialistic implications of the ethical directive to proficient work at the expense of its larger humanitarian or idealistic implications. To correct this imbalance, we shall point out that the operation of this directive in American life not only has led the whole nation along remarkable paths of material progress, but has also enkindled a "practical idealism" which has long distinguished our people and has infused the higher reaches of the spirit with the promise of the American Dream of what lies in store for men devoted to proficient use of their creative power. Furthermore, our historic commitment to proficient work, as tangible evidence of personal excellencies, includes commitments to distributive as well as commutative justice, both of which the plain man calls the "justice of equal opportunity." If we fail to counterbalance the economic import of our historic commitment to proficient work with its equally idealistic import, then with our own voice we convict ourselves of the common but false charge that America is the most materialistic civilization on earth. We shall seek to avoid this error.

Our fifth theme concerns the current status and prospects of the family farm, considered as a business in which the operating (management and risk-taking) family does most of the work of the business they operate.

Salient Features of the Medieval Landlord Civilization, Out of Which Our Farm and Nonfarm Society Emerged

Of all our major economic institutions, the family farm alone has a life span which connects our atomic age with the ancient and medieval landlord civilization out of which America and modern Europe emerged. To know and understand the belief-forming role of the family farm in the life of the nation is to be wiser concerning our whole American civilization, especially since our fathers were mostly family farmers throughout our two-hundred-year settlement era when our national character and institutions were in their most formative period.

In keeping with this fact, Griswold found it necessary to step back into the older landlord civilization out of which our rural society and modern Europe emerged in order to find the proper point of departure in understanding the freehold ideal. We need to follow the same procedure, especially since in our judgment, there is more in this ideal than is disclosed in Griswold's analysis. There is no other way of determining which of the deeply motivating beliefs underlying this ideal were carryovers from the older landlord civilization of Europe, and which ones were profoundly revolutionary breaks with this older order. To make this clear, six strategic features of this older culture need to be identified. In substance, they are as follows:

1. Commonly called feudalism but more accurately known as a traditionalism, this precapitalistic landlord civilization was distinguished by a system of master-servant beliefs and correlative institutions that distinctly segregated the managerial and labor roles of life into separate classes, called lords and serfs. The nobility was viewed as essentially personifications of divinelike managerial wisdom and power to know and administer an impartial justice for all classes. Those saddled with the work of the world were viewed as so lacking in managerial intelligence and other virtues that they were essentially personifications of turbulent passions and labor capacities, fit only for producing subsistence for the whole community.
2. This master-servant hierarchy both nurtured and rested on the equation of proprietorship with the spirit of self-mastery and other virtues, and on the correlative equation of tenantry and wage status with subservience, turbulent passions, and mental and moral incompetence.
3. This older culture necessarily included a self-sufficiency ideal of freedom. For within the master-servant relationship, the free man can be envisioned only as the one who has absolute command over all the personal services and other resources required to meet his needs. To the feudal lords, for example, it was self-contradictory to say in one breath that one is a free man and in the next breath say that his living depends on market exchange between producers and consumers.
4. This older hierarchy of superiors and inferiors both generated and rested on the belief that exemption from economic employments is prima

facie evidence that one possesses the praise-deserving qualities of mind and character, and that dependence on such employments proves that one is so deficient in meritorious capacities that he deserves only the lower stations, even servility.

5. This aversion to economic work as a badge of disrepute went hand in hand with the further belief that capital accumulation should be limited to the amount of goods and services required to support the customary needs of people in their various social roles or life stations. Acquiring property in excess of these limits was equated with greed and miserliness. Thus, this older culture would have been shocked by James Madison's axiom, widely shared by the founding fathers, that the chief object of good government was the protection of men in their "different and unequal faculties of acquiring property" (Madison in *Federalist Papers,* no. 10).

6. Finally, this older landed hierarchy of superiors and inferiors was the outward institutional expression of the belief that natural inequality is the proper guide to use in relating man to man with reciprocal rights and duties in all spheres of life. In line with this belief, the good society was viewed as one which invests the individual with only those rights and duties which mark him as an instance of a particular class. In one man it lodges the rights and duties of all serfs; in another the rights and duties of all artisans; in another the rights and duties of all landlords, and so on. In this way, the common sense of this older civilization tossed aside as nonsense the democratic belief that all men have a common nature in virtue of which they deserve an equal status that entitles each to the same bundle of rights and duties, whatever his particular social (class) roles may be.

Proprietorship as a Badge of Good Repute and Tenantry as a Badge of Servility

The striving for personal significance is apparently so universal that we assume people the world over share a profound *aversion* to being identified with status symbols which their age and civilization deem sure proof of servile attitudes and other vices. From Plymouth Rock and Jamestown on, the driving power of this tremendous urge for personal significance was a potent generator of the dream of an agrarian America of freeholders long before it was named the Jeffersonian Dream. For, in permitting the hitherto separate roles of lords and serfs to be recombined within the same skin, the two billion acres of virgin continent gave working people the chance to escape the age-old equation of tenantry and wage status with servility through becoming freeholders and in this way to identify themselves with the dignity—the esteem and proud sense of independence—which the ages had always posited in the lords of the land. The emerging agriculture of family farms along the moving frontier of the New World thus generated within everyday people an envisioned realm of equal dignity and worth, which all America soon enshrined within her national self-image much before Jefferson's day. Jefferson did not invent the Jeffersonian Ideal; he merely identified his name with an ideal which freeholders themselves already had embedded in their very bones.

By enabling ordinary people to view themselves as equally lords of the land, our earlier expanding agriculture of freeholders generated the aspiration for democracy long before even the brightest minds of the age were able to liberate themselves from the inherited equation of proprietorship with civic virtue, and nonproprietorship with civic vices. Two observations illustrate this fact. The founding fathers were still so guided by the carryover of this ancient equation that they were unable to envision the possibility of a democratic society in terms of a predominately wage and salaried population. At the constitutional convention, James Madison expressed this state of mind in these blunt words:

> In future times a great majority of the people will not only be without landed, [*sic*] but any other sort of, property. These will either combine under the influence of their common situation; in which case, the rights of property and the public liberty, will not be secure in their hands: or which is more probable, they will become the tools of opulence and ambition, in which case there will be equal danger on another side. (In Tansill 1927, 489–490)

In similar vein, Daniel Webster declared:

> A republican form of government rests no more on political constitutions, than on those laws which regulate the descent and transmission of property. Governments, like ours could not have been maintained, where property was holden according to the principles of the feudal system; nor on the other hand, could the feudal constitution possibly exist with us. Our New England ancestors brought hitherto no great capitals from Europe. They left behind them the whole feudal policy of the other continent.

> They came to a new country. There were, as yet, no lands yielding rent, and no tenants rendering service. The whole soil was unreclaimed from barbarism. They were themselves nearly on a general level in respect to property. Their situation demanded a parcelling out and division of the lands, and it may be fairly said that this necessary act *fixed the* future frame and form of their government. The character of their political institutions was determined by fundamental laws respecting property. (Webster 1851, 1:5–6)

As our earlier expanding agriculture of freeholders was the prime generator of an increasingly strong demand for political democracy, so this political objective in turn generated powerful demands for public land policies that would strengthen democratic government through enabling people with little or no capital except their labor to become freeholders. This strategy was expressed with uncommon clarity and eloquence by Thomas Hart Benton in the great land policy debates of the 1840s:

> Tenantry is unfavorable to freedom. It lays the foundation for separate orders in society, annihilates the love of country, and weakens the spirit of independence. The farming tenant has, in fact, no country, no hearth, no domestic altar, no household god. The freeholder, on the contrary, is the natural supporter of a free government; and it should be the policy of republics to multiply their freeholders as it is the policy of monarchies to multiply tenants. We are a republic, and we wish to continue so: then multiply the class of freeholders; pass the public lands cheaply and easily into the hands of the people; sell, for a reasonable price, to those who are able to pay; and give, without price, to those who are not. I say give, without price, to those who are not able to pay; and that which is so given, I consider as sold for the best of prices; for a price above gold and silver; a price which cannot be carried away by delinquent officers, nor lost in failing banks, nor stolen by thieves, nor squandered by an improvident and extravagant administration. It brings a price above rubies—a race of virtuous and independent laborers, the true supporters of their country, and the stock from which its best defenders must be drawn. (Benton 1854–56, 1:103–104)

While this reasoning enabled early America to turn her inherited equation of proprietorship and civic virtues to the service of democratic ends, it did so at the high cost of an invidious distinction between proprietors and nonproprietors with respect to moral excellencies. The longevity of this distinction is astonishing. For example, in an address not many years ago the head of one of our more conservative trade associations spoke as follows:

> Today the greatest threat to democratic institutions . . . and ultimately to freedom itself, lies in our big cities. They are populated for the most part with the mass of man, devoid of intelligence, and devoid of civic responsibility. . . . He will vote for anyone who offers him something for nothing, whether it be subway fares for half-price or public housing at one-third price . . . Our one hope of survival as a free country is that rural and semi-rural areas will dominate most of the state legislatures through their representative and still dominate the House of Representatives at Washington. (Cited by Hacker 1962, 84)

Jefferson's Contribution to the Freeholder Ideal

Jefferson is distinguished by his sharp differentiation of all proprietors with respect to civic virtues. He did so with a neat logic which sets up a causal relationship between freeholders and democracy. Here is the logic:

1. "Corruption of morals . . . is the mark set on those, who . . . depend for it on the casualties and caprice of customers. Dependence begets subservience and venality, suffocates the germ of virtue and prepares fit tools for the designs of ambition" (Jefferson 1984, 291) Holding precisely the same belief, the feudal lord would have declared that Jefferson took these words right out of his mouth. In both cases, the conclusion follows from a self-sufficiency ideal of freedom that is actually incompatible with market dependence.
2. Owing to its relative noncommerciality in Jefferson's time, farming enabled people to produce most of their own subsistence and in this way liberated them from dependence on "the caprice of customers" for a living; thus generating a proud spirit of personal independence which brooks no outside interference with the sense of self-mastery. Thus, the

really significant output of family farming is not food and fiber, but the brick and mortar of democratic society. In Jefferson's words:

> Those who labour in the earth are the chosen people of God . . . in whose breasts he has made his peculiar deposit for substantial and genuine virtue. It is the focus in which he keeps alive that sacred fire, which otherwise might escape from the face of the earth. (Jefferson 1984, 290)

3. From the premise that farming and nonfarm employments produce opposite types of character, Jefferson deduced his third premise of political science. Here it is: The possibility of a democratic society diminishes directly with the increase of the ratio of nonfarmers to farmers. He expressed this deduction in these words:

> [G]enerally speaking the proportion which the aggregate of the other classes of citizens bears in any state to that of its husbandmen, is the proportion of its unsound to its healthy parts, and is a good enough barometer whereby to measure its degree of corruption. . . . The mobs of the great cities add just so much to the support of pure government, as sores do to the strength of the human body. (Jefferson 1984, 291)

Throughout this reasoning, Jefferson, as Griswold observed, claimed a virtual monopoly of "good morals for farmers" (in Griswold 1948, 31). So completely different is the spirit of his disquisitions on farmers and nonfarmers from the spirit of the Declaration of Independence that it is difficult to realize that they were both composed by the same man. Lincoln, a farm boy and as matchless a politician as Jefferson, referred to the declaration as containing "the true axioms of democracy." But in speaking to a large gathering of farmers at the Wisconsin State Agricultural Society at its annual fair on September 30, 1859, he referred to the equation of farmers with superior excellencies as a device for flattering farmers. Said he:

> I presume I am not expected to employ the time assigned me in the mere flattery of farmers as a class. My opinion of them is that, in proportion to their numbers, they are neither better nor worse than any other people. In the nature of things they are more

> numerous than any other class; and I believe there really are more attempts at flattering them than any other, the reason for which I cannot perceive, unless it be that they can cast more votes than any other.

There is no question but that Jefferson served the already indigenous devotion of rural America to an agriculture of freeholders with singular distinction. But other than creating an ingenious device for flattering farmers, no useful purpose is served through imputing to them a virtual monopoly of good morals. Yet, there is no mystery concerning the survival power of this myth. For any group is most happy to be singled out and assured by high authority that they are "The chosen people of God" in "whose breasts he . . . keeps alive the sacred fire which otherwise might escape from the face of the earth." It would be less than human to expect farmers to stand up and deny it, and its oratorical potential is obviously too great to be overlooked by alert politicians and others seeking the good graces of the countryside.

Revolutionary Interpretation of the Ethical Significance of Proficient Work

Attention is now directed to the fact that the carryover of the ancient aversion to tenantry and wage status as the badge of inferior character was by no means the only directive to a freeholder agriculture throughout the settlement era. Equally potent directives were a revolutionary interpretation of the ethical significance of proficient work, which stemmed from the sixteenth- and seventeenth-century religious reformers, and a correspondingly revolutionary interpretation of the natural or moral right to acquire property, which stems from John Locke, the greatest of the natural rights philosophers.

The revolutionary interpretations of the ethical significance of proficient work, as sure evidence of highest personal worth, came into Western society in substantially the following manner.[2] In their breakaway from the Mother Church, the religious reformers of the sixteenth and seventeenth centuries had to face up to the question of what occupation is most truly appropriate for the upright man. According to the traditional view, such employment was boxed up in the monasteries where every moment of the twenty-four-hour day was organized into a systematic series of routines that were known as the Holy Callings. As long as the reformers left this view unchallenged, it was impossible for them to complete their breakaway from the Mother Church. They completed the break by taking the position that all employments, whether composing sermons, painting pictures, making mousetraps, or growing corn, are equally appropriate ways of showing that one possesses qualities of mind and character that deserve his own highest respect and esteem and the respect and esteem of others as well.

The following are typical illustrations of this revolutionary belief. Richard Steele, an exceptionally able minister, put it this way:

> God doth call every man and woman . . . to serve him in some peculiar employment in this world, both for their own and the common good. . . . The great Governour of the world hath appointed to every man his proper post and province, and . . . he will be at a great loss, if he does not keep his own vineyard and mind his own business . . . . [Therefore] Be wholly taken up in diligent business of your lawful calling, when you are not exercised in the more immediate service of God. (Cited by Tawney 1926, 240)

Again he clothes the same belief in these poetic lines:

> How is it that ye stand all the day idle? . . . Your trade is your proper province . . . Your own vineyard you should keep. . . . Your fancies, your understandings, your memories . . . are all to be laid out therein. (In Tawney 1926, 245)

As a guide to the use of time, Cotton Mather expressed the same belief in these words:

> There should be . . . some *Settled Business,* wherein a Christian should for the most part spend the most of his Time; and this, that he may Glorify God, by doing of *Good* for *others* and getting of *Good* for *himself.* (Cited by R. B. Perry 1944, 312)

And Baxter, with his usual lucidity, emphasized the same point in more detail:

> Keep up a high esteem of time and be every day more careful that you lose none of your time, than you are that you lose none of

> your gold and silver. And if vain recreation, dressings, feasting, idle talk, unprofitable company, or sleep be any of them temptations to rob you of any of your time, accordingly heighten your watchfulness. (Cited by Weber 1930, 216)

Under the guidance of this new belief, the profound need of achieving proofs of personal significance required ceaseless action, and an end to "taking it easy." In the words of Baxter:

> It is for action that God maintaineth us and our activities; work is the moral as well as the natural end of power . . . It is action that God is most served and honoured by . . . the public welfare or the good of the many is to be valued above our own. (In Weber 1930, 260)

This expanding concept of God's work to include all occupations released an avalanche of productive aspirations that literally reshaped the world. Vast energies that hitherto found release in building great cathedrals now found new expressions of the heavenward urge in sailing the seven seas, turning deserts into gardens, conquering pests and disease, breeding scrub stock into fine herds, transforming hovels into firesides of good cheer, and building new social worlds, new churches, new schools, new governments—new ways of living and of making a living in all spheres of human endeavor. These were the new songs of salvation.

No amount of riches ever exempts one from the responsibility for further expression of his powers in productive employment—a fact which the able Baxter put this way (cited by R. B. Perry 1944, 314–315):

> If God shew you a way in which you may lawfully get more than in another way (without wrong to your soul, or to any other), if you refuse this, and choose the less gainful way . . . you refuse to be God's steward . . . you may labour to be rich for God, though not for the flesh and sin.

Then he continues:

> Will not riches excuse one from labouring in a calling: No: but rather bind them to it the more; for he that hath Most wages from God, should do him most work. Though they have no

> outward want to urge them, they have as great a necessity of obeying God, and doing good to others, as any other men have that are poor.

In thus placing proficient work in any employment within the category of tangible evidence of character which most deserves emulation and esteem, this new breed of minister was obliterating the sharp and clear line which the ages had hitherto drawn between secular and religious employments. No one has ever expressed this new attitude toward the ethical significance of work more truly than did Hiram Goff, a simple shoemaker, in a conversation with his minister, John Jessig. To strike up pleasant chatter Goff remarked:

> I believe in honest work. Work is the law of nature and the secret of human happiness.

His minister replied:

> I am glad to see a man who can use the humblest vocation to the glory of God as you are doing.

This made the shoemaker's hair stand on end. Said he:

> There ain't no such thing in this wide world, pastor, as a humble vocation. Listen, you are a minister by the grace of God . . . I am a shoemaker by the grace of God. You'll carry up to the judgment seat a fair sample of the sermons you preach, and I'll carry up a fair sample of the shoes I've been making. If your sermons are your best, and my shoes are my best, He'll say, John and Hiram, you have used your talents about equally well. It's just as necessary for people to have good shoes as it is good sermons.[3]

No greater incentive to proficiency is conceivable than this identification of the services of any employment with sure evidences (proofs) of character which most deserves emulation, respect, and esteem for its own sake. For, whether it takes a religious or a secular form, this belief diverts the insatiable striving for evidence of personal worth from the "easy ways" into a ceaseless striving for being proficient, even far beyond one's actual abilities. Any amount of earnings one can ever generate from his work will always fall short of the amount he needs for the sake of showing he has all the initiative, genius, imagination, knowledge of right policy, and other virtues which entitle him to as much approbation and esteem as he would like to deserve and enjoy. Thus no achieved level of proficiency, however high, can ever release him from the felt obligation to strive for a still higher level. If he succeeds in making two blades of grass grow where only one had grown before, his thirst for a still finer image of himself then obliges him to find a way of making three blades grow where only two had grown before.

This deeply motivating belief not only directs people in their working activities, but also in their leisure employments—activities done without pay, such as vacations, club activities, and the like. As many writers have pointed out, we work as hard to generate evidence of personal worth through use of our "vacation time" as we do of our "work time" (see, for example, Dempsey 1958; Abrams 1961). Energized by this directive, people seldom find any rest and would be bored if they did; always they are on the move and never are so taut as when everything gets quiet and there is nothing to do except sit. In a passage previously cited, the great minister Baxter declared, "It is action that God is most served and honored by." By and large, American people, both farmers and nonfarmers, have long since tended to lose the older feeling that the proper purpose of proficient action, whether in "leisure" or "work" time, is to serve and honor God, but this has scarcely weakened their devotion to proficient action as the best way to serve and honor themselves and others.

Conflict between Proficient Work and the Precapitalistic Concept of Right to Acquire Property

Attention is now directed to the fact that the revolutionary commitment to proficient work, as a badge of character deserving highest emulation and esteem for its own sake, threw people into head-on conflict with the precapitalist (traditionalist) concept of the right to acquire property.

There were two aspects of this older concept. The first was that a man had a natural or moral right to "the full product of his hands." This belief was tied into the further belief that one's natural or moral right to acquire property is limited to what he can work with his own labor and use the products from. The justifying ground for this limitation was the belief that if one acquired more property than needed for this purpose, he deprived others of that land and other property which they needed for making a living.

Both aspects of this concept harmonize wonderfully well as long as people limit their work to the amount required to support their customary needs. But they were thrown into sharp conflict with each other as soon as people took to themselves the judgment that proficient work is sure evidence of character that most deserve respect and esteem for its own sake. For under the guidance of this belief, the striving to justify the highest possible valuations of one's personal worth becomes a motivation to work as much as one is able and to receive the full product of his hands as the just reward of his proficient work.

In line with this directive, people found themselves producing in excess of their customary needs. To put this excess product of their work to proficient use, they exchanged it for money which they invested in additional land, shops, stores, and the like. In this way, they expanded their possessions much beyond the amount they could work with their own labor, and also beyond what they needed for their own subsistence. This subjected them to the charge of depriving others of their natural and moral right to as much property as they needed for making a living.

Here was a conflict of ethical beliefs that was the very heart of the serious policy problem of seventeenth-century England. Government was on the spot. If it took the position that men could not acquire more property than they could work themselves and use the products from, it would thereby deny the individual his natural or moral right to the full product of his work whenever he was proficient enough to produce more than he needed for his own living. This didn't make sense. On the other hand, if government protected the individual in his natural right to the equivalent of his productive contributions whenever he produced more than he could use, then the government enabled some men to acquire more property than they needed for a living through depriving others of their right to acquire as much property as they needed. This alternative made as little sense as the first.

These conflicting meanings of the older concept of property generated the fundamental policy question of what is the primary responsibility of government to the governed, anyway. This kind of "value problem" can easily lead to civil war.

John Locke to the Rescue

The answer which John Locke worked out for this question was no less a revolutionary departure from the past than was the religious reformers' identification of proficient work with sure evidence of character deserving of emulation and esteem. For, in using widely accepted presuppositions, Locke demonstrated that the traditionalist concept of the moral right to acquire property held true only within exceedingly primitive societies where a money economy was totally absent. But whenever men consent to treat money as the exchange equivalent of all other forms of property, their nature-given right to acquire property, as fair reward for proficient work, expands to the limit of their abilities to contribute an equivalent amount of goods and services to society. This revolutionary concept worked hand in hand with the equally revolutionary belief in proficient work as the sure evidence of supremely praiseworthy character.[4]

Presuppositions

Locke derived his radical theory of property from the following widely held presuppositions concerning the prepolitical societies, which he called the "state of nature":

Postulate 1: In negative terms, the state of nature is characterized by a total lack of central political authority. This means three things. First it lacks:

> an established, settled, known law, received and allowed by common consent to be the standard of right and wrong, and the common measure to decide all controversies between them. (Locke 1690, sec. 124)

Again it lacks:

> a known and indifferent judge, with authority to determine all differences according to the established law. (Sec. 125)

Finally it lacks:

> a central "power to back and support the sentence when right, and to give it due execution." (Sec. 126)

Expressed positively, the state of nature is characterized by men who are "equally kings," each being:

> absolute lord of his own person and possessions, equal to the greatest and subject to nobody . . . without a common superior on earth with authority to judge between them. (Secs. 19 and 123)

Postulate 2: In the state of nature (prepolitical society) men are directed by reason to observe the "law of nature." This rational or moral directive

> teaches all mankind who will but consult it, that being all equal and independent, no one ought to harm another in his life, health, liberty, or possession; for men being all the workmanship of an omnipotent and infinitely wise Maker . . . are His property . . . sharing all in one community of Nature, there cannot be supposed any such subordination among us that may authorize us to destroy one another. . . . Everyone as he is bound to preserve himself . . . so, by the like reason when his own preservation comes not in competition, ought as much as he can, to preserve the rest of mankind, and not . . . take away or impair the life, or what tends to be the preservation of the life, liberty, health, limb, or goods of another. (Sec. 6)

Not only is each man obligated to observe this law of nature but also to use force if necessary in requiring others to observe it.

> that all men may be restrained from invading other's rights, and from doing hurt to one another, and the law of nature be observed, which willeth the peace and preservation of mankind, the execution of the law of nature is in that state put into every man's hands, whereby everyone has a right to punish the transgressors of that law to such a degree as may hinder its violation. For the law of nature would . . . be in vain if there were nobody that in the state of nature had a power to execute that law, and thereby preserve the innocent and restrain offenders; and if anyone in the state of nature may punish another for any evil he has done, everyone may do so. For in that state of perfect equality, where naturally there is no superiority or jurisdiction of one over another, what any may do in the prosecution of that law, everyone must have a right to do. (Sec. 7)

Postulate 3: Transcending "the law of nature," in case of conflict, men in the state of nature are directed by the law of "self-preservation."

> the first and strongest desire God planted in men, and wrought into the very principles of their nature, is that of self-preservation. (Cited by Strauss 1953, 227)

Rightful concern and responsibility for self thus takes precedence over concern and responsibility for others, in case of conflict. This means that nature endows men with absolute rights but only relative duties, as duties stand on the same footing as rights only when their observance requires no sacrifice of one's own right to life and pursuit of happiness. (In Strauss 1953, 226–227)

Postulate 4: By the very act of birth, nature and not civil society confers on all men a rightful ownership claim to their labor or productive capacities:

> every man has a property in his own "person." This nobody has any right to but himself. The "labor" of his body and the "work" of his hands . . . are properly his. (Locke 1690, sec. 27)

Postulate 5: Earliest societies were characterized by an absence of money as well as a central political authority. Locke describes this condition as:

> the first ages of the world, when men were more in danger to be lost, by wandering from their company, in the then vast wilderness of the earth than to be straitened for want of room to plant in. (Sec. 35)

Though characterized by an abundance of natural resources—potential plenty—this sparsely settled, highly undeveloped, primitive society is actually a state of "penury" in terms of usable goods (secs. 31 and 32): "Nature and the earth affords only the almost worthless materials as in themselves": such as acorns, berries, leaves, or skins. Existence is thus close to the bone; and no one can secure more than the bare necessities for self-preservation. From these postulates Locke deduces his theory of property.

The Right to Only Limited Property in a Pre-money Economy

In the first part of his theory, Locke demonstrates that within a pre-money economy, the "law of self-preservation" invests each man with a natural or moral right to acquire as much property as he needs for making a living, and that the "law of nature" (reason) denies him the right to acquire more. For if he did, he would be depriving others of their right to enough resources to support their survival needs.

Locke felt obliged to demonstrate this concept of limited right because traditional reflections on the "beginning ages" of the world had long raised doubt as to whether there can be any justification of the so-called right to private appropriation of natural resources. This doubt had its roots in the biblical dictate that the earth and its fruits were originally given to mankind in common (sec. 25). As Locke says:

> this being supposed, it seems to some a very great difficulty how anyone should ever come to have a property in anything.

Then he continues:

> But I shall endeavor to show how men come to have a property in several parts of that which God gave to mankind in common, and that without any express compact of all the commoners.

His proofs are as follows:

> every man has a "property" in his own "person." . . . Whatsoever, then . . . he hath mixed his labor with, and joined it to something that is his own, [he] thereby makes it his property . . . it hath by this labor something annexed to it that excludes the common right of other men. For this "labor" being the unquestionable property of the laborer, no man but he can have a right to what that is once joined to, at least where there is enough, as good in the common for others. (Sec. 26)

Locke then applies this labor method of rightful appropriation to the "parts of nature which God gave to mankind in common."

The first part are the perishables such as acorns, berries, wild fruit, and wild game. When and why do such "fruits of the earth" become the rightful and exclusive possession of anyone? He answers:

> 'Tis plain, if the first gathering made them not his nothing could. That labor put a distinction between them and the common. That added something to them more than nature . . . had done, and so

> they became his private right. . . . And taking this or that part does not depend on the express consent of all the commoners. . . . The labor that was mine removing them out of that common state they were in, hath fixed my property in them. (Sec. 27)

Locke next shows that the same principle applies to precious metals and wild lands (secs. 32 and 37). For if men did not have some means of appropriating to their private uses that which "God gave mankind in common," they would soon perish. The only fair way of doing so is through their own work; otherwise one would have to get his subsistence by robbery.

Locke also explains that within a pre-money economy, the amount of property one may rightfully acquire is limited to the amount required to support his necessary needs. First, a man may appropriate only as much as leaves "enough and as good" for others (sec. 26). This limitation is imposed by the similar right of others to have enough resources to support their survival needs. Second, with respect to perishables, one's right to appropriate is limited to the amount he can use before it spoils. If he picks more wild berries than he can use before they spoil, he wastes what others need for their subsistence. Third, in a pre-money economy, the right to appropriate land is limited to what one can use with his own labor by the fact that any larger amount is worthless. In Locke's words:

> As much land as a man tills, plants, improves, and cultivates, and can use the products of, so much is his property. He by his labor does as it were enclose it from the common. Nor will it invalidate his right to say everybody has equal title to it, and therefore he cannot appropriate, he cannot enclose without the consent of all his fellow-commoners, all mankind. God, when he gave the world in common to all mankind, commanded man also to labor, and the penury of his condition required it of him. . . . He that, in obedience to this command of God, subdued, tilled, and sowed any part of it, thereby annexed to it something that was his property, which another had no title to, nor could without injury take from him. (sec. 31)

Enlarging one's possession beyond this point is a sheer waste of effort.

> For supposing an island, separate from all possible commerce with the rest of the world . . . what reason could anyone have there to

> enlarge his possessions beyond the use of his family, and a plentiful supply to its consumption, either in what their own industry produced, or they could barter for like perishable, useful commodities with others. . . . What would a man value ten thousand or a hundred thousand acres of excellent land . . . where he had no hopes of commerce with other parts of the world, to draw money to him by the sale of the product? It would not be worth the enclosing, and we should see him give up again to the wild common of nature whatever was more than would supply the conveniences of life, to be had there for him and his family. (Sect. 48)

Introduction of Money Removes Older Limitation on the Right to Acquire Property

Having shown why the right to acquire property in a pre-money economy is limited to the amount each needs to support his survival requirements, Locke next shows that this limitation ceases to hold true as quickly as a money economy emerges. The major steps in his reasoning are as follows. Prior to consent to the introduction of money, "labor gave a right to property" (sec. 45) but not afterward. For in the process of introducing money, men agreed to divide among themselves all the hitherto unappropriated lands and other resources, thus wiping out the earlier natural right to acquire property from what "God gave all in common" by merely mixing his labor with raw resources.

> in some parts of the world (where the increase of people and stock, with the use of money, had made land scarce, and so of some value), the several communities settled the bounds of their distinct territories, and by laws, within themselves, regulated the properties of the private men of their society, and so by compact and agreement, settled the property which labor and industry began. And . . . either expressly or tacitly disowning all claim and right to the land in the other's possession, have by common consent, given up their pretenses to their natural common right . . . and so have, by positive agreement, settled a property amongst themselves in distinct parts of the world. (sec. 45)

Wherever this agreement to use money has not been made, much of nature's resources remain unappropriated and in a state of waste:

> there are still great tracts of ground to be found, which the inhabitants thereof, not having joined faith with the rest of mankind in the consent of the use of their common money, lie waste . . . and so still lie in common; though this can scarcely happen amongst that part of mankind that have consented to the use of money. (sec. 45)

Again, according to Locke, the consent to the introduction of money included consent to unequal possessions:

> since gold and silver . . . has its value only from the consent of men . . . it is plain that the consent of men have agreed to a disproportionate and unequal possession of the earth. . . . I mean out of the bounds of society and compact; for in governments the laws regulate it; they, having by consent found out and agreed in a way how a man may rightfully and without injury, possess more than he himself can make use of by receiving gold and silver, which may continue long in a man's possession, without decaying for the over-plus, and agreeing those metals should have a value. (sec. 50)

Again, in Locke's view, consent to the introduction of money included consent to the wage relationship, based on the contract of individuals involved:

> a free man makes himself a servant to another by selling him for a certain time the service he undertakes to do in exchange for wages he is to receive; and though this commonly puts him into the family of his master, and under the ordinary discipline thereof, yet it gives the master but a temporary power over him, and no greater than that is contained in the contract between them. (sec. 85)

Again, in consenting to treat money as the exchange equivalent of all other properties, men intended that money was not merely a medium of exchange but capital in that it rightfully earns its owner interest just as land earns rent. Its earning power stems from mutual agreements among those having unequal possessions.

> by compact (money) transfers that profit, that was the reward of one man's labor, into another man's pocket. That which

> occasions this, is the unequal distribution of money; which inequality has the same effect too upon land, that it has upon money. . . . For the unequal distributions of land (you having more than you can, or will measure and another less) brings you a tenant to the land . . . the same unequal distribution of money (I having more than I can, or will employ, and another less) bring me a tenant for my money. (Locke 1691, 22–23; cited by MacPherson 1951, 557)

Locke Turns the Tables on the Traditionalists

In this reasoning, Locke neatly turned the tables on the precapitalistic concept of appropriation according to which the individual had a natural or moral right to the full product of his labor, but a right which was limited to only as much property as was necessary for his customary needs. The advent of money removed this restriction for two reasons.

First, it removed all limits on the amount of possessions individuals may seek to acquire. For, if one can acquire more land than he can work himself, others will pay him rent for the privilege of using it, or he can exchange it for money and then loan out the money for interest, or invest it in a business and reap a profit from it. Thus as Locke said:

> Find out something that hath the use and value of money amongst his neighbors, and you shall see the same man will begin presently to enlarge his possessions. (Locke 1690, sec. 49)

Second, in addition to thus releasing each man's desires to expand his possessions without limit, money provided society with the means of returning the individual the full equivalent of all he could possibly produce, however much this might exceed the amount required to support his customary needs. Therefore, if for any reason the individual was motivated to produce more goods and services than required to support his customary wants, then the ancient justice of receiving "the full product of his labor" gave him a natural and moral right to more money than required for his customary needs. Since the difference between his total product and his subsistence needs (savings) was his to use as he pleased, he had a perfect right to exchange it for more property than he could work with his own labor. Thus, the more proficiently each man works, the more he gets riches for himself in return for equivalent riches he gives to society—a fact which Locke pointed out through a comparison of the pre-money economy of the Indian tribes with that of his native England. The former, said he,

> are rich in land and poor in all the comforts of life . . . for want of improving it by labor, [they] have not one hundredth part of the conveniences we enjoy and a king of a large and fruitful territory there feeds, lodges, and is clad worse than a day laborer in England. (Locke 1690, sec. 41)

Locke's Theory of Property as Reinforcement of the Radical Ethical Interpretation of Work

In thus using the traditionalist's own presuppositions as a means of proving that the natural right to the full product (or money equivalent) of one's labor also included the right to acquire more property than required to support one's customary needs, Locke rescued the religious reformers from devastating attack. For, while ministers might use the Scriptures in supporting their radical belief that men fail in their moral responsibility if they choose the less gainful way when "God shows them more gainful ways," they were intellectually defenseless against the charge that they were actually teaching people to enrich themselves through exploiting their brothers. Locke relieved them from this embarrassment by showing how it follows from the money linkage of commodities with each other that society is most enriched when each man receives the full product (or money equivalent) of his labor even though he may acquire more property than he can work with his own labor.

Commitments to Proficient Work and Right to Acquire Unlimited Property as Dominant Directves of Settlers

The older landlord civilization could provide no home for radicals who were committed to the beliefs that proficient work is the true mark of upright men and that there is no justifiable limit on the amount of property which any individual has a natural or moral right to acquire from the earnings of his work, except the limit of his abilities to produce goods and services for society. These presuppositions pointed to a new destiny and were a different faith from that which the older landlord civilization had trusted and guided its feet by for a thousand years. So, the new radicals had no choice except to tear up this older order or get out. They did both. American settlers stemmed mainly from those who got out.[5]

Attention is now directed to the main ways in which their heritage of radical commitments led them to revolutionary departures from ways of life and social organization generated by older traditionalist beliefs, some of which they also shared.

In line with their directive to proficient work, settlers demanded a freeholder agriculture. For within their virgin continent of cheap land, a system of debt-free, full-owner operated businesses gave each individual a better chance to earn and accumulate more property from proficient work than could any other system. Settlers also demanded a freehold agriculture because full-owner operatorship enabled them to escape the onus of "subservience and venality" which the ages had imputed to tenants and wage workers.

This traditionalist motivation reinforced the proficiency directive for a freeholder agriculture, but it was not the dominant factor. For the desire for freeholder status as means of escaping the onus of tenantry could be met by public land policies which limited farm sizes to what the usual family could handle with its own labor and management, except perhaps for relatively short seasonal labor peaks. But such policies were incompatible with the Lockean belief that each man's natural or moral right to a fair reward for his industry is violated by any state-made limit on the amount of possessions he may acquire.

To be sure, in their long struggle for equitable land policies, settlers did limit their *request* for land from the government to the amount a man and his family could handle with their own labor. But this limited request did not arise from any belief that this was all the land one had a moral right to acquire; it stemmed from the fact that under any alternative policy, settlers were fleeced by absentee speculators. If, for example, the government sold only large tracts, it gave monied men a monopoly of first opportunities to acquire title to public land, which they could turn to a speculative profit by resale in small tracts to settlers. As a protection against such exploitation, settlers demanded policies whereby the government would give the man with little or no capital, except his labor, the first opportunity to acquire at least as much public land as he and his family could work themselves, but would place no limit on any additional amount which one might acquire through his own initiative and industry. In this spirit, the settlers, like John Locke, were a highly capitalistic-minded people.

Griswold argues that Locke's concept of natural right to only limited property was "confirmation, if not inspiration for his [Jefferson's] ideal community of small land holders" (Griswold 1948, 41). This may be correct. But it is appropriate to add that Locke's whole aim was to show that this traditionalist concept of right to only limited possessions applied only to pre-money societies and not to money economies like those of the settler. Under such conditions, the heart of Locke's theory is that each man has a natural and moral right to acquire as much property as his initiative and industry will enable him to acquire, however much his possessions may exceed the amount he can work with his own labor. In this belief, Locke never had more kindred disciples than the settlers themselves.

With respect to improving prevailing ways of living and making a living, the settlers' directive to proficient work comes into clear dominance over traditionalist biases in favor of old methods handed down from father to son. To be sure, there are many instances of their resistance to change. A classic example is the often-cited hesitation of pioneering prairie farmers to use the steel plow for fear it would "pizin the land." In reasoning from such cases, and their number is legion, the impression is sometimes given that settlers were fundamentally "tradition bound," and that present-day agriculture resulted from the fact that scientific and technological leadership of the land grant colleges and commercial pressures of modern industry simply *engulfed* our older rural society with a tidal wave of proficiency beliefs and values which were never indigenous to farm people.

This puts the cart before the horse. It overlooks the fact that long before the rise of agronomists and agricultural engineers, the settlers, by and large, were well known for their belief that one fails in his duty to earn the respect and esteem of himself and others unless he is on the alert for improved ways of removing drudgery and relieving want and privation from his own household, his country, and even the whole world. From earliest times, this belief bore good fruit. For however much the settlers may have been slow to change their ways, the fact remains that innovators were their heroes. In emulating such heroes, they become "tinkerers" long before the rise of agricultural specialists. Out of their tinkering, prior to the 1860s, came most of the basic machines and implements which entered into the mechanization of agriculture during the late nineteenth century. Widespread agricultural fairs centered around their interest in improved farming. It is thus no accident that the so-called captains of industry who guided the building of industrial America from the Civil War on to around 1900 were mostly migrants from family farms of premachine America (L. Hacker 1950, 265–266). Viewed in this light, our machine age, including modern scientific agriculture, is the cumulative expression of the prior commitment of farm and nonfarm people alike to increasingly proficient endeavor as the hallmark of praiseworthy character, and not the other way around.

To be sure, the spectacular advance of the last one hundred years in either industry or agriculture would be inconceivable without the cooperation of scientists and technological leaders with businessmen and farmers alike. But it does not follow that they were a sufficient condition. From the very beginning, they couldn't have gotten their foot in the door, as it were, except for the fact that farmers and businessmen were already seeking ways of working more proficiently. Both yesterday and today, biases of farmers for the old ways have often impeded a faster rate of progress than was actually achieved. But we know of no reason for supposing that such biases have been more pronounced among farmers than nonfarmers, especially so in view of the fact that for a century the gain in output per worker has been nearly as rapid in agriculture as in industry, and in the last decade it has been appreciably faster.

The dominance of the proficiency directive among settlers is especially evident with respect to what is commonly called "quietism" and "activism." *Quietism* is the technical term used in denoting such traditionalist commitments as the belief that work is the badge of inferior worth, and that no sensible person will do no more of it than is necessary to support his customary needs so as to have ample time for good fishing, picnics, festivals, wholesome leafing, delightful companionship of coon dogs, and the like. *Activism* is the technical term used in characterizing people in whom the proficiency directive is so urgent as to be a well-nigh insatiable need for ceaseless action as means of earning favorable valuations of their worth (De Grazia 1948, 59–71).

In these terms, it is questionable if any people were ever so completely energized by the proficiency directive as American farmers and others of the settlement era. Historians picture this fact in many ways. For example, in their account of America from 1820 to 1850, Morrison and Commager observed that this period

> was America's busy age. . . . Each Northern community was an ant hill, intensely active within, and constantly exchanging ants with other hills. Every man worked . . . the few who wished to

> idle, and could afford idleness, fled from the opprobrium of "loafing" to Europe where they swelled the chorus of complaints against democratic institutions. . . . The Northern American had not learned how to employ leisure; his pleasure came from doing things. (Morrison and Commager 1937, 1:391–392)

Such zeal for industry is far in excess of that which we would expect of a people cherishing ancient ways and whose summum bonum was a freehold big enough to support their customary needs and provide escape from the ancient onus of tenantry as a badge of subservience. This traditionalist motivation characterized French peasants who finally succeeded in achieving their freeholder ideal through the French Revolution. But for this very reason they differed from the American settlers as night from day. Tocqueville pointed out this fact in a memorable passage:

> In certain remote corners of the Old World. . . . The inhabitants, for the most part, are extremely ignorant and poor; they take no part in the business of their country and are frequently oppressed by the government, yet their countenances are generally placid and their spirits light.
>
> In America I saw the freest and most enlightened men placed in the happiest circumstances which the world affords; it seemed to me as if a cloud habitually hung upon their brow, and I thought them serious and almost sad, even in their pleasures.
>
> The chief reason for this contrast is, that the former do not think of the ills they endure, while the latter are forever brooding over advantages they do not possess. It is strange to see with what feverish ardor the Americans pursue their own welfare and to match the vague dread that constantly torments them, lest they should not have chosen the shortest path which may lead to it. A native of the United States clings to the world's goods as if he were certain never to die; and he is so hasty in grasping at all within his reach, that one would suppose he was constantly afraid of not living long enough to enjoy them. . . . A man builds a house in which to spend his old age, and he sells it before the roof is on; he plants a garden and lets it just as the trees are coming into bearing; he brings a field into tillage, and leaves other men to gather the crops; he embraces a profession, and gives it up; he settles in a place, which he soon afterwards leaves to carry his chargeable

> longings elsewhere. If his private affairs leave him any leisure, he instantly plunges into the vortex of politics; and if, at the end of the year of unremitting labor, he finds he has a few days' vacation, his eager curiosity whirls him over the vast extent of the United States, and he will travel fifteen hundred miles in a few days to shake off his happiness. Death at length overtakes him, but it is before he is weary of his bootless chase of that complete felicity which forever escapes him.
>
> At first sight, there is something surprising in this strange unrest of so many happy men, restless in the midst of abundance. The spectacle is, however, as old as the world; the novelty is, to see the whole people furnish an exemplification of it. (Tocqueville 1898, 163–164)

Thus energized by an insatiable hunger for proficient action, the settlers were likely just as eager for innovations as are modern farmers. The safe conclusion appears to be that the settlers' quest for novelty could not be gratified nearly as rapidly as that of modern farmers since they had no vast system of research institutions and agricultural specialists to feed them an ever hastening stream of new farm know-how. But this handicap does not obscure the fact that from earliest time until now the driving power of the typical farmer's proficiency image of worthwhile life so keeps him on the move that what he most dreads is the arrival of the day when he must retire and "take it easy."

The same belief in the ethical significance of work which thus energized settlers with a boundless activism also generated a peculiar brand of practical idealism which has always distinguished our people. In virtue of this belief, the settlers were seeking a work bench on which to prove themselves. This they found in a virgin continent of potential plenty. But this realm of potential plenty was a hard world, especially so for a people with little or no capital except their bare hands. But however severe the privations and cruelties of the new continent, it would nonetheless turn into marvelous shapes and forms under the touch of patient industry; and soon there emerged the inspiring vision of a whole wide wilderness transforming into farms, homes, and thriving cities in response to diligence and creative toil. As men saw the oak in the acorn, so they envisioned farms in swamps and thickets, ports and thriving cities on river bends, paths of commerce along the wild-game trails. In this way the poetry of the spirit joined the sinews of the hand with the stuff that dreams are made of. Thus was born the American Dream as the felt assurance of nature and Providence alike that, in the capacities for superior industry, men have ample means to bring their actual circumstances increasingly in line with their aspirations. As Santayana observed, the typical American is "an idealist working on matter" (Santayana 1924, 175). This fact accounts for his skittishness toward either visionary idealism or crass materialism, and of no one is this more true than the usual farmer. Few things so get on his nerves as the preaching of "ideals" that are without tangible promise of such materialistic outcomes as conquering disease or unlocking the secrets of photosynthesis. Thus, his "idealism" and his "materialism" are like the sides of a coin. The one is inconceivable without the other, although neither is identical with the other. Accordingly if one calls him a materialist, he scowls. Call him an idealist, and he wonders if you think he is soft headed. But call him a practical idealist, and he dilates with good feeling. Add that he is a self-made man, and he bursts with pride. His "practical idealism" is thus only another name for his work ethic faith that, in their capacities for proficient industry, people have ample means for bringing their actual conditions increasingly in line with their dreams or visions through an ever greater mastery of nature, both human and physical.

Finally, the same directive for proficient work which generated the settlers' demand for a freeholder agriculture, his thirst for innovations, his ceaseless activism and practical idealism, also included unique concepts of equity. For the belief that the key responsibility of the individual to himself and society is to earn high standing through increasing competence in any useful employment of his choice obviously includes the further belief that society owes three reciprocal debts to individuals. These debts are the obligations to (1) provide all its members with opportunity or access to the means (for example, public schools) necessary for developing their potential to the fullest extent possible; (2) offer opportunities for productive roles in keeping with their abilities; and (3) give each a fair return for their contributions. Thus the directive to proficient work places society under duties to the individual which are no less binding than those which it places on the individual to himself and society. Accordingly, it is as impossible for the individual not to resent the unfairness of a society which fails to do its best to discharge all of these debts to the individual and at the same time expect him to earn good repute through proficient work, as it is for society not to resent the unfairness of the individual who seeks a living and a favorable valuation of himself, but is unwilling to earn these goods through superior industry.

These three concepts of equity are all caught up in what is commonly called "the justice of equal opportunity." The first two debts are called distributive justice, and the third is called "commutative justice." That is, distributive justice includes the belief that society owes its members (1) access to the means necessary for developing their potential as fully as possible and (2) the opportunity for a productive role in keeping with their abilities. Commutative justice includes the belief that society is obliged to return a fair reward for their contributions.

There is no "natural harmony" between these beliefs. Individual capabilities are themselves largely the function of goods and services that are within society's power to extend or withhold. Therefore, distributive justice may require severe limitations on income inequalities that many might regard as incompatible with the right of each to a fair return for contributions made.

It is interesting to observe that the settlers rejected the medieval method of resolving this conflict; yet, thanks to the abundance of cheap land, the resolution they actually achieved closely approximated the one called for by the medieval theory of natural right to only limited possessions. As previously stated, the heart of this theory is the belief that each person has a natural or moral right to the "full product of his labor" but that his right to acquire property is limited to the amount which he can work with his own labor in supporting his customary needs. Obviously this limitation precludes the possibility of inequalities of income ever becoming so great as to throw the just demand for equal opportunities for personal development and creative roles into conflict with the equally just demand for each to receive the equivalent of his contribution.

But the settlers held this limitation would rob the industrious of their just deserts. They did so because of their Lockean belief that, if the individual produces more than he needs for a living, he has a natural and moral right to exchange the surplus for money, which he may invest in additional property that yields additional money, which he has a right to invest in still more income-producing property. There is no conceivable limit on the income inequalities which this process may generate. For there is no assignable limit to the diverse and unequal capacities of men for proficient industry and wise investment.

But, as Madison ably stated, this means that protection of "different and unequal faculties of acquiring property" is the first object of government (Madison in *Federalist Papers,* 1961, no. 10), however great may be the income inequalities which are generated by their unequal capabilities. But no government can be totally committed to this objective and also totally committed to (1) providing all its citizens with access to the means necessary for developing their potential as fully as possible, and (2) offering them opportunities for productive roles in keeping with their abilities.

This contradiction was resolved in early America by the abundance of land. For with relatively inexpensive farm technologies and enough "dirt cheap" land available for everyone, no one would work another person's land because he could earn more by acquiring title to a farm big enough to fully utilize his own labor. Thus, the abundance of land held income inequalities within very narrow limits and in this way prevented any conflict that would have otherwise arisen between the competitive requirements of commutative and distributive justice.[6]

But today the conflict is serious, made so by high-priced land, highly productive farm technologies, limited outlets for farm products, and a national economy with 5 to 6 million unemployed. Under these conditions, price-depressing surpluses siphon off to the rest of society a disproportionate share of the cost-reducing benefits of the farmer's increasingly superior industry. Thus, commutative justice is disturbed. Seeking to rectify this inequity through programs designed to cut back production to the level at which total supply balances total demand at reasonable prices improves incomes for agriculture as a whole. However, such programs may worsen the inequality of opportunity to productive roles and fair incomes among the relatively few families who are on proficient farms and the great bulk who are on inadequate farms. Seeking to rectify this inequity through credit programs designed to accelerate growth in the total number of proficient farms generates additional price-depressing surpluses, thus worsening the lack of fair returns to farmers as a whole. Seeking to get out of all these boxes through programs designed to move surplus workers out of agriculture expands the national pool of unemployed. This further infringes on the natural and moral right of all citizens to productive and fairly rewarding roles in keeping with their abilities.

Thus, all our genius and wit are far less effective than was the abundance of cheap land of the settlement era in limiting income inequalities to the point of harmonizing the otherwise seriously conflicting demands of distributive and commutative justice.

The Widening Gap between the Freeholder Ideal and Proficient Practice

As previously explained, the freeholder ideal was an agriculture of businesses in which the will and interest of all participants are totally excluded except those of the operating families who do most if not all the work of their businesses. In such self-sufficient firms, the managerial and risk-taking role of the operator is that of absolute lord and master over all conceivable operations of his business. Divested of all commitments to any outsider, such as creditor, landlord, government, or others, the freeholder was truly the monarch of all he surveyed.

Furthermore, until the late nineteenth century, the actual status of farm families and their ideal status as freeholders were identical. The actual and the ideal were not hooked together by an "agricultural ladder" in which a would-be freeholder started out as a farmhand, worked a few years until he saved enough from his wages to buy a line of equipment, became a tenant until he could save enough to buy a farm with the help of creditors, worked and saved a few more years until he could pay off all his creditors, and then finally live out the evening of his life in the proud feeling that he was absolute lord and master of all that lay within his fence lines. Farmers of the settlement era would have regarded this separation of the ideal from the actual (theory and practice) as a monstrosity, for their ideal was an agriculture in which the young man began the race of life as a freeholder—a debt-free, full-owner operator—and the ladder of his dreams extended from that point on up.

Thus the concept of the "agricultural ladder" is wholly the product of the postsettlement era and stemmed from the need for assurances that continual departures from dying ideals are in fact merely stepping stones to their fulfillment. Four observations bear out this point.

First, as cheap land gave out and farm technologies became increasingly expensive, the amount of land and other resources required for proficient farming quickly exceeded the amount which the usual farm family could acquire with its own limited assets. In line with this fact, increasing numbers of farmers and family operators departed from their freeholder status by seeking the help of creditors who would supplement their limited resources with real estate as well as capital loans on reasonable terms. Both public and private lenders responded to this invitation. But the price of their doing so was a debt contract that limited the farmer's absolute managerial power over his own equity until creditor liens were satisfied. In this way, a wedge was driven between his actual status and his ideal status as a freeholder. But sufferance of the gap was made easy by the faith that, within a few years of diligent industry and thrift, debtor operators would accumulate enough savings to pay off their creditors and wind up their later years as the absolute sovereigns of a Jeffersonian freehold.

Two innovations in practices of farm lenders in recent years indicates that the essential function of agriculture credit today is to enable operators to acquire operating control over proficient farms through perpetual credit. First this tact is implicit in the establishment of forty-year real estate loans. For the lifetime of such loans substantially exceeds the productive life span of the usual farmer. Second, the same fact is implicit in the use of loan renewals and lengthening the life span of chattel loans. It is also implicit in a small increase in incorporated family farms.

Second, as increasing numbers of family farmers sought the help of creditors in purchasing farms, so they likewise sought the help of landlords who would supplement their working capital by lending them the use of their land and buildings. Landlords responded but the price of their response was a rental contract that limited the operator's absolute managerial power. For example, to protect his contribution to the operator's business, the landowner might require commitments of the operator to follow a specified rotation system, and make no alterations in fixed improvements without the landowner's approval. In this way, a second wedge was driven between the actual status of increasing numbers of family farmers and their ideal status as freeholders. But again the sufferance of this gap was made easy by the faith that within a few years of diligent industry and thrift the usual tenant operator would eventually save enough to achieve a debt-free, full-ownership status.

Third, since the 1930s farmers have sought government assistance in helping them to balance their output to total demand at fair prices. Government has responded with price-support guarantees. But the cost of this response has been production contracts which again limit the hitherto absolute managerial power of the farmer to produce whatever amounts of commodities he chooses. In this way, a third wedge has been driven between the farmer's actual managerial status and his ideal status as a freeholder.

Finally, in recent years merchandisers of a very limited number of foods like poultry, table eggs, and certain fruits and vegetables began bypassing wholesalers and warehouses and going to first processors in the marketing chain with contracts which specified the prices which they would pay for specified volumes, delivery dates, and quality characteristics of given products. First processors in turn translated their contracts into corresponding production contracts with farmers—the first producers. Thus, in addition to creditors, landlords, and the government, contractors, including feed dealers, have entered into the farmer's business with contractual arrangements that further limit the historic ideal of himself as the absolute master of a self-sufficient firm. Today, the farmer is only a relative master of a contractual firm.

It is often said that "the family farm as we have known it is on the way out." Right. But this has been true for nearly one hundred years. It first became true when the family operator allowed creditors, especially real estate creditors, to be participants in his business, thus limiting his absolute managerial power with contractual commitments to comply with specified rules which his creditors deem necessary for the protection of their interest in the farmer's business. Again the family farm, "as we have known it," disappeared when the operator consented to the landowner becoming a participant in his business, limiting his absolute managerial power with commitments whereby the landowner could send out the sheriff if he didn't live up to this agreement.

Still again, the family farm as we have known it disappeared when family operators by the millions sought for their government to become a participant in their business by guaranteeing them price floors in exchange for commitments which limited the farmer's hitherto absolute power to produce any amount of whatever products he pleased. Finally in some sectors of agriculture the family farm as we have known it has disappeared because the operator has allowed contractors to become participants in his firm through contracts which specify both the price and the product he may produce. In return the farmer limits his absolute power to produce as he pleases with commitments to comply with certain specified practices which will meet the specifications of the contract.

In following their three-century directive to be proficient farmers, actual family operators over the last one hundred years have steadily shifted their operations from self-sufficient firms to contractual firms. I see no possibility of turning the clock back. If this be true, the Jeffersonian Dream in the strict sense of the word is dead. But is it realistic to be this strict?

Prospect

An agriculture in which operating families do most of the work and have varying degrees of risk-taking and managerial power is very much alive and on the go. Otherwise an increasing proportion of total farmwork would be accounted for by hired labor, and also an increasing proportion of total farm output would be accounted for by larger than family farms, that is, farms using more hired labor than can be supplied by an ordinary farm family, which is approximately 1.5 person-years. Unless both of these changes were definitely underway, an agriculture of predominantly family farms could not be going down the drain. They do not appear to be underway.

Actually the proportion of farmwork done by hired workers declined rapidly, but both declined at about the same rate, although the rate decline in hired work was somewhat faster (Nikolitch 1962, fig. 6, p. 23, table 21, p. 40). Again, family-operated farms accounted for over 74 percent of total farm marketings in 1954 as compared with nearly 67 percent in 1949 (table 16, p. 29). Preliminary indications are that approximately the same relationships were obtained for the 1954–59 period as for the 1949–54 period. (The figures cited were unaffected by changes in farm wage rates or prices received by farmers.)

But the fact that the family farm is not going down the drain does not mean that the road ahead is clear of danger. My chief apprehension stems from two areas; one pertains to the nature of future farm technologies, the other to the increasing capital requirements of proficient farms.

In the visible future will scientists and engineers develop farm technologies that make possible the widespread introduction of the factory process into food and fiber production, as was done in industrial production at the beginning of the Industrial Revolution?

Before the rise of the machine process, agriculture and industry were characterized by a sequential pattern of operations in which the various production steps were separated by time intervals, so that the same individual could perform one production step after another until the final product was the embodiment of his own planning and effort. The shift from hand to machine methods quickly wiped out this premachine similarity of agriculture and industry. But the shift to machine production in farming, by and large, does not transform the age-old sequential pattern of operations into a simultaneous pattern. If it did, plowing, cultivating, and harvesting would need to be done concurrently. But nature does not permit this; it separates these operations by very wide time intervals.

Were this not the case, the shift to machine farming would quickly multiply the number of things that must be done at the same time far beyond the number of workers in any family. In this way, it would destroy the premachine institution of family farms (or businesses) and replace them with factories. A factory is simply a production organization in which the many steps involved in making a product, such as an automobile, are divided into different places along a production line and done concurrently with the input-output rate of each operation controlled by the flow rate of materials between operations. No family could possibly operate a factory. The number of things that need to be done concurrently is far greater than the number of workers in many, many families.

In some agricultural processes, nature separates production steps by very short intervals. This is true of the time spans between animal feedings, for example, and milking cows. Approximation of the factory process in farming has been achieved to some extent in these situations. In such instances, the family farm becomes a physical impossibility for the same reason that a family factory is unthinkable in industry. In the main, however, the "Industrial Revolution" in agriculture is mainly a spectacular change in the implements with which operations are performed, whereas in industry it is a further revolution in the sequence of productive operations.

The age-old sequential pattern of operation is the physical basis of the family production unit, in either industry or agriculture. It is long since gone in industry. Can it be expected to continue indefinitely in agriculture? There is reason to suspect it will not. The family farm would go out like a light if the secrets of photosynthesis were unlocked, as it seems they surely will be. It is said we are now closer to this achievement than we were to the atomic bomb in 1905 when Einstein opened up the possibility of it with his famous formula for the equivalence of matter and energy.

Even much before this event, scientists and engineers may go far in wiping out the physical possibility of the family farm through discovering profitable ways of separating the production of crops and livestock on the same farm throughout the Corn Belt and elsewhere. Should this happen, there probably would not be much left of family farming. From a physical standpoint, livestock feeding can now be organized into an approximate factory process. Although my foresight is very limited, the day may not be far off when specialists will bring forth profitable farm technologies that will wipe out the wide time intervals between the many operations on given farm products.

In addition many fear the family farm is on the verge of being wiped out by its expanding investment requirement. It is not uncommon for proficient farms to represent real investments from $75,000 to over $200,000, and non–real estate investments of an additional $35,000 to $75,000.

How can individual families with assets of this magnitude pass them on intact to the next generation? The ready answer is that the farm real estate owners may combine into relatively large corporations, convert their assets into stock shares, and then protect the value of their assets and assure reasonable dividends by appointing boards of directors who will hire professional farm managers to see that farming is carried on in a proficient manner. By withholding rental privileges, and offering stock shares in exchange for renters' investments in working capital, the corporation would be able to induce family farmers now on a rental basis to become shareholders in the corporation. At their option they might remain on the same farms as before, but as wage or salaried workers of the corporation under the direction of professional managers.

This may happen, but there is an obstacle that may prevent it from happening very soon. As I see it, the obstacle is this. The relationship between prices received and prices paid by farmers has seldom been sufficiently favorable to make possible a return to labor comparable to that of workers of similar capacities in nonfarm employment. If agriculture had to compete with industry for workers in the same general labor market, then to take over in great numbers in any given area, large farming corporations would have to pay workers as much as they can earn in other employment, else they would choose industrial employment. Furthermore, from early times up to now very large agricultural businesses have become the dominant institution in American agriculture only in those regions, such as the far West and the plantation areas of the South, where peculiar institutional arrangements have enabled agriculture to secure large supplies of relatively cheap labor outside the general labor market and in this way escape the necessity of competing with industry for labor services. No similar institutions prevail throughout all the major regions in which family farming has long been the dominant business unit of American agriculture, and as I see it there is little possibility of such institutional developments in these areas. Therefore, my guess is that investment requirements of proficient farms will fail to proceed to the point of wiping out a predominantly family farm agriculture in the visible future.

Expanding investment requirements may become incompatible with the family farm without leading to big stock-share corporations. The trend in farm mechanization is in the steady direction of larger capacity machinery and equipment which decreases labor per unit of output. Fuel, oil, and other machine costs might be cut appreciably by larger business that enables operators to purchase supplies in wholesale rather than retail lots. Savings might also be achieved in farm marketing costs. Such factors suggest that we may be reaching a point where increasing numbers of operators, by expanding their business to four- or five-man farms may be able to achieve enough savings in costs to compete with industry for labor. This eventuality might or might not involve large stock-share corporations, but in either case it would wipe out the family farm. The main obstacle to this development, as I see it, is high risk on large investments which stems from uncertainty of farm prices and incomes. I do not expect these obstacles to be overcome in the near future, but I may be mistaken.

Conclusions

Considered as a business in which the operating family is a risk-taking manager of an undertaking in which it does most of the work, the family farm is no less a dominant institution of American agriculture now than it was in Jefferson's day. The difference lies in the fact that the typical family farm of his day was a highly self-sufficient firm in which the will of the operator was absolute; his power to run his business as he pleased was not limited by outside participants, such as his creditor, his landlord, his government, or buyers of his products, or his suppliers, such as the feed dealer. Furthermore, in Jefferson's day of cheap land and low capital requirements per farm there was a high correlation between this concept of a self-sufficient firm, in which the operator's will was absolute, and the requirements of a proficient farm—that is, a farm with sufficient resources and productivity to utilize a "full line of equipment" and generate at least enough income to meet expenses for operating costs, fixed charges, family living expenses, and enough capital growth for new investments to keep in step with technological advance and rising levels of living. Under modern conditions, this older concept of the self-sufficient firm has long ceased to coincide with the requirements of a proficient farm. In the actual life of farm people, requirements of proficient farming have taken precedence over values which may attach to their image of themselves as absolute masters of their businesses. As a result, the typical family farm of today is a contractual firm in which the operator's managerial power and risk taking are limited in varying degrees by his commitment to follow certain rules under which others are willing to extend him services that enable him to operate a more proficient farm than he could otherwise do.

In Jefferson's time no one did more than he to advance the cause of scientific agriculture. From this fact, we may infer that he sought above all else an agriculture of proficient family farms. Such an agriculture is not possible today except as the usual family seeks and secures the help of other owners of production and marketing services. In light of this fact, we may infer that Jefferson would expand his agricultural ideal to include contractual firms in which families consent to limit their otherwise absolute managerial power with commitments under which others are willing to extend them production and marketing services they need for becoming operators of proficient farms.

In these terms, the traditional ideal of a predominantly family agriculture of proficient farms remains as relevant to present-day policy considerations as in the settlement era. In that era, implementation of the ideal merely called for equitable land policies that gave families with little or no capital except their labor first opportunities to acquire fee-simple ownership to enough public lands for a proficient farm. Today the lines of implementation are quite different.

Achievement of an agriculture of proficient family farms calls for continual improvement in agricultural credit policies and programs to enable competent operators of inadequate farms to expand their resources and productivity to farms of sufficient size to effectively utilize a full line of equipment. As this is done, we approximate an agriculture of farms which are able to produce foods and fibers for society at minimum cost per unit of output. Limited available data on the economies of farm size indicate that, under conditions of reasonable prices for farm products, such an agriculture will enable operating families to have earnings for their labor and management which are comparable to earnings of people of similar capacities in nonfarm employments. The data also indicate that in most types of farming, farms which are larger than proficient family farms would not be able to produce foods and fibers at less cost per unit of output, although they would return greater profits to management because of their larger volume of output.

Again, achievement of the traditional agricultural ideal of proficient family farms calls for continual effort to establish and maintain equitable contracts between tenants and landlords; between operators and contractors seeking products of specified grades and quality; and between government and farmers seeking ways to bring total farm output in line with total demand at reasonable prices. Third, in connection with new land developments, such as large reclamation projects, implementation of the family farm ideal calls for policies that keep the door of opportunity open to families seeking to become established as operators of proficient farms. Finally, implementation of the ideal calls for continual scrutiny of tax laws to see that these do not take unfair advantage of proficient family farm operators.

But the striving to achieve this Jeffersonian agriculture of proficient family farms runs into conflict with another historic ideal of ample opportunities for farm people for productive work in return for which they earn decent incomes and the high standing they seek in their own eyes and in the eyes of others. When land was so abundant as to be "dirt cheap" there was room enough in agriculture for every family to have a proficient farm that wanted one.

Today, this is far from true. In 1959, there were over 2.4 million commercial farms. Depending on what assumptions are used, one may reach somewhat different estimates of the number of proficient family farms that would be needed to provide society with all the foods and fibers it wants at reasonable prices. But, all the "educated guesses" I know of indicate that about 1 million proficient farms would be enough to do this job.

Achieving an agriculture of proficient family farms comes at a very high price when the cost of it is the uprooting of well over half the present commercial farm population. It will not do to say they will shift into higher paying nonfarm employments. This alternative is limited by a lack of over 5 million opportunities for the needs of the national labor force. It also is severely limited by the fact that farm families commonly lack the training and skills required to take advantage of nonfarm opportunities, and also by the fact that anyone's occupational mobility is definitely circumscribed once he has trained and geared himself for a particular career such as farming, mining, or teaching and research.

Above all, the clash between the drive for proficiency and the need of opportunities for farm families is not basically a knock-down drag-out struggle between an agriculture of family farms and an agriculture of larger farms. For, as previously explained, in terms of the proportion of total farmwork and total farm output accounted for by farms on which operating families do most of the farmwork, family farming is not losing out. This means that the drive for achieving the historic agricultural ideal of proficient family farms is in fundamental conflict with the equally historic ideal of providing ample opportunities for farm people; and this conflict cannot be resolved by "programs to save the family farm."

This does not mean surrender of the family farm ideal, but it does mean that policies and programs to achieve it must be incorporated within the larger objective of finding opportunities for farm people who are otherwise sacrificed by an overwrought dedication to the agricultural ideal of proficient family farms. The overriding concern of Jefferson was with policies that would provide abundant opportunities for people; and his interest in an agriculture of proficient family farms was incidental to his larger concern with the need of people for opportunities.

If this be a correct estimate of the man, then the first requisite of Jeffersonian principles bids us turn attention more directly to ways of expanding educational opportunities for rural areas and of creating more new nonfarm employment opportunities in both rural and urban sectors of the nation. Only as we move toward this objective of more abundant opportunities for all people do we open a realistic road to more rapid achievement of an agriculture of proficient family farms for all who remain in farming. That, as I understand it, is the true relevance of the Jeffersonian Dream today.

Two on Jefferson's Agrarianism

GENE WUNDERLICH

> There is no name in American history so intimately associated with the twofold theme of agrarianism and democracy as that of Thomas Jefferson.
>
> —A. Whitney Griswold, *Farming and Democracy*

PAUL THOMPSON defines the agrarian tradition by way of the many forms it has taken through the centuries (see "Agrarianism as Philos- ophy," this volume). One of the more enduring forms is associated with Thomas Jefferson, a founding father of America and a distinguished agriculturist.[1]

Two among the legions who have searched the agricultural origins of American character are A. Whitney Griswold and John M. Brewster. Both have drawn upon Jefferson in their writings. Griswold's *Farming and Democracy* (1948; hereafter FD) was written on the two hundredth anniversary of Jefferson's birth.[2] It was Griswold's only entry into agriculture and its concern was largely the relation of democratic institutions to agrarian ideology and policy. Brewster, a long-term employee of the U.S. Department of Agriculture, wrote extensively about the agrarian ideoogy underlying the family farm. Brewster's perspective of Jeffersonian agrarianism can be drawn from one of his best-known works "The Relevance of the Jeffersonian Dream Today" (1963; "The Relevance of the Jeffersonian Dream Today," this volume; hereafter RJDT).[3]

Griswold's *Farming and Democracy*

Griswold (1906–1963) was, by discipline, a historian. His undergraduate and graduate education, indeed virtually all of his professional career, was tied to Yale University, of which he was president from 1950 to the year of his death (Anonymous 1964). His study of the idealisms of agriculture in a political environment was supported by the Guggenheim Memorial Foundation, the Social Science Research Council, and Yale University. It was undertaken after World War II, a time of great concern about the nature and future of democracy in the world.

Griswold introduces FD with the proposition "that the nature of the American state requires the preservation of the small, owner-operated, family-size farm . . . that the fate of democracy is somehow or other bound up with the fate of the agricultural community whence it emerged" (1948, 4). In six chapters, he examines in detail the origins and forms of agrarianism in the United States, Great Britain, and France, the role of Jeffersonian and agrarian ideals in policy, and prospects for agriculture and democracy in America. He concludes that family farming had neither created, nor will preserve, democracy, but that democracy depends upon widely dispersed control of power, economic and political. He writes: "The lesson is plain in history. Family farming cannot save democracy. Only democracy can save the family farm" (1948, 204).

Griswold conventionally, but perhaps incorrectly, attributes much of Jefferson's philosophy of government to John Locke.[4] He also attributes the rationale of property to Locke, with one significant difference:

> "Locke's ideas on property were construed in the interests of the rich as well as the poor. . . . If Locke was more interested in defending the right to private property than in promoting its equal division, to Jefferson both were important." (1948, 41)

From Griswold, his contemporaries, and his followers, the Jeffersonian and American ideal of the family farm tied the entrepreneurial, decision-making farmer to freehold landownership. Jefferson led the attack against entails and primogeniture in Virginia on grounds that they created unequal shares among the heirs of estates (Edwards 1943, 54). The politically pragmatic Jefferson, Griswold acknowledged, recognized that equal division of land was an ideal, not an exact practice (Griswold 1948, 43). The Jeffersonian Ideal, interpreted realistically, means that land should be widely, if not equally, distributed.

The Griswold Omission

Given the general orientation and conclusions of FD, it is a bit strange that Griswold would fail to complete a famous quote from Jefferson, a quote that first seems to unreservedly support small farmers as a singular source of virtue, but in full text supports their disappearance. The quote is from Jefferson's letter to John Jay, August 23, 1785. (I have italicized parts commonly omitted.)

> *We have now lands enough to employ an infinite number of people in their cultivation.* Cultivators of the earth are the most valuable citizens. They are the most vigorous, the most independent, the most virtuous, and they are tied to their country, and wedded to its liberty and interests, by the most lasting bonds. *As long, therefore, as they can find employment in this line, I would not convert them into mariner, artisans, or anything else. But our citizens will find employment in this line, till their numbers, and of course their productions, become too great for the demand, both internal and foreign . . . As soon as it is, the surplus of hands must be turned to something else.* (In Koch and Peden 1944, 377)[5]

The selective quotation from Jefferson seems to ally Jefferson with the agrarian fundamentalism Griswold aims at proving has no connection to democracy. The reason for omitting portions of the larger Jefferson statement is not obvious, but perhaps Griswold wanted to round out an image of extreme agrarianism. Jefferson was a supporter of agriculture and rural values, so it would not be difficult to caricature his position by omitting his qualifications and differing views. Jefferson's personality, interests, and perspectives are reputed to have occupied many, sometimes contradictory, compartments (Malone 1981, 146; and Ellis 1997).

Had Griswold emphasized the omitted portions of the Jefferson quote, he could have used the Jefferson insight to support his own views of economic sector changes of the mid-twentieth century. Indeed, Griswold a few pages later in FD cites Jefferson as saying, "we must now place the manufacturer by the side of the agriculturist" (1948, 35) and terming Jefferson's position a "conversion." But the omitted passage reveals that recognition of economic restructure was a part of Jefferson's thinking all along; no "conversion" took place. Why Griswold omitted the critical passage is somewhat mysterious, and mischievous for scholars who get their Jefferson from FD.

Brewster's Jeffersonian Dream

John Brewster (1906–1965) was a contemporary of Griswold. Brewster had a farm background and graduated from the College of Emporia (Kansas) in English and psychology (Brewster 1970). In graduate school, he studied philosophy at the University of Chicago under George Herbert Mead, who died during Brewster's graduate studies. Brewster left Chicago to complete his Ph.D. in philosophy at Columbia University. Brewster's writings grew largely out of his career in the Department of Agriculture, where he emphasized farm values and attitudes, agricultural organization, and related farm policy. He frequently employed his particular concept of family farm in his articles.

Brewster opens RJDT ("The Relevance of the Jeffersonian Dream Today," this volume) with the statement, "It is timely to take a current look at Griswold's *Farming and Democracy*" (he did not explain why he thought it was timely). Brewster stated Griswold's concern was the validity of Jefferson's belief in the "causal relationship between farming and the political system of democracy" (RJDT).[6] Within the first ten lines of RJDT, Brewster dismisses Griswold with: "We shall not belabor this issue further; a more fruitful question is why such a belief became so much a part of our folklore in the first place" (RJDT).

Brewster's discourse in RJDT, as in many of his other writings, centers on his concept of the "family farm . . . an agricultural business in which the operating family does most of the work and is *manager* of ongoing operations of the business as well as a *risk taker* in the outcome (financial returns) of the business venture" (RJDT). He elaborates the family farm idea:

> "In this definition, *operatorship is equated with varying degrees of managerial power and with risk-taking involving management and production inputs, including labor*. . . . This definition applies to the Jeffersonian freeholder and the modern family farmer alike." (RJDT)

The distinction, if any, that Brewster makes in the structure of agriculture in modern and Jeffersonian times was in "self-sufficiency," by which he meant "a firm from which the will and interest of all conceivable participants in the business, except those of the operator, are totally excluded" (RJDT). He connected this organizational form with the value set of "the

Jeffersonian Dream . . . Jefferson's affection for and desire to establish and preserve an agriculture of freeholders—full-owner operators, debt-free, unrestricted by any contractual obligations to anyone" (RJDT).

From these definitions and concepts, Brewster evaluated the agriculture of the day (ca. 1962). He employed two strands of argument in his explanation of the beliefs carried by Americans about the structure of agriculture. One strand was in terms of managerial dominance by a sole entrepreneur, presumably the head of household (family). The other was ownership of property of the farm, which then and now means primarily land. In RJDT and generally, Brewster avoids distinguishing land from other assets or property.

For the first strand, Brewster found ethical roots in "proficient work as the hallmark of praiseworthy character" (RJDT). He devoted a section to the religious and moral support of proficient work, from holy callings to the craft of shoemaking. Agriculture was a source of honest, honorable work. The ethical dimensions of labor were extended to earnings and the accumulation of wealth. In general, the foundations of the family-farm work ethic and the earnings therefrom appear to be more or less the Calvinistic Protestant work ethic (see also Brewster 1970, 7–66).

For the second strand, he found ethical roots in "the natural or moral right of men to acquire all the property they can from the earnings of their work" (RJDT). Here, as elsewhere, Brewster calls upon his champion, John Locke, to whom he referred twice in RJDT as "the greatest of the natural rights philosophers." The Lockean ethic for property is based on the mixing of one's labor in the accumulation of wealth, a feature attractive to Brewster in his concept of a family farm. A large part of Brewster's article, to the point of distraction, is an argument on the Lockean interpretation and justification for holding property. By avoiding a penetrating distinction of land as a resource, Brewster avoids awkward questions about how land is created by mixing with labor, and thus how land can be property except by a claim through contractarian (unnatural rights?) principles.

The Lockean natural rights view that justifies property by mixing with labor provides Brewster with a foundation for his concept of the family farm. The family farm of Brewster's time was a package of labor/management and property held by a sole proprietor. In his view, the (Jeffersonian) freeholder agriculture was the ancestral model of that package. Because the model saw the two strands of entrepreneurship and ownership as inextricably connected, interactions with the market, lenders, landlords, and government meant that "the Jeffersonian Dream in the strict sense of the word is dead." (RJDT). But then he relaxes the strict sense and turns to a simple criterion of farm size, based on 1.5 years of family labor, and returns to a rosier scenario: "an agriculture in which operating families do most of the work . . . is very much alive and on the go." He expressed apprehension about the effects of technology and capital requirements, but arrived at less than startling conclusions, consistent with most agricultural economists of his and the present day. Farm size will continue to rise, farm numbers will decrease, farms will be managed within more complex contracts, farms will function within a lessening agricultural economy, society, polity. Brewster reads the increases in farm size and structural changes in agriculture as consistent with his definition of family farm but with continued out migration of the farm population. Farm programs should not interfere with this process:

> This means that the drive for achieving the historic agricultural ideal of proficient family farms is in fundamental conflict with the equally historic ideal of providing ample opportunities for farm people: and this conflict cannot be resolved by "programs to save the family farm." (RJDT)[7]

Brewster and Griswold hold similar positions on market and migration solutions to the farm size and number issues, positions similar to that of Jefferson, if one restores the Griswold Omission. Agricultural economy will have to transmute to an urban-cybernetic-industrial form.

Both Griswold and Brewster, when describing the state of agricultural economy in their respective days, were consistent with agricultural economists at the time. Their historian and philosopher perspectives and vocabularies added special interest to generally accepted explanations of the trends toward fewer farms, out migration, increased productivity, below-average incomes (the lag of economic change), and reduced influence (but still proportionately strong) on national policies. Their retrospective, and sometimes selective, use of Jeffersonia in FD and RJDT, however, reflected an inadequate distinction between entrepreneurship and landownership. After all, in the time of Jefferson the distinction did not matter, at least in terms of the distribution of economic and political power, and the future of democracy.

Griswold devoted an entire chapter to the questions: "Are they [farmers] still the most precious part of the state? Were they ever?" His answer was that, in reality, no. As a myth, however, Jefferson's "ideal of democracy as a community of family farms has lived on to inspire the modern lawmakers and color the thoughts of their constituents when they turn their minds to rural life" (1948, 46).

Brewster opens RJDT by stating (incorrectly) that Griswold only addresses the issue whether the Jeffersonian Ideal as an essential ingredient of democracy squares with reality. Brewster, therefore, goes on to examine whether the Jeffersonian Ideal or "dream" was itself reality (RJDT). He concludes yes, building his case almost entirely on Locke. But Locke was concerned more with the rationale for property—that is, mixing one's labor with the property object—than with a particular occupation. Locke's primary concern was freedom, and in his time freedom meant freedom from government. He was less concerned with freedom from other individuals or entities. Brewster in defining his family farm in terms of proprietor labor and management and debt-free landownership (RJDT) merges the Lockean view of property with his own view of farmer control. He was observing a massive substitution of mechanical, biological, and chemical capital for labor. Farms, by earlier standards, were swallowing huge areas of land. In order to reconcile the system of Lockean ideals with what he observed, he defined and emphasized a labor-based concept of family farm. He acknowledged tenancy but dismissed further implications of landownership and control by focusing on the physical or management dimensions of labor. An exception was the inclusion of a tenure section in his chapter in a 1961 National Basebook of Agriculture, which I coauthored with him (Brewster and Wunderlich 1961). A moment's reflection about economic conditions of Brewster's and our time reveals the inherent conflict between farmer freeholder and widespread ownership of land.

If agricultural land is to be held by operators, it will be highly concentrated because farmers are a small percent of the population, because their number is shrinking (one-third the number of the 1930s) and because the rest of the population is increasing. Land acres are relatively constant, declining slightly, but value is increasing while percent of real estate value or total asset value is decreasing. Structural arithmetic tells us that wide distribution of farmland and land to the tiller is incompatible. If the small number of operators own the land, landownership will be concentrated. Policy must decide, therefore, either freeholder farmers *or* widespread ownership, not both. This proposition is implicit in Griswold, Brewster, and most observers of agricultural development in the United States. It is simply necessary to state it clearly and simply.

From the discussion above, what can be said about the practical, pragmatic problem of policies on the organizational structure of agriculture? How many farmers? How much land to the tiller? Because farmers are now a small proportion of the population and of enterprises, ownership of farmland by farmers is inconsistent with an objective of widespread ownership of farmland. Indeed, farm tenancy today is more closely related to Jefferson's ideals than freehold, full-owner operators. In the meantime, land as property and use issues arise from new concerns and sources. Land, including farmland, issues rely partly on an ethic separate from agriculture as an industry.

The American Philosophy

How is Jefferson's agrarianism linked to pragmatism? Clearly, Jefferson has no direct connection to the philosophy of pragmatism formalized by Peirce and James, who succeeded him by two generations. Yet Jefferson was a man of his time, and his time was one of blossoming science and invention in America and in Europe. In America, science tended toward "useful experiments and improvements whereby the interest and happiness of the rising empire may be essentially advanced" (Greene 1984, 417). Practicality and utility directed science and development in the New World. And, "there was Thomas Jefferson, no great scientist himself but a tower of strength and encouragement to all who labored to advance the cause of science in the rising American nation" (419).

Although Jefferson was a learned man, his forte was less in creating profound ideas than expressing those ideas in the language of common people. But perhaps that *is* the essence of pragmatism. In the Declaration of Independence, for example, "it was not Jefferson's task to create a new set of politics or government but rather to apply accepted principles to the situation at hand" (in Perry and Cooper 1978, 318). The Declaration of Independence was virtually the single-handed work of Jefferson. He had sound reason to have personal attachment to the draft. However, when his strongly worded opposition to slavery threatened the acceptance of the Declaration of Independence, he eliminated the paragraph from the historic document as a practical political move (317).

On the agrarian policies of Jefferson, Griswold comments: "Jefferson was too shrewd a politician to believe that a perfectly equal division of the land was possible. It was another of those ideals that he tried to realize pragmatically" (Griswold 1948, 43). Griswold quotes a letter from Jefferson to

Du Pont that he says, "provides the key to his [Jefferson's] whole character." In that letter, Jefferson writes, "what is practicable must often countrol [*sic*] what is pure theory, and the habits of the governed determine in a great degree what is practicable" (Griswold 1948, 32). Griswold then writes, "he displays here that pragmatic empiricism which has governed Anglo-American political thought and action through history, confounding European critics in our time as it perplexed Du Pont in his" (32).

Jefferson's credentials as a pragmatist in politics were demonstrated throughout his career. When there was a question about the legality of the Louisiana Purchase, for example, Jefferson declared that "to lose our country by a scrupulous adherence to written law, would be to lose the law itself" (*Annals of America* 1968, 173).

Jefferson's agrarianism was distinctly rationalist but occasionally sentimental. His pro-rural sentiments are infused in his correspondence such as the frequently quoted letter from Paris to John Jay referring to "cultivators of the earth" as "most valuable citizens" (in Koch and Peden 1944, 377). The rationalist agrarian, but pro-rural, Jefferson is present throughout his only book, *Notes on the State of Virginia*. With his agrarianism he joins a distinctive array of intellectuals averse to one or another features of cities or urbanism, among them Henry James, brother of pragmatist William James (White 1962). Jefferson was not so averse to cities, as, say, Thoreau, that he rejected the arts and refinements of urban living or the necessity of manufactures and commerce; in these compromises, Jefferson was pragmatic. In a response to an inquiry from a Dutch businessman, Hogendorp, Jefferson wrote:

> You ask what I think on the expediency of encouraging our States to be commercial? Were I to indulge my own theory, I should wish them to practice neither commerce nor navigation, but to stand, with respect to Europe, precisely on the footing of China. We would thus avoid wars, and all our citizens would be husbandmen. Whenever, indeed, our numbers should so increase as that our produce would overstock the markets of those nations who should come to seek it, the farmers must either employ the surplus of their time in manufactures or in navigation. (In Koch and Peden 1944)[8]

In his letter to the Dutchman, Jefferson presents the rationalist side of the agrarianism expressed more romantically in the Jay letter quoted partially by Griswold. Together the letters to Jay and Hogendorp capture Jefferson's agrarian philosophy: farming may be the vessel of virtue, but necessity may dictate other occupations. They reveal what might be termed Jefferson's pragmatic agrarianism.

But a cautionary comment about Jefferson's pragmatism: the enigmatic Jefferson was, foremost, an intellectual, an academic. The intellectual's view of problem and solution is that of confusion and explanation.[9] A problem is solved when its cause and effect are explained. Jefferson is at his best when writing on the principles of government, his worst when he was governor, his best as a learned agriculturist, his worst as a farm manager, his best as an explorer when Lewis and Clark did the exploring. If one were to call Jefferson a pragmatist, one must extend the idea of pragmatism beyond just doing the expedient. Pragmatism, to accommodate Jefferson, must be extended to accept general principles, even as those principles are expressed as a particular time or particular circumstance. Thus, Jefferson opens his argument for the political independence of the American colonies with: "When in the course of human events . . . ." Pragmatism as a philosophy is consequentialist, utilitarian, and empiricist, all features of the Jefferson lifestyle, if not philosophy.

While Griswold and Brewster add useful insights to the Jeffersonian sources of agrarian ideals, dreams, and myths, and the relation of those ideals to the structure of agriculture and agricultural policy, they did not address the relation of America's agrarian roots to America's philosophy, pragmatism. First, in FD and RJDT, the purpose of the authors was to examine the structure of farming either as its role in the economy or as a product of beliefs only generally connected to Jefferson. For Griswold the issue was politics, and Jefferson's agrarianism was instrumental in shaping democratic values. For Brewster, Jefferson was entrance to Locke's system of ideals. Second, explaining the development of American pragmatism was not their agenda. "Despite Mead's strong influence, Brewster was not himself a pragmatist. He thought highly of this school of philosophy for the insights it provided into the way individuals work together in society. . . . What impressed Brewster about the pragmatists was . . . their social awareness and social commitment" (in Brewster 1970, 2).

Griswold clearly established the pragmatic qualities of Jefferson's politics (FD op. cit., 32). Those qualities may account, in part at least, for Jefferson's lasting esteem in American history. In Jefferson's time, agrarianism was not an abstract ideal. Farming was a tangible, real life aspect of democracy, and the goals, problems, and solutions of the new democracy were never far from agriculture. By Griswold's time America was no longer an agrarian economy so democracy's role for farming was historical, idealistic. Hence, his proclamation, that the family farm will not save democracy but democracy may save the family farm, reflected his mid-twentieth-century assessment of politics. More recently, Thompson and associates, under the heading "agrarian traditionalism" (1994, 248–258), draw ethical essences from agrarianism that affect public policies, illustrating extensively from Wendell Berry's writings. Griswold said, "for democracy to save the family farm, those who believe in the latter must find it in more than a romantic symbol or tradition" (1948, 206–207). Thompson and associates suggest that these symbols and traditions may be agrarianism's strengths.

In 1936, Brewster joined an administration and agency with action orientation, where programs were framed in terms of practical ends and alternatives. He accepted and adapted to a "pragmatic" environment surrounded by programmatic personalities. Brewster drew on Lockeisms for ideals and ends, but drew on pragmatics when he dealt with the practical issues of the structure of agriculture. In this, Brewster was Jeffersonian, or, in other words, able to compartmentalize his worldview to avoid uncomfortable incompatibilities of thought.

Brewster, in his adoption of the family farm as a central concept of his agrarianism, revealed the "instrumentalist" form of pragmatism, perhaps adopted, consciously or unconsciously, from Dewey, with whom he was well acquainted. Brewster's concept of family farm became the focus of his interpretation of the structure of agriculture, at times problem, at times solution, but never long out of sight. His analyses of farm issues were subject to the hazard of instrumentalists: marriage to the instrument, requiring definition stretching and reason bending. It meant that Brewster by centering on farm family management devoted insufficient attention to landownership and land tenure issues that took on new forms and new importance following *Silent Spring*.

The New Agrarianism: Environmentalism and the Attention to the Land

What may one conclude about the current status of agrarianism in the American value system? Notwithstanding the massive changes in the food and fiber industry, generally, and in production agriculture, a legacy of positive feelings about farming, ranching, and country life remains in the American psyche, nurtured by politicians, playwrights, and poets (Howarth 1995). Since Jefferson, and even since the latter-day interpreters Griswold and Brewster, the shape of agrarianism has changed. The virtues of the farm are found less in the independence and enterprise of management and more in the country lifestyle, a relation to nature, a nurture of the environment. Such matters are better left to others; here the change in agrarianism is described as a shift from farm to land. The reasons are, well, pragmatic.

Jefferson wrote at a time when the abundance of land was such that the human uses of land were two, agriculture or no use. He wrote: "we have now lands enough to employ an infinite number of people in their cultivation" (in Koch and Peden 1944, 377).

Griswold wrote in 1945–46, a time when agricultural land was fully expanded and grossly overpopulated in terms of the available technology. He premised FD with the observation: "at the moment we are dedicated to a goal that promises equal opportunity to everybody—full production and employment in a free society . . . If farmers are to realize the full measure of either, it is more than likely that many will have to give up farming for employment in other industries" (Griswold, 1998, viii)

Brewster wrote during the immense wave of off-farm migration that eventually cut the number of farms to one-third of its peak of 6 million in the 1930s. At the time Brewster wrote RJDT, the 1959 census showed 3.7 million farms, of which he noted 2.4 million were "commercial farms." He then cautiously concluded, "all the 'educated guesses' I know indicate about one million proficient farms would be enough to do this job" (RJDT).

The problem for the nation is no longer how to settle vast territories with Americans to prevent their settlement by other Europeans, or how to increase production of raw materials to exchange for processed goods from Europe. The problem is no longer supporting democracy by converting hillbillies into construction workers. The problem is no longer saving family farms by increasing their capacity to own, control, and inherit ever larger amounts of capital and land. The problem, for many at least, is an exploding, consuming population equipped with a technology of unlimited destructive capacity.

The structural changes in agriculture anticipated by Jefferson, and later noted by his two commentators, Griswold and Brewster, were forthcoming and continue still. But some other significant changes have more clearly separated the economic occupation of farming from the stewardship of the land by farmers and people generally. Perhaps most significant is the environmental awareness beginning in the early 1960s, often associated with Rachel Carson (Carson 1962). Environmental concerns brought forth regulations to prevent destructive behavior on land. Many of those regulations affect farming practices simply because such a very large portion of private land is in agricultural use.[10] Conflicts between land quality protection and agricultural production were inevitable. Then, partly as reaction to environmentalism, a "property rights" movement arose espousing libertarian values.

The last decade of the twentieth century saw a renewed interest in the concepts of property.[11] Noted cases, *Nollan, Lucas, and Dolan,*[12] focused on the issue of "taking" private property without just compensation. Politically conservative groups have sought to reduce regulations on land use. Many of the property rights arguments have a Lockean-like quality, which, as indicated earlier, shows much concern about inviolability of rights of property holders but little concern for the way property wealth is distributed. Evidence of the popularity of this perspective is revealed in legislation introduced into the U.S. Congress and passed in a number a number of states.[13]

The old, occupationally oriented agrarianism carried well into the twentieth century (Griswold, Brewster), but it no longer applies to today's agriculture or the public's view of agriculture. Evidence of the passing of the old agrarianism is found in the Agricultural Market Transition Act (P.L. 104–127), containing amendments that treat farming more like other businesses, without special consideration, yet continues to support conservation activities (Title III). The new agrarianism sees farmers as stewards of the land.[14] The new agrarianism still wants to support agriculture as a form of macro landscape architecture—less concerned with the individual farmer, more with the total scene. The new agrarianism is concerned, for example, with incursions of residential development on farmlands.

Pragmatism of the New Agrarianism

Few would doubt the importance of agrarianism in the American worldview, or "philosophy." The scholars Griswold and Brewster probed the origins and essence of Jeffersonian agrarianism for its role in the American psyche—Griswold for the role in post–World War II policy, Brewster for a defining quality of the family farm. Both found aspects of agrarianism to criticize; neither questioned agrarianism's presence or power. Neither Griswold nor Brewster claimed a mantle of pragmatism, yet both reveal elements of pragmatism in their works.

Did the two scholars provide any special insight about a connection between agrarianism and pragmatism? Probably not directly. However, their historical and philosophical writings about the value systems and practical consequences of those value systems in policy definitely had pragmatic qualities. They reflected the distinctive American outlook. That outlook extols the virtues of purposeful endeavor (work), practical solutions to immediate problems, and utility achieved by pursuing tangible goals, even as those goals change over time.

The Griswold/Brewster era ended about the time environmental awareness began. Organic farming prepared the way for sustainable agriculture. Urban land use touched farmland preservation on the fringes of metropolitan areas and beyond. Wildlife and endangered species merged with agriculture. Farming became an aspect of America's new, greening agrarianism, and stewardship reentered the vocabulary of both agriculturists and conservationists.

The shift of agrarianism from the occupation of farming and toward land will affect some value emphases: the family farm as a element in the structure of agriculture will decline. Greater concern will focus on the ownership of land and the way contracts for use are created. Distributive equity issues will not be confined to farm operators. The nature of property, already an important issue, will require the increased attention of citizens and policy makers. Stretched by the new agrarianism, property in land will extend to rural landscapes, cultural artifacts, communities of species, and spatial awareness. A constantly changing, emerging property system will accommodate people's needs and intentions.

And how might Jefferson view the land today, two hundred years from the eve of his presidency? Who knows? Let us speculate with a quote from Richard Rhodes's *The Inland Ground:*

> He [Jefferson] organized the American wilderness, before its details were even mapped, on a grid borrowed from Euclidean geometry: so laid it out despite the fact that he was a countryman who understood curving fields and wandering streams and the vagaries of plantings: laid it out to our eternal complication, his grid the woe of his fellow citizens ever afterward, impossible to find those damned cornerposts, impossible to track those inhuman

squares. A Dionysian land, a Dionysian continent, plains and tortuous rivers and sharp mountains and wandering shores, and he overlaid upon it a Euclidian grid as if Apollo truly reigned and the Furies were forever banished . . . this man caused a continent to be penetrated, then bluntly annexed, then measured out in squares. He embodied all the American contradictions: extended his fantasies and his fear of those fantasies out onto the land itself: loved it and hated it, and because of that love and hate saw it uncertainly. (Rhodes 1970, 13,14)

Steinbeck and Agrarian Pragmatism

RICHARD E. HART

Maybe the challenge was in the land; or it might be that the people made the challenge.

—John Steinbeck, Afterword, *America and Americans*

THROUGHOUT HIS long and illustrious career as writer, journalist, war correspondent, and chronicler of the American soul, John Steinbeck consistently eschewed the label "philosopher," or more specifically "philosophical writer." Dating back to the 1930s he forcefully denied ever having a philosophy or being influenced in his thinking and writing by philosophy. Deeply skeptical of abstracted intellectuals—what he considered the suit-and-tie crowd—he spoke disdainfully of those who make a career out of the mind detached from the world and experience. As one illustration, in 1938 Steinbeck, then hard at work on *The Grapes of Wrath,* was sent a lengthy questionnaire by Merle Danford of Ohio University, who was preparing the first graduate thesis (MA) on Steinbeck's early fiction. Given his natural shyness and resistance to interviews, Danford had to prod Steinbeck for quite a time, even getting her thesis director (C. N. Mackinnon) to exercise his powers of persuasion. After "lamenting the academic decay such a thesis indicates," Steinbeck finally conceded and gave written answers, most imbued with a sense of irony, flippancy, self-effacement, exceptional candor, and evasiveness. Upon completing his answers, he appended a personal note to Ms. Danford in which he stated, "And as to the questions as to what I mean by—or what my philosophy is—I haven't the least idea. And if I told you one, it wouldn't be true. I don't like people to be hurt or hungry or unnecessarily sad. It's just about as simple as that" (Fensch 1988, 27).

I

As simple and profound as this may sound, the matter of Steinbeck and philosophy is not so simple at all. Clearly, Steinbeck was not a philosopher in any of the typical senses we have come to accept today. He was not an abstract systematizer, professional ethicist, high-minded theorist, hair-splitting logician, philosophical visionary, logic chopper—none of the above in any pure or easily identifiable sense. Yet in curious ways his writings perhaps invoke and articulate a little something of each of these descriptions, a tentative claim I hope will be clearer by the end of this essay. While critic Stanley Edgar Hyman, in a review of *Sea of Cortez* for the New Republic, once accused Steinbeck of being preoccupied with "rambling philosophizing" for its own sake (namely, a kind of intellectual game), biographer Jackson J. Benson writes that Steinbeck was "almost alone among important fiction writers in this country of his generation in his interests in formal philosophy" (Benson 1988, 233). Ella Winter, a reporter who interviewed Steinbeck for a 2 June 1935 article in the *San Francisco Chronicle,* observed, "Steinbeck loves to read physics and philosophy and biology" (Fensch 1988, 5), but as an artist he wanted his ideas worked out and expressed in his fictional creations.

Benson amply documents that the young Steinbeck was greatly influenced at Stanford University by the history of philosophy courses he took with Harold Chapman Brown as well as the ideas of scientist William Emerson Ritter. This suggests the notion of philosophy and science working together rather than at odds with one another, a pervasive theme in Steinbeck's later work. But if there was any pattern of early and sustained philosophical development it was very private for Steinbeck. He shared such ideas with no one until his now-famous adult friendship with marine biologist Ed Ricketts. Benson, however, regards Steinbeck's philosophical interests seriously enough to describe him, given his unrelenting openness to new thought and experience, as a "philosopher in progress." From my perspective, scholar Gloria Gaither summarizes the matter correctly, albeit in broad strokes, when she proclaims, "Any discussion of Steinbeck the social reformer, Steinbeck the artist/writer, Steinbeck the journeyer, Steinbeck the marine biologist, remains inconclusive without a deep appreciation for and genuine understanding of Steinbeck the philosopher" (Gaither 1992, 43).

Adding to the confusion Steinbeck himself perpetrated, the secondary literature is divided on the subject of philosophical influences and basic orientation. For example, Benson argues that Steinbeck was not (as some have thought) essentially a philosopher in the American strain, to wit, a transcendentalist or idealist in the tradition of Emerson and Thoreau. Rather he contends Steinbeck was most influenced by the ancient Greeks, particularly the Greek notion of "nature" as living, continuously evolving and companionable, with man as a unique living being within the whole of nature. If true, this may provide interesting clues to our later consideration of Steinbeck and agrarianism. In a different, but I believe compatible vein, Frederick Carpenter, in a famous 1941 essay, "The Philosophical Joads," praised The Grapes of Wrath for what he terms its Aristotelian sense of philosophical wisdom (its criticism of life). But chiefly he claimed (without much elaboration) that Steinbeck's epic novel serves as a synthesis, a culmination of var-ious strands of American philosophical thought. Steinbeck is, thus, situated squarely within the clas-sic traditions of American philosophy, including pragmatism. Carpenter asserts that Steinbeck created the character Jim Casy ("the preacher") to interpret and embody the philosophy of the novel. Carpenter alleges that, beyond the basic story line, the ideas of Steinbeck and Casy have an extratextual significance all their own:

> They continue, develop, integrate, and realize the thought of the great writers of American history. Here the mystical transcendentalism of Emerson reappears, and the earthy democracy of Whitman, and the pragmatic instrumentalism of William James and John Dewey. And these old philosophies grow and change in the book until they become new. They coalesce into an organic whole. . . . Jim Casy translates American philosophy into words of one syllable, and the Joads translate it into action. (709)

For Carpenter, Jim Casy and the Joad family quite literally think and do all these philosophical things. This anticipates our later reflections on Steinbeck and the spirit of pragmatism.

I think it plausible *both* Carpenter and Benson are, to a degree, right. They are on the right path inasmuch as the ancient Greek and American philosophical influences (and themes) can, with some work, be uncovered in Steinbeck's writing. While each probably exaggerated certain claims about Steinbeck, the philosopher, I think, following their lead, it most appropriate to conceive of Steinbeck in an experimental, experiential Socratic sense of philosophy and philosopher. To me this is in no way discontinuous with the thrust of American pragmatic thought, but rather blends and harmonizes the Greek and American influences. He is, first and foremost, a writer, a literary artist, one who, like Socrates of the *Euthyphro* or *Apology*, is a constant inquirer, explorer, discoverer. Through his work he examines ideas and values and presents his exploration and struggle to others through the texts themselves. Steinbeck, like Socrates and the writer Plato, is at the very intersection, the subtle yet powerful interface, between philosophy and literature. Steinbeck, like Socrates, is not inclined to posit definitive philosophical solutions. Like all great artists, he does not provide answers, but rather asks questions, poses dilemmas, points up contradictions, exposes untruth (falsity) in search of better ways of seeing and living. His work is motivated by a unique assembly of ideas and theories, values and experiences, and, to this extent, moves beyond the parameters of fiction, beyond the pure telling of an interesting story.

With this as background, I simply add that while Steinbeck cannot be called an agrarian or pragmatist philosopher per se (the dual subjects of the present volume), one does nonetheless sense strains of both agrarian and pragmatist thought and sensibility woven throughout his writings, especially works from his early, so-called California period. Arguably, his early life experience in rural California, his fundamental American values, the reading he did, the close friendships he formed with people like Ed Ricketts, all somehow coalesced to move him in these philosophical directions. Without claiming that his work is consciously grounded, in a formal sense, in such philosophies, agrarian and pragmatist concepts and perspectives are present in a number of his fictional works and nonfictional logs and journals. Put differently, the convergence of these two American skeins of thought helps elucidate important aspects of Steinbeck's philosophical worldview. As Paul Thompson points out in his introduction "Agrarianism as Philosophy" (this volume), agrarian literature and art offers imagery that may rest on philosophical principles or may seek to cultivate an agrarian mentality or experience. In this respect, Steinbeck, like the Vanderbilt agrarians, produced literary works wherein philosophical principles are often masked or subdued, but nonetheless always present.

In what follows I shall briefly sketch some particular overlaps and parallels between the two schools of thought as found embedded and operative in Steinbeck's writings. This will be approached in two ways: one involving the examination of nature and Steinbeck's unique sort of "naturalism," the other through a focus on experimental method and nonteleological thinking. I then conclude with a brief consideration of some fictional illustrations from the early California writings.

II

Students of American intellectual history know that both agrarianism and pragmatism place decided emphasis on nature. On the most basic biological level, both modes of thought tend to conceive of human beings as but one species of natural organism, alongside and integrated with all others. Humankind is subject to natural laws and natural explanations much as apply to the process of photosynthesis or the movement of planets. Humanity is, for sure, unique within nature (as a species) but never removed from the clutches and possibilities of nature, never somehow beyond nature. Conceptually, we hear resonating in this notion echoes of evolutionary biology, as present also in Dewey's version of pragmatism, as well as recollections of the early ecological thinker, Aldo Leopold, and his notion of "the land ethic" founded on the "biotic community." For Leopold the "biotic community" is the community of all living things, including soil and water, what he refers to simply as "the land." Leopold writes that "the individual is a member of a community of interdependent parts" (Olen and Barry 1999, 461). His notion of "the land ethic," moreover, enlarges the boundaries of the community to include soil, water, all plants and animals. Beyond enlarging empirical and social boundaries, "the land ethic" presents ecological imperatives in the form of ethical obligations to preserve the integrity, stability, and beauty of the "biotic community." Such obligations parallel our obligations to the human community and its individual members. This sort of early ecological consciousness (well prior to any organized environmental movement) was characteristic of Steinbeck's appeals from his earliest days as a writer. Biographer Benson summarizes: "As early as the mid-1930s he [Steinbeck] was talking about man living in harmony with nature, condemning a false sense of progress, advocating love and acceptance, condemning the nearly inevitable use of violence, and preaching ecology at a time when not even very many scientists cared about it" (1988, 196).

Steinbeck was a writer of the land and all its inhabitants. He wrote of simple people and transformed them into memorable characters whose struggles have become part of American folklore. He sought always to preserve their ecological and natural integrity as functioning units within the biotic community. In this respect, his work is firmly ensconced within the tradition of naturalism. Accordingly, Steinbeck fully accepted that man was a discrete though integrated cog in the vast machine of nature, and thus writes in the *Log from the Sea of Cortez,* summarizing his cosmic worldview: "All things are one thing and . . . one thing is all things—plankton, a shimmering phosphorescence on the sea, and the spinning planets and an expanding universe, all bound together by the elastic string of time. It is advisable to look from the tide pool to the stars and then back to the tide pool again" (257).

But Steinbeck's is no simple-minded, univocal, purely scientific naturalism, as was the sort of "literary naturalism" of his time that sought to conceive of man with detached scientific objectivity. According to this view, man is controlled by his passions, his social and economic environment, to the extent that he has no genuine free will. Steinbeck's "modified naturalism," which I regard as a complementary dimension of philosophical naturalism fully compatible with the American pragmatists, holds that while, scientifically speaking, the natural universe, including man, operates according to neutral, indifferent forces of cause and effect, man is also unique as a value-bearing and creating being who witnesses his nature through deliberation and action. Humans, an integral part of, but never fully reducible to the mechanisms of, nature are simultaneously the locus of free choice. Humans, through acts of will and valuing, rise up against the indifferent determinism of nature, thus exercising contextualized critical intelligence and moral choice in a manner so compellingly explicated by pragmatist philosophers like Dewey. Steinbeck's expansive rendering of naturalism thus harmonizes determinism and freedom, individual and group person, biological necessity and responsibility. This is the Steinbeck who, in chapter 14 of *The Grapes of Wrath,* writes of man's singular status within nature, so inextricably related to his will that he, unlike any other species, is willing to "die for a concept." This is the familiar Steinbeck who constantly affirms the possibility of moral grandeur, through choice and action, mainly on the part of common people like gamblers, ranch hands, and prostitutes. Inasmuch as Steinbeck's characters can and do help themselves purely through acts of will, thereby affirming human dignity, he shares with American naturalists and transcendentalists like Emerson, Thoreau, and Whitman the idealistic vision of the human as simultaneously individualistic and selflessly altruistic. In this vein, Steinbeck, ever the realist/optimist, often alludes to the infinite perfectibility of the human, another theme one can see interwoven with the brighter side of transcendentalism and pragmatism. Steinbeck's fiction thus articulates, through the agency of art, the contextualized, lived experience of a multilayered, at times paradoxical, naturalism. His work is thereby shot through with a complex appreciation for, and understanding of, nature common to both agrarian and pragmatist thought.

Inseparable from Steinbeck's embrace of a multifaceted philosophical naturalism is his pursuit of so-called nonteleological thinking as a method of analysis and literary presentation. As Steinbeck's way of seeking the truth in an inductive, scientific manner, the tie to the present themes of agrarianism and pragmatism is intriguing. As commonly known, the particular sort of agrarianism arising out of the American context overlaps with pragmatism's emphasis on experimental or scientific method. Here Steinbeck presents yet another unique twist on such basic method with the now well-known theory of nonteleological thinking. Significantly, this theory allows Steinbeck to wed his "scientific mechanism" (evolutionary biology) and "humanistic vitalism" (ethics) while following out the logic of an empirically driven theory.

This variation of scientific method, experimental by its very nature, is elaborated by Steinbeck (and marine biologist Ed Ricketts) in the book *Log of the Sea of Cortez* (compare "Coming Full Circle?," this volume). Readers will recall that the book offers a day-to-day description of a 1940 expedition undertaken by Steinbeck and Ricketts in a sardine boat to collect marine invertebrates from the beaches of the Gulf of California. In the pages of the log, particularly the Easter Sunday sermon, the two men debate philosophically the sort of method appropriate to accounting for everything from their beach discoveries to the movements of heavenly bodies to the behavior of men. The essence of nonteleological thinking is that, in taking account of the whole, it focuses on what actually "is" the case rather than what "should" or "could" or "might" be (the stock in trade of theologians, ideologists, and normative philosophers). As Steinbeck writes in the *Log*, "The truest reason for anything's being so is that it *is*" (176). Such method seeks to describe reality (including social reality) as it is rather than prescribe how it should be, to see the thing disinterestedly rather than through the lens of personal, or interested, utility. It seeks to answer "what" or "how" (for example, how things work) rather than "why" (the speculations of theorists).

On first blush this method may seem decidedly nonphilosophical insofar as it deliberately avoids the traditional "why" questions of philosophy. Importantly, however, I suggest that this sort of thinking, in part, reflects and extends (through application) central tenets of America's unique, home-grown philosophy, namely pragmatism. I have in mind here two related points: pragmatism's commitment to the primacy of thinking embedded in a particular, concrete context, and all that follows philosophically; and the central role of immediate, brute experience of the here and now rather than inchoate reflection on some speculative or visionary future. Steinbeck and Ricketts considered their version of "is" thinking to be logically associated with evolutionary biology and natural selection as Darwin understood it. For them the attempt to find truth (whether in a scientific travelogue or a work of fiction) must be through "abandoning popular beliefs, making observations firsthand, getting the facts together, and achieving the inductive leap to discover a great principle," a message one also gleans from pragmatism. As with Darwin, this method allows for a contextualized view of the whole of nature, "with humans a part of all natural interrelations," while denying "traditionally religious or romantic notions which exalt our species by setting it apart" (Railsback 1995, 27). The conceptual circle back to Steinbeck's sophisticated philosophical naturalism is thus completed.

When creating his fiction Steinbeck often claimed that he simply wrote about what happened. His goal, like a good reporter, and perhaps reflecting the influence of his early forays into journalism, was to tell the story of people and events without taking sides or espousing doctrine. I submit, however, that his contextualized way of approaching the story, his way of rendering experience through the conceptual and philosophical lenses here articulated, reveals universal aspects of the human condition and instantiates themes central to the heart of both American agrarianism and pragmatism.

III

Space permits but a few suggestions of how philosophical naturalism and nonteleological method appear in some of Steinbeck's early California fiction. In speaking about his 1937 classic novella, *Of Mice and Men,* Steinbeck characteristically denied any philosophizing or drawing of moral judgments. He claimed to be simply retelling, with some literary license, an interesting story based on actual events he once witnessed while working as a ranch hand. He simply reported the facts as experienced. In nonteleological fashion, dramatic events happened as such, and the reader is left to draw his own evaluations. Among the reader's considerations are the ways in which nature and naturalism permeate the story. One simple example is the persistent longing of George and Lennie, throughout the story, to return to the comfort and solace of nature, to acquire their couple of acres so that they can "live off the fatta the lan'" (14). Their pastoral dream interestingly reflects a fundamental agrarian sense of property rights in which they would, if only the dream could be realized, be liberated from typical dependency on a broad network of economic relationships. Consider further that at story's end, in the throes of impending disaster, George and Lennie find themselves back along the quiet banks of the Salinas River (where the story began), symbolizing their refuge in comforting nature. Similarly, at one point boy Jody, in the story "The Red Pony," becomes frightened by the tall, dark Santa Lucia Mountains while also ironically longing to be absorbed into their mysterious comfort. Nature, in the same story, further represents man's great impediment, as illustrated by Jody's grandfather describing the line of old men along the Pacific shore hating the ocean for stopping their onward push to the west, their "westering," as Steinbeck terms it.

Of Mice and Men also illustrates, succinctly and penetratingly, the two sides of philosophical naturalism, both objective determinism and the freedom of humans' will. George and Lennie can be apprehended, on one level, as constituted products of nature and society, Lennie through his inherited mental incapacity and George through his passions, unrelenting fantasizing, and even overpowering love for Lennie. But they are not entirely helpless or hopeless in the throes of nature. They are not just fated, pathetic characters who simply go down in defeat. In my reading, George especially becomes the exemplar of responsibility and free choice. At the end of the story, George has to make a life-altering choice and quickly. There is nothing to rely on except his own free will. The forces of nature (Lennie's inability to ever change) and society (the approaching lynch men) are closing in. George can abandon Lennie to the mob or take Lennie's life in a merciful, painless act of euthanasia. Clearly, all they have left is each other. And then,

> through an heroic act of will, George demonstrates, with simple, unparalleled clarity, the transcendent power of love. In killing Lennie, thus saving him from indignity and torture, George

> morally reverses (for the moment) the forces of fate and nature that have both of them in its grip. Nature controls *and* man is free. Hopelessness and courage exist side by side. (Hart 1997, 51)

While one could elaborate further on the courage and moral conquests embedded in the land and labor struggles of *In Dubious Battle* or the commodification of land and people, of virtue resident in the breasts of farming people, in *The Grapes of Wrath*, all important themes to American agrarianism, I trust the preceding suggestions offer at least some illuminating fictional examples in support of my claims regarding Steinbeck's philosophical outlook.

In sum, it is clear to me that agrarian imagery and rhetoric, as well as basic pragmatist assumptions and tendencies, run through the thought and literary art of John Steinbeck. While we should not try to render him a philosopher in any formal or professional sense, Steinbeck was, on my account, a vehicle, indeed forerunner in some instances, of rich philosophical concepts and dilemmas. Unlike many American writers, his was a refined and mature intellect. His thoughts ran deep and clear. His serious reading, his working friendships, his experiential pursuit of the sciences, of human spirituality and perfectibility—all led to the appearance in his work of philosophical themes deeply ingrained in the American experience. I have not sought here to label Steinbeck as an agrarian or a pragmatist in any strict sense. Rather through initial reflection on two themes—philosophical naturalism and nonteleological thinking—I have tried to succinctly demonstrate how strains of American pragmatism and agrarianism, how sensibilities and perspectives common to each, evolve, merge, and reveal themselves in the work of one of America's greatest writers.

Coming Full Circle?

Agrarian Ideals and Pragmatist Ethics in the Modern Land-Grant University

JEFFREY BURKHARDT

After sixty or so years, agrarianism has begun to resurface in land-grant universities in the United States, especially the colleges of agriculture, which constitute their historical and philosophical core. This agrarianism, which will be referred to herein as *neo-agrarianism* since it differs in some basic ways from older American agrarian philosophies and movements, can be seen in three subtle but important "wind changes" in colleges of agriculture: (1) an increasing interest in agricultural sustainability or alternative agricultural techniques and technologies; (2) an attempt at the "humanization" of the research and instructional agenda of agricultural (and natural resource) schools or colleges, especially in the form of courses, symposia, workshops, etc., in agricultural ethics; and (3) a growing belief that agricultural colleges, and land-grant universities in general, be socially accountable. While I will not overestimate the extent to which these new agenda concerns have permeated the land-grant system, they do represent a potentially profound shift from a half-century trend toward the land-grant university's being the handmaiden of positivistic science and large-scale agribusiness. Indeed, with considerably more work on the part of those involved in sustainability and agricultural ethics, neo-agrarianism may become the guiding force in the land-grant mission into the twenty-first century, just as traditional agrarianism undergirded the mission of the land-grant system near the turn of the twentieth century.

Philosophically, what makes this potential ideological change significant is that the neo-agrarian vision, or rather visions, are pragmatist at their core: without explicit reference to American pragmatist philosophers, much of the neo-agrarian vision is best understood in terms of pragmatist, in particular, Deweyan notions of science, ethics, education, and democracy. This new agricultural philosophy, most explicitly developed in disciplinary philosophy-based agricultural ethics, is in fact a revisitation of traditional American agrarian themes in modern American pragmatist ethics garb. Admittedly, there are differences among various neo-agrarians in focus or intent. However, what ties together the various neo-agrarianisms in a unified critique of a land-grant mission purportedly gone astray is a vision or ethical philosophy which is fundamentally Deweyan: to humanize and democratize the land-grant institution. As I will discuss below, there is much work to be done on a variety of fronts for this to be successfully accomplished, and there are many obstacles to face. I will argue, nevertheless, that is under way.

The Call for Reform

Twenty years ago, in *The Unsettling of America*, Wendell Berry argued that the land-grant universities in the United States are indictable for the perversion—if not near destruction—of agrarian ideals and sustainable farming practices. Though intended to serve agrarian interests and a larger social-ethical purpose, the land-grant system had been captured by the ideology of scientism—the worship of things scientific (Rosenberg 1976)—and what is now referred to as industrial agriculture. Berry wrote:

> The tragedy of the land-grant acts is that their moral imperative came finally to have nowhere to rest except on the careers of the specialists whose standards and operating procedures were amoral: the "objective" practitioners of the "science" of agriculture, whose minds have no direction other than that laid out by career necessity and the logic of experimentation. They have no apparent moral allegiances or bearings or limits. Their work thus inevitably serves whatever power is greatest. That power at present is in the industrial economy, of which "agribusiness" is a part. (Berry 1977, 155–156)

There is ample evidence in the social-scientific literature on the negative impacts of agricultural science and technology generated by the land-grant complex on smaller farming operations, farm labor, and the environment to suggest that Berry's critique has been proven correct (see, for example, Berardi and Geisler 1984). It is, however, interesting to note that for at least fifty years, various actors within the land-grant system have been pointing to potential negative impacts and suggesting, for pragmatic reasons if for no other, that the anti-agrarian tone and direction of the research and development agenda (and perhaps teaching and outreach [extension] agenda) need to be redirected. Interestingly, as early as the 1940s some agricultural economists noted that if farming becomes just a business like any other, the lack of uniqueness leads to the lack of a distinct identity for agricultural economics and many of the other agricultural sciences and, therefore, their lack of relevance or necessity in the land-grant university (Choe 1977). Nevertheless, from at least the 1940s, and probably well before that, the positivistic science and agribusiness orientation of the land-grant system progressed unabated.

It is not by coincidence that in the early 1980s a number of the leading U.S. land-grant universities, especially their colleges of agriculture and natural resources, began to change. There were many precipitating events for the development (or at least incipient development) of the new agenda for agricultural research and education, and many of these were the result of concerns of individuals and groups outside the land-grant system. For example, the environmental movement increasingly targeted agriculture as a major contributor to problems in public health and environmental safety as early as the 1960s. The farm financial crisis of the late 1970s, brought on in part by official USDA and land-grant system incentives for farmers to "get bigger," led to the loss of large numbers of small farms—and attendant major protests by small farmer activists. And the negative impacts of land-grant-developed agricultural technology on small vegetable farmers and farm laborers in California (for example, the infamous tomato harvester [see Hightower 1973]) resulted in a major lawsuit against the land-grant university there. More directly, throughout the 1970s and early 1980s, a number of academic critiques and even high-level "official" reports targeted the land-grant complex for (a) lacking appreciation of the larger social and environmental context in which farming is practiced; and (b) essentially abandoning the practical (and political) mission-orientation of the land-grant system in favor of a basic science-oriented research agenda (Mayer and Mayer 1974; Rockefeller Foundation 1982).

Nevertheless, the initial impetus for changing the research agenda and curricula of colleges of agriculture came in the form of attempting to make colleges of agriculture more relevant to modern agriculture—the world of high technology, heavy chemical and industrially dependent inputs, global markets, and the like. Ironically, however, one of the means of addressing the concerns about big picture, relevance, and critical thinking was the introduction of teaching, research, and extension programs devoted to sustainable agriculture, sustainable development, and human resource development, and most significantly, agricultural (and natural resource) *ethics*. Almost despite itself, the land-grant system began to re-agrarianize its orientation. But I am getting ahead of the story; a brief history of the land-grant mission gone astray and its incipient modern return to its roots is in order.

The Land-Grant College: Education for Democracy

The land-grant university (LGU) concept is a uniquely American one, though the land-grant model has now been exported around the world. The origin of the LGU was a gift, from the federal government to each of the states of the U.S., of federal land for the establishment of a public college. The reason, or "mission," as it is called, of the LGU system, as stated in its 1862 enabling legislation, the Morrill Act, was politically pragmatic but philosophically Jeffersonian. The purpose of the land grant was

> the endowment, support, and maintenance of at least one college where the leading object shall be, without excluding other scientific and classical studies . . . to teach such branches of learning as are related to agriculture and the mechanic arts . . . to promote the liberal and practical education of the industrial classes in the several pursuits and professions of life. (*U.S. Code Annotated* 1964)

In essence, the Morrill Act set out that the federal government provide to each state and territory a gift of 30,000 acres of federally owned land, per senator and congressman, that would provide the basis for a continuing revenue stream for the support of the colleges. Later (1890), a second Morrill Act granted an additional $15,000 per year for further support. Additionally, a series of federal and state legislative actions broadened the scope and deepened the public commitment to serve the larger good through serving agriculture and rural people more generally. These are the Hatch Act of 1867, which established the Agricultural Experiment Station as the research component of the LGU system; the 1890 Morrill Act, which also provided for separate colleges of the agricultural and mechanical arts (A&M) for black constituents in the former Confederate states; and the Smith-Lever Act of 1914, which formalized LGU outreach programs in the Cooperative State Extension Service. Subsequent legislation, usually parts of so-called major farm bills, enacted by the U.S. Congress every five years (since 1935) with minor modifications in the intervening years, have attempted to fine tune LGU activities through funding initiatives and reorganization schemes among USDA's various sub-agencies, and various state initiatives have had impacts on the direction and scale of programs within their respective LGUs. Nevertheless, in essence, the LGU system remained in the 1990s structured as it was intended in those early enabling pieces of legislation (Works and Morgan 1939; Kunkel et al. 1997).

As mentioned above, the mission of the LGU is generally regarded as fundamentally pragmatic. Each of the legislative acts establishing what are referred to as the "function areas" in the LGU system—teaching, research, and extension—were intended to respond to social and economic needs, although the Morrill Act, establishing the colleges of agriculture, was also linked to a political agenda. It has occasionally been viewed as a ploy or a not-so-veiled bribe to each of the states in the Confederacy to give up their allegiance to the Confederacy, insofar as the land grant was withheld from Confederate states until after the Civil War ended (Ross 1969). Even so, the act was a straightforward policy tool whose intent was to improve agriculture through the education of the next generation of farmers to the latest, science-based (though still primitive by our standards) farming methods, and provide those new farmers with knowledge in the liberal arts and sciences which would make them more active and functional citizens. Regarding the improvement of agriculture, it need only be noted that although the majority of the U.S. population still resided on farms or in smaller rural communities, increasing industrialization and urbanization—together with the reduction in the labor pool available for labor-intensive agriculture, due to factory opportunities and deaths during the Civil War—entailed increasing the productivity of each farm and farmer who remained in agriculture. The hoped-for consequence of agricultural or engineering education would be stable food supplies and stable food prices for the population as a whole.

The idea of educating young members of the industrial classes—especially young farmers—was clearly Jeffersonian in tone, and likely in intent. The availability of an education for every citizen, rural or urban, wealthy or poor, undoubtedly reflects the idea that democracy best thrives when literate citizens engage in productive work on the land. Though it is doubtful that the authors of the college of agriculture envisioned a "happy yeomanry," they did hold prominent the Jeffersonian notion of the freeholder and the importance of bringing science to all productive work. That college-educated, literate farmers would be willing to mount serious challenges to the urban, monied, Eastern Establishment in the form of the Populist movement was undoubtedly beyond the comprehension of Congress in passing the Morrill and Hatch Acts (although, post-Populism, the Smith-Lever Act explicitly addressed this possibility). Rather, with a pragmatic and Jeffersonian faith, the founders of the LGU saw their efforts in roughly utilitarian terms—service to the greatest good of the Republic.

The Hatch Act, which established the Agricultural Experiment Station (AES), carried on this utilitarian, pragmatic, and Jeffersonian direction. The idea here is that these institutions—in conjunction with the college of agriculture—would generate useful knowledge right in the locales where it would be put to use. That is, rather than rely on the results of scientific investigations at the older, more established research institutions—Harvard, Johns Hopkins, the University of Chicago—applied research and technology development would be engaged in near the farm; in fact, most early experiment stations were actually experimental farms, a practice which continues today, though to a much lesser extent. In many cases, the education of a student in the college of agriculture consisted of hands-on research on the very crops or animals he would be managing once back on the farm. Ultimately, useful knowledge in the service, again, of the larger public good through service to farms was the ultimate driving consideration.

The Smith-Lever Act, which institutionalized outreach programs in the form of the Extension Service, most deliberately and explicitly addressed a social, economic, and political problem. As mentioned above, the Populist movement of the 1870s through the first decade of the twentieth century was a challenge to the (sometimes real, sometimes perceived) increasing dependence of farmers on banks and manufacturers. President Theodore Roosevelt's Country Life Commission (1908) found that poverty, unrest, and, indeed, reduced productivity continued on farms and in rural settings despite the "successes" of the LGUs and AESs. What was needed, it was concluded, was a more dramatic and systematized delivery system of LGU/AES knowledge and expertise to all rural residents, including an updating of farming practices as newer, science-based alternatives arose. The Cooperative State Extension Service (ES) (1914) was to be the vehicle for (a) delivery of information and technology to rural constituencies; and (b) feedback from the countryside as to farmers' needs and desired new knowledge. As implied above, there are more sinister interpretations of the actual intent of the ES: to forestall any new Populist stirrings. In all likelihood, however, this was an unintended (though not entirely successful) consequence of the work of extension personnel. In fact, the "Cooperative State" part of the ES was to place knowledgeable individuals in locations throughout each state. These people would be become mentors, friends, and trusted experts of farmers, and hardly agents with missions to subvert antiestablishment or antigovernment sentiments.

The point of this brief (and somewhat official) overview of the origins of the LGU is to suggest that from the outset, a philosophy of service to agriculture as an indirect way of serving the goals and interests of a democratic market system underlay congressional and state legislative actions establishing the LGU. And insofar as farming at that time, although beginning to become industrialized, was essentially a matter of Jeffersonian freeholders and small family farms, an indisputable agrarianism was also part of that commitment. Although it is far from clear that the authors of the land-grant system explicitly endorsed pragmatism as either a metaphysical, epistemological, or ethical system, the idea of "useful science for the public good" is certainly apparent in their actions, and in outcomes of their efforts to improve young farmers and American farming in general—although, as I will suggest, that notion has been, as Berry wisely points out, somewhat corrupted over time.

Positivist Technoscience vs. the Land-Grant Mission

The "middle history" of the LGU can be read as an allegory for the success of science over humanities, of capitalism over community, of humans over nature. For in the over seventy years beginning around 1920, the LGU contributed in significant ways to the industrialization of agriculture—characterized by larger-sized farms and their dependence on ever-increasing and ever-more sophisticated technology in the service of increased productivity or the ideology of productionism (Thompson 1995). Along the way, the LGU retreated from democratic ideals in terms of the actual practices and outcomes in education, science, and technology development and business practices. And indeed, with the general urbanization of society, even agriculture began to take a back seat in terms of its role in the larger land-grant university. Until the fairly recent rekindling of agrarian ideals, and the pragmatist and Jeffersonian stirrings in the LGU, the LGU found itself basically in the service of the industrial establishment the agrarian populists had so vigorously opposed at the beginning of the twentieth century.

To give a precise date to the change in LGU practice, if not to its stated mission, is difficult, as are the complete range or reasons or causes for this change. Busch et al. (1991) link it at least in part to the successes of scientific agriculture emerging out of research conducted on agricultural chemicals by Justus Liebig at the end of the nineteenth century. Also significant is the development of a functional internal combustion–powered tractor around 1920. Politically, an important turning point came when, under considerable pressure from the American Seed Trade Association, the state AESs halted their practice of distributing new varieties of grain seed freely to farmers, in essence forcing farmers to purchase seed from seed manufacturers (Busch and Lacy 1983). Of note is also the 1930s development of hybrid corn by a private firm, Pioneer, a considerably improved corn variety but one which required farmers, again, to purchase new seed yearly instead of storing corn from a previous year's harvest for the next year's planting. Despite the hardships of the 1930s depression, U.S. government policy directed at securing adequate food production through such measures as the Agricultural Adjustment Act (AAA) of 1933 also contributed to the change in the farming—and LGU—landscape. The details of this policy are beyond our scope here, but one of the results of the AAA was to tie farming ever more closely to big business and big government. As most observers of agriculture now readily acknowledge, by the 1930s the industrialization of agriculture was in full swing. It took only the "chemical revolution" of World War II agriculture to lead agriculture to its current state: most Americans' food is produced by large farms which are heavily dependent on financial, mechanical, and chemical inputs, and essentially driven by agribusiness concerns rather than agrarian or freeholder interests.

It is difficult to assess whether the LGU was a deliberate or unintentional party to the transformation of agriculture, but LGU research and extension efforts were nevertheless an important factor in this change. As

Busch and Lacy (1983) demonstrated, over the period in question (roughly 1920 to 1980), the composition of LGU faculty changed from predominantly farm raised, agrarian in outlook, and mission oriented to more urban (or at least nonfarm rural), disciplinary-science oriented, and less concerned about agrarian goals—or even the impact of their research on smaller farms. These characteristics, and related dispositions toward education, moved the teaching component (both on campus and in the form of extension efforts) of the LGU decidedly toward a basic-science and technology-delivery set of practices, if not explicit philosophy. Indeed, prior to the critical reports alluded to previously, little serious consideration was given to educating the whole person, with even less attention devoted toward educating students to be active citizens in a democratic society—one of the original land-grant ideals. Interestingly, even to this day the vast majority of students in colleges of agriculture have no idea what the original idea of "land-grant" means, and, while not faulting them, they do not seem to care. With but a few exceptions they are not preparing to go back to the family farm to improve it, or to contribute something of value to their rural communities. Rather, they are preparing for careers in finance, high-technology business, and veterinary medicine, among other agriculture-related but decidedly nonfarm occupations.

With less than 2 percent of the American people actually involved in commercial agriculture, it is not surprising that the LGU has changed in terms of its research agenda, teaching practices, and extension programming. Indeed, many historical colleges of agriculture have changed the title of the college—"Life Sciences" or "Food, Natural and Human Resources." One might argue that this in itself is a pragmatic move. It is just that along with the near disappearance of the objective instantiation of agrarianism—the family farm—the ideals of agrarianism were approaching extinction. It may be inevitable that in a highly urbanized, high-technology society, with an ever-increasing world population, the dominant form of agriculture (and business in general) would not be smaller, family-owned-and-operated concerns (see Burkhardt 1988). It is not, however, inevitable that the *ideals* of agrarianism, including *concern* for its objective instantiation, become extinct.

Neo-agrarian Ideals and Pragmatist Ethics

History may not always repeat itself, but it is interesting to note that the reemergence of agrarian notions in the college of agriculture and out in the countryside parallel the developments which led to the Populist uprisings of a century ago. Populism, it should be recalled, was essentially a reaction on the part of farmers and other rural people to their perceived—and in some respects very real—powerlessness in the face of strong political and economic forces. Indeed, U.S. agrarian Populists felt threatened that their livelihoods were at the mercy of banks and other financial institutions, farm technology/implements manufacturers, and various government policies which could negatively affect farmland values and farm commodity prices. In many respects, the Populists simply demanded that their opportunities (or even rights) to farm be validated and that they retain some political and economic control over their communities and enterprises (Goodwyn 1978).

Similarly, many modern-day agrarians and populists are united in the view that the industrialization of agriculture threatens smaller family farms, and that this is a bad thing, for a variety of reasons. And the causes of those fears, anxieties, and ultimately philosophical if not political responses are the same: the system of industrial, financial, governmental, and now educational institutions (especially the LGU) has "conspired" to undermine opportunities for family farms. The forces of science and technology, high-level financial maneuvers, and the government-fostered globalization of the agricultural economy have put smaller farmers in the United States (and even more so in developing nations) in an increasingly precarious position. A more contemporary addition to this concern is that the (complete) industrialization of agriculture has also put ever-increasing strains on natural resources and the environment, ultimately raising the specter of an agricultural crash—our simply being unable to produce enough food because of a depleted and despoiled environment (Worldwatch Institute 1997). In any event, the call for an agrarian new agenda for government policy and LGU practices parallels nineteenth-century Populism in the urgency of its cry and the locus of its critique.

The nineteenth-century populists were, however, driven by political and economic self-interest. "Saving farms" was literally their sole pursuit, and their agenda was directed toward specific public policy changes. For example, they rejected a return to the gold standard for the dollar, which they maintained would further impoverish and disempower them. In contrast to that self-interested, political movement, neo-agrarianism embodies a number of connected philosophical, metaphysical, epistemological, and ethical critiques, only one "wing" of which explicitly calls for the political-economic protection of family farms.

Although farming, or the family farm, is the central unifying concept in neo-agrarianism, it is the source of differing kinds of critiques and/or proposals for change—in agriculture per se as in LGU teaching, research, and extension directions. As mentioned above, new agenda concerns range from sustainable agriculture to humanization of agriculture (and agricultural education and research to the democratization of science and technology) to the protection of farming opportunities for small family farms. Let us begin with the last point of focus.

Jim Hightower and Marty Strange are central among neo-agrarians whose main intent is literally saving family farms. Hightower takes a more traditional, Jeffersonian-laced agrarian Populist stand in arguing for the complete remaking of agricultural policy and the research and technology-development agenda. Hightower makes no case for any aesthetic or symbolic or ethical value to be associated with family farms as does, for example, Berry. Rather, family farms are in essence the last bastion of individualism, self-determination, and populist democratic power. As such, his neo-agrarianism is decidedly political in intent and focus (Hightower 1973, 1976). Strange's view is similar insofar as he argues that saving family farms is as much a matter of economics as it is of politics. In his *Family Farming: A New Economic Vision* (1988), Strange maintains that family farms are in fact more "efficient" than industrial agribusiness farms and more likely to achieve long-term economic goals we have set for agriculture generally: steady yields, quality commodities, a secure and sustainable food source. He argues for incentives and protections for family farms, much in the same vein as Hightower, but with more "utilitarian" reasons in the forefront, in contrast with the near-libertarianism often associated with Hightower's views.

Less directly concerned with saving the family farm per se, some proponents of sustainable agriculture or alternative agriculture nevertheless find in smaller, family-type farming operations a greater likelihood that farm production will be maintained at steady levels, that nonrenewable resources such as oil will be depleted at slower rates (if at all), and that farming practices in general will be able to keep up with population growth (see, for example, NRC 1989; Redclift 1987). (In fact, family farms in the developing world are considered one potential solution to overpopulation there.) Though there is no logically necessary connection between family farms and sustainability, there is, according to many sustainability advocates, enough evidence to suggest that saving or preserving or encouraging family farms will solve the problem of the "unsustainability" of high-tech, high-input, high-environmental impact industrial agriculture (see Jackson 1984). Sustainability advocates who do not take the "official" USDA/LGU line, which maintains that "sustainability" simply means long-term agricultural productivity and profitability (see Thesig 1989), are part of the neo-agrarian fold in virtue of that fact alone. That they advocate "wise use" of the land and the virtues associated with being a good farmer (and good member of the local community) only lends further credence to their "membership" in neo-agrarianism (Berry et al., 1990).

Wendell Berry's views are analyzed elsewhere in this volume, so I will not go into much detail here regarding them. Suffice it to say that at one level, Berry shares the political and economic perspective of Hightower and Strange, as well as the belief that family farms are the best sources of sustainable agriculture. Beyond that, however, Berry's poetic-philosophical-ethical vision, and the source of his critique of industrial agriculture and the LGU, is that the family farm *means something*. That is, saving the family farm may be a morally obligatory thing to do, but equally of moral importance is that the values and virtues that are part and parcel of a "good family farm" and farming community are virtues and values which should themselves be preserved and fostered. Community, self-reliance, respect for nature, and harmony of spirit are fundamentally human virtues and values which modern agriculture and modern agricultural research and instruction have sought to undermine if not destroy. Perhaps not everyone can farm; everyone, nevertheless, can and ought to practice these virtues. A system of production and education that does not foster these and in fact is opposed to or myopic in regard to these virtues is inhuman, and morally reprehensible for that fact. That denigrating those virtues has accompanied the systematic undermining of the actual family farm only makes it more indictable (Berry 1977).

Taking Berry's vision and critique to its logical conclusion, and in sympathy with Hightower, Strange, and sustainability proponents, are a final group of neo-agrarians who find the ultimate source of the threat to the actual family farm, as well as to the *ideal* of the family farm, to be the turning away of the LGU mission from its democratic, agrarian underpinnings. Papers and books by Thompson and Stout (1991), Thompson (1995), Burkhardt (1988), and Busch and Lacy (1983), among many others, express the view that the demise of the family farm, the threats to sustainability, and the dehumanization of agriculture and agricultural science and education are at root the results of the "takeover" of agricultural production and the LGU mission by positivist science, large agribusiness concerns, and government bureaucracies less concerned about people than about productivity—"productionism" again. The key focus here is on rehumanizing agriculture and LGU activities, opening up the agenda to all interested and affected parties, and simply allowing the shape of agriculture and agricultural research and education to be shaped by the democratic process. In some respects, modern agriculture and modern LGU activities are only symptomatic of larger social problems or issues: environmental threats, scientism, technology-mania, race, ethnic and gender inequities, and the like. Where this orientation within neo-agrarianism ultimately stands is with the view that our ultimate goal in agriculture, in LGU research and education, and indeed, in all of our institutions and practices, should be to serve people. In ethical terms, the LGU should be serving (and serving up) the "whole person" for active participation in a just, democratic society. Neo-agrarians, all, see us having lost our way in this regard; they differ only in terms of where to focus the energies for change: on the farm, in the classroom, in the laboratory, in the larger public forum.

These final points lead me to suggest that, if there is a common philosophical or ethical thread in neo-agrarianism, it is Deweyan pragmatism. This is perhaps as it should be, given the arguments of previous chapters in this book. If American pragmatism owes its fundamental vision and many of its themes to older agrarian notions and many of their problem sets, it makes sense that a reworked, contemporary agrarian philosophy will find a special affinity with the most developed American pragmatist philosophies. Few, if any, direct references to Dewey (or even to pragmatism itself) are found in the written works of neo-agrarians. Nevertheless, Dewey's philosophy, especially his ethics, provides just the philosophical completeness that neo-agrarianism in any of its manifestations needs in order to self-consciously assert its critical perspective on agriculture and especially the LGU system.

Dewey's Moral Philosophy

Perhaps the best point of departure in making this connection is to suggest that it is the conclusions of Dewey's ethics, more clearly than his method, which connect Dewey with neo-agrarianism. Neo-agrarians all appear to embrace the idea that "right conduct"—on or off the farm—must be a matter of fostering individual-in-community, that community which includes both the family and the larger ecological community, and even the nation or global community. The family farm either is or represents the context in which right actions so understood are given the best opportunities to be played out. The call for public policy to help preserve or enhance the family farm is a straightforward call to allow that ideal moral context to continue to exist. The idea that a family farm provides the context for morally obligatory sustainable agriculture is a related notion. And visions or images of the family farm can serve as ciphers for what the truly moral life is about. In each case, the idea that ethics is about self-development and community-enhancement remains fundamental. As such, neo-agrarianism appears ethically Deweyan. Let us briefly explore Dewey's perspective in more detail.

John Dewey's pragmatics ethics contains two main themes: first, Dewey viewed ethical thinking—perhaps metaethical is the better characterization—as a "science" complete with a "logic of discovery" and a "logic of justification," to use Popper's terms. Second, regarding substantive moral or ethical principles, Dewey held that they would emerge out of the scientific process in a dialectical or organic way. Rather than apply a priori principles or rules to cases at hand, any real-world opportunity for ethical reflection would itself suggest the appropriate, rational, ethical response. Dewey's developed ethical thought appears in his (and J. H. Tufts's) *Ethics* (1908), which, though revised in 1931 and again in 1938, contains his "settled thoughts" on metaethics and ethical theory (Welchman 1995). The overarching theme of his argument and analysis is that moral theory (in philosophers' terms, the set of justified ethical prescriptions) is not, and cannot be, derived "in a vacuum." Indeed, a pragmatic set of ethical norms can only be developed through the *process* of actually making decisions. Reminiscent of Hegelian dialectic, whereby Reason "unfolds itself" through a process of revealing the incompleteness or inadequacy of any given instantiation of "reasonableness," "The Good" will be discovered only in a scientific-like and systematic "testing" of ethical "hypotheses" (Dewey 1938).

Like the physical sciences, ethical science requires definitions and procedural rules. For Dewey, a first principle governing reflection on ethical conduct is that any and all concepts, axioms, and even assumptions we bring to such reflection and analysis have fundamental practical value. By this he means that even such basic ethical notions as "moral agency," "responsibility," "duties," and even more substantive concepts as "rights" and "respect for persons" are inappropriate for ethical thinking if they do not or cannot fit into the range of people's actual experiences in the actual world. Thus, for example, while a utilitarian's notion of "satisfied preferences" may provide some explanation for the justifiability of a particular action or policy, the fact that "satisfied preferences" may not have been explicitly (and self-consciously) a part of the person's reasons for choosing that course of action or policy renders "satisfied preferences" a useless ethical concept. (Interestingly, this is analogous to one criticism of neoclassical economic theory; namely, that the claim that all rational agents try to maximize utility is either a priori true or trivial or both! (Hollis and Nell 1975). In short, ethical concepts must have some significance to actual people, as they actually make decisions, for such concepts to have any *analytic* value.

Welchman (1995) and Campbell (1995) are essentially correct in characterizing Dewey's approach in this regard as "naturalistic." Over and above the demand that the concepts and definitions we use to analyze and characterize ethical decisions and actions be meaningful to moral agents themselves (and not just to philosophers), the actual discovery of "first principles," if there are any such things, proceeds "naturalistically" as well. It is clear that no one enters into moral agency without moral baggage. As Peirce had argued, the process of discovering truth is essentially the process of disposing whatever falsehoods we have been taught by our significant elders, essentially testing them in that quasi-scientific, quasi-dialectical way. Dewey himself had thought that the real business of coming to know truths should proceed accordingly: "what revisions and surrenders of current beliefs about authoritative ends and values are demanded by the methods and conclusions of natural science?" (Dewey 1929, 252). Similarly, in ethics, the only real metaethical imperative is that we position ourselves to reflect continually on our own ethical experiences in real life cases. One can do so, at least in part, in hypothetical space—by analyzing, in what we might now call a Rawlsian reflective equilibrium, how our currently settled ethical principles or beliefs would stand up, given a new test (Rawls 1971). But the real test is how we achieve such an equilibrium in our whole lives.

As for substantive ethical principles, then, the proof, as the adage goes, is in the pudding. This, however, should not be construed as a situational ethics, where "whatever works" in ameliorating conflict (inner or outer) is the right thing to do. Situational ethics (though Dewey did not use this term) degenerates into egoism or relativism as easily as a prioristic ethics (Kant), and even a simple naturalistic ethics (Bentham) degenerates into authoritarianism. The key to Dewey's substantive ethics lies in his belief that our conclusions be liberating, on the one hand, and community-enhancing, on the other. In this respect, ethics is fundamentally political, in Aristotle's sense. I will discuss the democratic orientation of Dewey's ethics below. Suffice it to say here that principles which do emerge out of the scientific or dialectical process of "rightness-seeking" are likely (though not inevitably) to advance the inner strength of the whole person while at the same time fostering fundamental respect for people (and perhaps nature itself), a necessary condition of community. Both outcomes, it should be noted, are themselves rooted in the "natural condition of humankind"—individuality, yet connectedness and interdependence. Finally, it should be noted that Dewey's notion that principles or guideposts for human conduct should be based on an understanding of what is natural to human life experience is also reflected in the notion that the mechanization, routinization, commmodification, and alienation of industrial society is antagonistic to a fully human life.

On this last point alone one can begin to see most clearly how neo-agrarians' critiques—and perhaps even their methods—closely parallel Dewey's thought. Yet, there are other, more direct connections. Although Berry's notion of the connection between family farms and certain truly human virtues may have an a prioristic appearance, Berry's point is that those markers for "right conduct" developed in a historical and cultural context: they worked because they served the whole person and the whole community, including nature. Moreover, as Swedish agricultural scientist and avowed agrarian Ulrich Nitsch has argued, family farmers of necessity make decisions in an organic fashion (not to mention they farm organically): they are continually striving to find the right combination of mechanical, biological and managerial ingredients in order to strike a balance between producing crops or livestock and maintaining the integrity of the farm family and farm community (Nitsch 1984). Dewey's notion of the organic development of ethical goods rings true here, as does his metaethical notion that ethical or normative concepts must have meaning to the actors themselves. Similarly, neo-agrarians concerned with sustainable agriculture have stressed that the truth regarding right methods of farming emerges out of the unique soil, climatic, ecosystemic, and cultural context of the farm (Altieri 1987). Though of course Dewey did not address questions of site specificity—as proponents of context-driven as opposed to so-called universal farming methods refer to their goal (see O'Neill 1986)—the idea that concepts and or principles should reflect the context and experiences of actual people is clearly reflected in these strands in alternative agriculture's and sustainable agriculture's philosophies.

The most important and final connection I wish to draw between Dewey's thought and the neo-agrarian vision is this: in both perspectives, the moral imperative of democratic empowerment of individuals and communities, and the notion of the whole person as central to the success of political democracies, are fundamental. Hightower's self-reliant "freeholder" and Strange's (individually and socially) functional family farm are clearly in line with Dewey's position here. As mentioned, Dewey's ethics is political, through and through. Just as Aristotle had maintained that the person of character must live in a just society and cannot exercise the virtues without a prior condition of justice, for Dewey the essence of the individual's moral development depends in large part on there being strong democratic traditions and institutions. Thus it is that neo-agrarians, especially of the last orientation discussed above, look to the establishment of a larger context in which farming—or any other self-determined way of life—can be made possible. This leads to the agreement with Dewey by neo-agrarians that political disempowerment, just like economic or psychological disempowerment, can stifle ethical conduct. Thus, it is incumbent on individuals, making ethical decisions and growing in terms of self and community, to support and promote those institutions which undergird democratic society. Education is one such institution, and in Dewey's political-ethical thinking as well as in the minds of many if not most neo-agrarians, establishing—or in the case of the land-grant university, reinstating—education of the whole person for participation in his or her local community, the life of the nation, and as a citizen in a global community ultimately is the key. In this regard, all neo-agrarians, especially the last perspective discussed above, find kinship with Dewey's position here. It is indeed in concert with Dewey's political-ethical philosophy of education that we come full circle to the mission of the land-grant university: can it genuinely reflect pragmatic, ethical, and agrarian methods and conclusions once again?

Agricultural Ethics and Neo-agrarianism in the LGU

I stated at the outset of this chapter that the neo-agrarian-pragmatist agenda is reflected in the LGU and in work on sustainability, on social accountability, and on humanizing the curriculum and research orientation through agricultural ethics. Sustainability and accountability are indeed important components if the new agenda concerns are to be fully institutionalized and applied in the activities of LGUs. However, without the third component, agricultural ethics (or agricultural philosophy more generally) and the transition to a more humanized, democratic orientation and set of institutions would probably not be possible. This is because agricultural ethics, of all the neo-agrarian impulses finding their way into LGUs, is explicitly and self-consciously concerned with the neo-agrarian-pragmatist vision—to make possible the actualization of the whole person in the context of LGU education and research, all the while promoting the democratization of LGUs.

Unlike much other philosophical work in the area of applied ethics, agricultural (and natural resource) ethics *emerged* from a pragmatist-agrarian impulse (and movement). As mentioned at the outset, critics of the land-grant system, both inside and outside, identified a problem set as well as offered land-grant administrators and scientists a challenge to which they were forced to respond in a variety of ways. In a number of cases a land-grant university responded by seeking out philosophers or theologians concerned with ethics. This is in contrast to many of the other areas of applied ethics, where philosophers and/or theology-oriented ethicists approached a problem in, for example, law, business, medicine, or journalism simply as an intellectual puzzle to which methods and concepts from philosophical ethics could be directed. This is not to say that other areas of applied ethics are less authentic; rather, it is that the proximate cause for the existence of agricultural ethics *inside* the LGU research and instructional agenda was the decisions of scientists and administrators inside the system itself.

It is interesting to note here that agriculture was rarely if ever considered an area where philosophical ethicists might have something to say, except in the context of already-established scholarly work in, say, environmental ethics. Thompson (1995) argues that for years (continuing to this day) most applied ethicists either ignored agriculture or casually identified agriculture as a contributor among many to the environmental problems environmental ethicists sought to analyze. Specific practices associated with agriculture, such as the use of animals or the larger public policy issues, such as the equitable distribution of food, may have been of interest or concern to a small number of applied ethicists. However, agriculture per se was not a matter of separate, or unique, focus for scholarly ethical research or teaching.

Nevertheless, because of the emerging problem set and the interest of a small number of philosophers in agriculture and agrarianism, agricultural ethics did begin to develop in the 1980s as a separate and in many respects unique application of ethics in the context of the LGU. As noted, a few philosophers were recruited into colleges of agriculture. Some retained primary appointments in philosophy or religious studies or even sociology or political science departments; some held joint appointments in philosophy and one or the other agricultural science disciplines (usually agricultural or resource economics); a few were employed full time (either temporarily or permanently) in a college of agriculture. With the support of like-minded (read: neo-agrarian) colleagues in agricultural economics, rural sociology, anthropology, and across the spectrum of physical and biological scientists in colleges of agriculture (and kindred spirits in liberal arts colleges or business schools), the field of agricultural ethics continued to develop. By the 1990s, professional journals and professional societies devoted to agriculture and ethics, agricultural and environmental ethics, and agriculture viewed from a humanistic (or human values) perspective have come into existence. Although the numbers of professional philosophers engaged in agricultural ethics (or agricultural philosophy more generally) remains relatively small, the coalescence of their professional expertise with the concerns of social and biological scientists, research administrators within the college of agriculture, as well as with activist critics outside the LGU system, has set the stage for the opening of entirely new instructional and research directions inside the LGU. Because many if not most of the issues critics and less impassioned analysts of agriculture and the LGU system have raised are matters of distributive justice, or obligations to future generations, or risks to people or the environment, or the control over science and technology, philosophical-ethical concepts and analysis have come to be seen as a key component in rethinking (and reinventing) agriculture and agricultural science. In large measure, philosophers engaged in agricultural ethics work have had an effect on setting the terms of discourse far in excess of what their numbers might suggest.

Origins aside, there is something which seems to be true of agricultural ethics which does not seem to be as strong, if it even exists, in much other applied ethics work. With some trepidation, I will suggest that agricultural ethics—as practiced by philosophers in agriculture or connected with agriculture, as well as when engaged in by rural social scientists or agricultural scientists, and even by activists outside the system—is driven by a neo-agrarian (and pragmatist) urge. In some cases, a commitment to pragmatist ethics, ethics for the people, preceded an individual's journey into agricultural ethics; in other cases, the neo-agrarian problematic forced individuals to engage agricultural ethics in a more hands-on and pragmatic way; in still other cases, specific ethical problems in agriculture led to a coalescence of neo-agrarian interests and pragmatist ethical commitments. This neo-agrarian-pragmatist ethical urge is a fundamental component of work in agricultural ethics.

By neo-agrarian-pragmatist urge as a keystone of agricultural ethics, I mean that nearly all contemporary work in agricultural ethics, including both research and teaching, is predicated on the critiques of neo-agrarians, especially Wendell Berry, and on the recognition that something must be done to ethically respond to his (and others') critiques. Regardless of the specific issue area under consideration—food safety, international development, farm animal welfare, gender-related problems—it is the critique of modern industrial agriculture and the land-grant system that establishes the first and fundamental problem set to which ethical analysis and argument must be directed. It is not a great exaggeration to say that were one not familiar with Berry—or at least with similar critiques others offer—one would not be engaging agricultural ethics in its most fundamental problematic. Rachel Carson's *Silent Spring* (1962) and its indictment of the use of DDT in agriculture certainly predates Berry (1977), and questions were certainly raised about industrialization's effect on farmworkers well before that (Williams 1984 [1939]). But it was Berry's *The Unsettling of America* that provided much of the terminology specific to that fundamental problem set. Though Berry is not a disciplinary-educated professional philosopher, and *The Unsettling of America* contains few arguments framed in traditional philosophical-ethical language (aside from discussions of character and virtues), the points and arguments lead directly to our asking questions about agriculture and people and the environment in standard philosophical-ethical language. I will return to this last point below.

The other element is that insofar as agricultural ethics grows out of and articulates a "movement," nearly everyone involved in agricultural ethics is involved in "hands-on" work—described by an extension specialist as work "where the rubber meets the road" (Olexa, pers. comm., 1995). That is, while philosophers and others engage in agricultural ethics research and present their work before colleagues in their disciplines or in interdisciplinary meetings such as those of the Agriculture, Food and Human Values Society, most professionals engaged in agricultural ethics work find their primary audiences to be the people who actually are (or will be) making decisions and acting in ways which directly affect farming, natural resource management, farm labor, and the like. In other words, agricultural ethicists are engaged in a sort of "missionary work"—though the point is not to convert to a particular point of view, but to demonstrate the legitimacy of considering points of view and practices alternative to the mainstream productionist bias in the LGU. The point, after all, of courses in agricultural ethics or published research on ethical issues in agriculture, or even extension programs taking ethics to the larger public, is to educate for the common good.

As I state above, agricultural ethics is self-consciously one means for the reinstitutionalization of the public interest, social responsibility, and a whole-person-oriented education for students, faculty and administrators, and the general public: in short, a revitalization of the spirit of the land-grant institution. Practically speaking, undergraduate courses in agricultural and natural resource ethics are at many institutions the only exposure students receive to "big picture" questions. As mentioned above, most are so locked into career tracks in their major fields of study in the sciences that they do not have the luxury of taking a set of ethics, or sociology, or anthropology courses to alert them to questions about industrialization, environmental problems, or what social justice means.

Philosophers and others who work in agricultural ethics have concerned themselves, in their research and teaching activities, with a whole range of issues. Among these are certainly the ethics of sustainability and our moral obligations to save family farms (see Comstock 1987). However, larger questions of technology and industrialization, and specific issues concerning biotechnology, food safety, international aid, trade and development, gender issues, animal rights and welfare, community development, and farm labor, are all part of the agricultural ethics domain. The point is that each of these real-world problem areas is a legitimate candidate for ethical reflection and analysis. More significant is that agricultural ethics, whatever the specific area one focuses on, makes a pressing demand on students, faculty, and administrators, as well as farmers and every member of the general public: we all *must* critically reflect on these issues. Not to do so constitutes a failure on the part of the LGU in its public mission, a failure made even more grievous by the fact that each of these issues—as ethical issues—are real concerns of real people in an increasingly interdependent world. In short, agricultural ethics matters to our ability to make sound decisions as participants in a system of which agriculture is still a fundamentally important part.

A final point, then, concerning metaethics. As Dewey maintained, ethical concepts must mean something to moral agents, and a priori ethical theory is illegitimate as we think about real-world ethical contexts and quandaries. It is true in large measure that agricultural ethicists—and perhaps neo-agrarians in general—approach real-world issues and conflicts with no predetermined commitment to a given ethical theory. The best work in agricultural ethics is that which, following Wittgenstein, "shows the fly the way out of the fly-bottle." This means helping students—whatever their station in life—to see how various ethical concepts and ethical theories shed light on the matter at hand. Most agricultural ethicists do reject an economic interpretation of utilitarianism as their guiding philosophy. Yet, economic utilitarianism, institutionalized in the decision-making process of the LGU at least for the past seventy years is, in many respects, the cause of the LGU mission gone awry. But apart from that more-or-less general orientation, agricultural ethicists cannot rule out of hand any ethical notion or theory that accounts for potentially ethically sound options, directions, and potential outcomes of our decision and actions. Once again, the good must emerge, organically or dialectically from dialogue and critical reflection on the myriad issues we face.

The above is not meant as an accolade for agricultural ethics, but rather as a characterization of the generally acknowledged foundation and mission of this area of applied scholarly endeavor. However, I do not think it is too strong a claim to state that work in agricultural ethics—whomever is engaging in it—is a key factor in reinstitutionalizing an agrarian vision, now in the form of neo-agrarian-pragmatist methods and ideals. We need only reflect, in conclusion, on neo-agrarian prospects in the LGU.

Conclusion: Coming Full Circle? A Parcel of Fears and Hopes

Despite the inclusion of teaching and research on sustainable agriculture, concerns related to social accountability and even in some LGU research and technology development explicitly addressed toward aiding small farms (usually in developing nations), detractors of the LGU system continue to have grounds for maintaining their criticisms. In some respects, the industrialization of agriculture and scientism and productionism in LGU teaching and research are as strong as ever—even growing in some places and in some fields and disciplines. Certainly since the late 1970s, with the emergence of biotechnology as a "tool" for the agricultural sciences and as a technology/commodity to be exploited by large agribusiness firms, the trend toward high-tech, large-scale farming remains firmly in place, and even growing worldwide. Within the LGU, the emphasis on scientific specialization and limited-option career tracks is continuing (see Burkhardt 1998).

Moreover, some of the work on sustainable agriculture has been co-opted by reductionist science and agribusiness concerns, such that sustainability often translates into technologies for long-term profitability for the large farms remaining in agriculture (Allen and Sachs 1992). Social accountability has been in some cases reduced to a matter of dollars and cents: making sure that public expenditures on LGU research and education are directed to just those areas where useful knowledge or commodities might be delivered. Some LGUs have even lost the agricultural ethics component of their teaching and research functions.

It might even be argued that the inclusion of sparks of the neo-agrarian vision as part of the official agenda of the LGU has been simply a public-relations or marketing ploy all along. In response to criticisms that LGU activities were irrelevant, a public policy course or ethics symposium was served up to show that the LGU system was actually responding in a sincere and proactive way. This would certainly not be without precedent: business ethicists have argued for years that business ethics runs the risk of being co-opted as another public-relations trick.

Nevertheless, an argument can be made that, despite these pitfalls and dangers, neo-agrarian ideals and agricultural ethics concerns may continue to have an impact, make further inroads, and ultimately give the LGU system one final push in the (re)direction of being "Democracy's Colleges."

Several years ago, I argued that applied ethics work in any field—business ethics was my particular focus at the time—always runs the risk of degenerating into ideology. By *ideology* I meant the standard Marxian notion of self-interest parading as morality, and I suggested that talk of business ethics by businesspeople or in business schools might ultimately be nothing more than a cover for self-justification. However, I also maintained that there can be something insidiously good about such talk (Burkhardt 1985). It can turn out that people who "talk the talk" can come to understand, actually act upon, the concepts, theories, intuitions, etc., that they are even deliberately using to undermine any real change. In an interesting twist on the Platonic notion that knowledge of the good impels people to do the good, the fact is that people forced in one manner or another to use ethical terms to account for their actions or even just discuss issues may come to see the world in those terms. It is a reverse image of what Jon Elster called *adaptive preference formation*—the phenomenon of sour grapes (Elster 1983). Having to use the language of ethics, even in defending one's actions against critics, can indeed lead to a (subtle) shift in perceptions, which may over time lead to even more significant alterations in perspective. This is most effective with some external "finessing," as it were: the existence of external critics, but more significant, internal gadflies. To the extent that someone is actually demanding ethics talk, ethics might seep its way into the worldviews of the targets of such talk. Of course, to return to Dewey's metaethical point, the ethics talk being foisted on the listener must at some level mean something to the listener or audience.

This is to suggest that as long as there are neo-agrarians who articulate their vision and ideals, advocates of sustainability who relentlessly pursue their goals and agricultural ethicists who continue to subject the issues and elements of agriculture and agricultural research to close ethical scrutiny, the prospects for reinventing the LGU as it has sometimes been called (for example, Kunkel et al. 1997) will remain in place. It has sometimes been claimed that the reason ethical issues get media play is because they *are real;* and in our interconnected and interdependent world they are made even more because of the media and rapid communications technology. Whatever the reason, I have found that in agriculture and in the LGU, neo-agrarianism and agricultural ethics talk are given a listen. Others involved in sustainability, in saving or at least respecting family farms, and other agricultural ethicists as well, have communicated similar experiences.

Dewey was criticized, after the publication of the *Ethics,* for suggesting that all philosophical ethics should be applied ethics—engaged in with an eye toward the improvement of the human condition (Welchman 1995). His notion that ethical reflection and analysis should direct itself toward the realization of more democratic institutions and better contexts for the self-realization of the whole person was rejected as wild-eyed liberal optimism, and hence unprofessional for a "disciplinary" philosopher. I suspect that Dewey, like many neo-agrarians and perhaps even the older agrarian populists, might embrace rather than cringe at the notion that they (we) are optimistic and not disciplinary in a rigidly and narrowly defined sense of the term. Optimism concerning the future of our democratic institutions, optimism concerning the future of agriculture and agricultural science, was certainly fundamental to the mission and vision of the founders of the land-grant system. Even more fundamentally, democratic faith and agrarian ideals were fundamental to the vision of one of our greatest agrarian thinkers—Thomas Jefferson. I would like to think that Jefferson would have approved of the nineteenth-century agrarians and American pragmatist philosophers and would actively support neo-agrarians all, even agricultural ethicists.

Notes

Agrarianism as Philosophy

1. Michael O'Brien describes the agrarian movement as an attempt to depose the previous generation of southern intellectuals, especially in the person of Edwin Mimms. Mimms was the chairman of English at Vanderbilt, and it was he who hired Ransom, Tate, and Warren, along with fellow agrarians such as Donald Davidson and Andrew Nelson Lytle. As portrayed by O'Brien, Mimms was quite aware of the talent under his charge and lobbied Vanderbilt Chancellor James M. Kirkland strongly on behalf of his faculty. Yet Mimms's own aesthetic aspirations were to bring European high culture to the provinces, and in this he represented a self-conception of the South that the agrarians were determined to unseat.

Michael Kreyling updates O'Brien's analysis of the Vanderbilt agrarians, also seeing them as essentially involved in constructing an ideology that would place persons of their ilk at the forefront of intellectual politics. Kreyling relies upon Karl Mannheim's *Ideology and Utopia* in portraying the agrarians as a self-formed intellectual elite intent upon providing an ideology and interpretation of the world for other groups. Like Mannheim, Kreyling is less interested in the content of this ideology than in the social networking it facilitates (Kreyling 1998, 7). Kreyling's discussion reveals much about that networking and further grounds the suggestions made by O'Brien and Bové. Yet there is an untheorized gap between the approach of O'Brien, Bové, and Kreyling, on the one hand, and the study proposed for this volume, on the other. The power-seeking dimensions of agrarian discourse clearly have moral and philosophical significance, but Mannheim's sociology of knowledge emphasizes the elements of social power that would affect *any* attempt to promulgate an "interpretation of the world." Questions such as whether the content of agrarian ideas are particularly disposed to false or morally objectionable interpretations, thus, do not arise.

Paul Bové reads Vanderbilt agrarianism from a post-Marxist, poststructuralist perspective and sees it as a surprisingly sophisticated material critique of capitalism. He believes that Ransom and Tate grasped the imperial state's capacity to subvert opposition by incorporating the opponent's critique within its own power structures. Bové compares them favorably to Gramsci, though he notes that the critical argument in *I'll Take My Stand* was launched from a conservative platform that prevented the Vanderbilt agrarians from achieving their objectives. O'Brien, Kreyling, and Bové deepen our understanding of the Vanderbilt agrarian phenomenon, but the extent to which they further marginalize agriculture is remarkable in the present context. Vanderbilt agrarianism is now rendered as having everything to do with power and influence, and nothing to do with agriculture at all.

Susan Donaldson also reads the Vanderbilt agrarian corpus as a discourse of power and ideology, though with less approval than Bové or O'Brien. She sees the agrarian movement as a calculated attempt to marginalize black and women writers. At one point, Donaldson characterizes Allen Tate's thought as "Agrarian and modernist" (Donaldson 1997, 493). Such a characterization may well fit Tate's views, but what is telling in this context is that Donaldson displays no sensitivity to the way in which agrarianism and modernism are, on the face of it, contradictory. What is crucial about her view of agrarianism is the extent to which its advocates promoted an ideal of the Old South as a white-male dominated society "rooted in time and place, and unified in sensibility" (Donaldson 1997, 493). This stereotype is criticized at length in Jones and Donaldson's *Haunted Bodies: Gender and Southern Texts* (1997). The agrarians are crucial to the critique because they both represent and in large measure articulate what these feminist critics find most objectionable about southern literature:

> The assumption that a white man can represent a universal or "representative" southerner here uncritically extends Wilbur J. Cash's white male "Mind of the South," perpetuating a southern myth deliberately constructed by (among others) southern agrarians like John Crowe Ransom, Allen Tate, and Robert Penn Warren." (Donaldson and Jones 1997, 4)

Donaldson herself documents the agrarians' antipathy toward women writers and racial equality, quoting especially from Allen Tate's shocking attempt to minimize the moral significance of lynching (Donaldson 1997, 504–505). However, Donaldson also finds that as the oppressive themes in agrarian discourse were more explicitly stated, its advocates (especially Tate) became less and less enamoured with what they had created (Donaldson 1997, 507). Yet the salient features of the Vanderbilt agrarian discourse that give rise to this critique may have little to do with agriculture as such. Donaldson does not examine the agrarian tradition of thought beyond the rendition Tate and Ransom gave to it in some of their most polemical writings. Like Bové, O'Brien, and Kreyling, she reduces agrarianism to a discourse of power that neglects its substantive claims.

2. Clearly, libertarian property rights are also broader than agrarian property rights in the range of use they afford. Agrarian property rights do not necessarily protect a property owner's disuse of land, for example, and in many agrarian cultures squatting on unused land is considered fair play. Furthermore, extending land-based property rights to intangible goods begs the key philosophical question: even if *agrarian* property rights are justified by an argument from liberty, why should we presume that *every* claim of property right is crucial to liberty. The material nature of agricultural production and the practical skills of smallholders link farming and liberty, especially in subsistence societies. But as already noted, the sense in which a right to capital gains or monetary interest secures a person's liberty is not, on the face of it, entirely clear. The most crucial difference, however, is that libertarians presume that property rights include not only rights of access and use, but also rights of transfer. If the right of transfer is included in the conception of property, the thinking of libertarians begins to come into focus. Forced transfers would be seen as coercive, hence as threats to liberty. But the libertarian analysis would not apply to Native Americans and Saxon freeholders, for example. They are but two cultures that established traditions where access and use were secure in protecting the agrarian right to apply labor in bringing forth sustenance. Yet these systems lack a "fee-simple" concept of property. Unused land reverts to nature, free for use by all. The coercion that threatens

the liberty of Native Americans or Saxon freeholders has to do with occupancy of the land and control over its produce. "Losing" land means losing access, not taxation or forfeiture of title. In these systems, the land itself was never something that anyone could own in the first place.

American Agrarianism

1. A more extensive account of Emerson's views on political economy and the critical role they accord to morality is to be found in his essay "Wealth" (1851).

2. Thoreau's "poetical farmer" is discussed in Liberty Hyde Bailey's *Outlook to Nature* (1911, 79).

3. A useful study of the Grange in the nineteenth century is D. Sven Nordin's *Rich Harvest* (1974).

4. Every social movement needs its villains. These, for the Grange, were basically the same individuals that agrarian movements—and thinkers like Cobbett—have blamed the farmers' difficulties on lawyers, Jews, and manipulators of commercial wealth.

5. On the Manichaean tendency of the Populists to see all history as a struggle between a small conspiracy of "robbers" and those whom they have robbed, see Richard Hofstadter, *The Age of Reform* (1963, 64).

6. As Liberty Hyde Bailey, the leading thinker of this movement put it, "How to make country life what it is capable of becoming is the question before us" (1913, 61).

7. On the diverse strains comprising the Country Life Movement, see especially David Danbom's *The Resisted Revolution* (1979, chap. 2). One can identify at least four components: those who gave relatively greater weight to the reform of city life; those social scientists concerned to radically alter traditional patterns of what they saw as excessive rural individualism as well as backwardness; those who advocated (either for economic or social reasons) the return of city folk to the land; and those who were primarily concerned with the spread of useful scientific knowledge, which was seen as fostering rural productivity.

8. A survey, for instance, of leading members of the Country Life Movement shows that almost 20 percent of them held degrees from such obscure rural institutions as Harvard, Columbia, and the University of Chicago. On this membership, see particularly William L. Bowers (1974, chap. 3).

9. On business support for the Country Life Movement, see Bowers 1974, 18–19. As one might expect, such industrial and transportation magnates as Hill were primarily concerned with the agricultural production practices, both in reference to conservation and the familiar complaint during these decades of rising farm prices that farm production needed to become more similar to industrial production.

10. Although Bailey certainly was a romantic and very much an agrarian, it is a mistake, I think, to pigeonhole him—as, for example Bowers does in chapter 4 of his book—as an urban agrarian; for there is hardly a single criticism of rural life raised within the Country Life Movement which Bailey does not endorse (even as he differs from many of the less agrarian inclined in the remedies he will accept). Bailey's views ultimately are a synthesis of those who criticized urban life and those who criticized rural life in turn-of-the century America.

11. Bailey sounds most like Bryan in his "Cross of Gold" speech some two decades earlier, in such passages as this:

> The city sits like a parasite, running its roots into the open country and draining it of its substance. The city takes everything to itself—materials, money, men—and gives back only what it does not want; it does not reconstruct or even maintain its contributory country. Many country places are already sucked dry. (1913, 20)

12. This is not to say, however, that Bailey was unconcerned with the economic problems of farmers or against all government actions. Certainly he did not oppose actions designed to limit the monopolistic powers of those "middlemen" who stood between the farmer and consumer (c. 1913, 149–164). Still, his views on farm activism were very much opposed to the Grangers and Populists of his time, inasmuch as he opposed concerted political action by farmers as a class.

13. Unlike Cobbett, though, Berry gives a special role to the importance of an aesthetic appreciation of nature—see especially page 30.

14. See Berry's own appreciative comments on Jefferson, pp. 143–144.

15. In a later essay, "Whose Head Is the Farmer Using? Whose Head Is Using the Farmer?" in Wendell Berry, Bruce Coleman, and Wes Jackson, eds., *Meeting the Expectations of the Land* (1984), Berry indicates more of his opinions on the current crisis facing agriculture. The emphasis in this essay is on the futility of the kind of "industrial" approach to agriculture counseled by many supposed experts (economists and assorted agricultural scientists). In the final analysis, Berry contends, we need an agricultural policy which will maximize not wealth, but the number of good productive farms and farmers (35ff.). This, in turn, will mean an agriculture which achieves three currently unsatisfied goals: a healthy agricultural product, a just concern for the rights of the agricultural poor and land hungry, and an ecologically sensitive concern with the earth "and its network of life" (37–39).

16. *Annals of Agriculture* 26:214; quoted in Raymond Williams, *The County and the City* (1973, 67).

Land, Labor, and God in American Colonial Thought

1. In my view, Perry Miller's classic, sweeping narrative *The New England Mind: The Seventeenth Century* is still one of the best introductions to Puritan and New England life. Miller's work has been challenged at many places but remains a wonderful point of entry into the colonies. For a fine bibliography of work supporting and challenging Miller, see Alan Heimert and Andrew Delbanco's *The Puritans in America*, 418–420. Not all colonialists were Puritans, but in this paper I am giving prominence to their contribution as a distinctive voice in the New World, more so than, say, Anglicans.

2. Richard Bushman's anthology, *The Great Awakening*, is an excellent resource for reviewing the salient early American literature on depravity, self-reproach, and spiritual rebirth.

3. I believe that much of the subsequent American critique of early American life is indebted to tenets that were in place in the colonial period. So, the appeal of divine ownership, God's calling, covenental ethics, and natural rights have all been employed in arguing against slavery, the destruction of Native American life, the oppression of women, and so on. As I argue in *Contemporary Philosophy of Religion*, theism undergirds and contributes to the judicial ideal of impartiality (chap. 7).

4. Lest one conclude that idealism is now dead, readers may wish to consult the contemporary works of John Foster and Howard Robinson.

5. I articulate a related philosophy of mind and God in *Consciousness and the Mind of God*. Other contemporary philosophers who defend the intelligibility and plausibility of such an overriding theistic account include Richard Swinburne, R. M. Adams, Alvin Plantinga.

6. See Bushman 1969 for interesting material on the nature of self-awareness. Also Miller 1939.

7. For my own modest defense of contemporary agrarianism, see "Family Farms" in Comstock's *Life Science Ethics.*

8. A wide range of philosophers have defended the thesis that intentional agency must be linked to intelligible goals; for example, A. MacIntyre, Peter Winch, J. Searle, Paul Grice.

9. For a treatment of how reason itself may be underwritten by theism, see recent work by Alvin Plantinga, Richard Creel, and Richard Taylor. The Cambridge Platonists would have been allied with these philosophers, as would Thomas Reid. "The Mind of Man Is the Noblest Work of God" (Reid 1990, 105). Moreover, I believe that it as a common commitment to theism that provided a core agreement among rationalist and empiricist colonial ethicists. The rationalists trusted reason while the empiricists trusted experience and reflection, and both parties trusted the Great Mind, who benevolently made them and the whole cosmos.

10. Eliot's warning to farmers has a familiar ring today: "We know indeed our present Ability, but we depend greatly on what we think may be hereafter. Thus many run into Debt without measure, and without end, hoping they shall be able to Pay next Year, when they have no visible means for a ground of their hope" (in Bushman 1967, 128; see Bushman's discussion of farming expansion, chaps. 4 and 8).

11. For comments and guidance, I thank Richard Bushman, Peter Field, F. M. Taylor, and Gary Comstock. I am especially grateful to Irve Dell and Kira Obolensky for helping me appreciate anew Winthrop's "bonds of . . . affection."

Franklin Agrarius

1. C. Vincent Buranelli: Franklin was a "remarkable man" but "not a philosopher." There were, however, "four colonial philosophers who deserve the title in its full meaning: Jonathan Edwards, Samuel Johnson, Cadwallader Colden, and John Witherspoon." The operative criteria that Buranelli is using—and with which I do not agree—are that these four thinkers "attacked the technical problems of being and knowledge, and who reasoned out solutions that still challenge interest" (Buranelli 1959, 353–354).

2. In this chapter, Benjamin Franklin's writings are cited from the Yale University Press critical edition (Labaree, Willcox, and Oberg, 1959–), indicating volume and page number [for example, "(10:129)" for volume 10, page 129]. The *Autobiography* (Labaree, Ketcham, Boatfield, and Fineman 1964) will be cited using an "A" in the place of a volume number [for example: "(A:129)"]. Material from those years not yet reached in the new edition reference (Smyth 1907) cited as "W" followed by the volume and page number [for example: "(W10:129)"].

3. I have developed the pragmatic and philosophical aspects of Franklin's thought in Campbell 1998.

4. Such simple equations have resulted in the distorted pictures of Franklin that are found in such authors as Max Weber 1930; D. H. Lawrence 1923; and Charles Angoff 1931.

5. C. Clinton Rossiter: "All that Franklin was trying to tell his fellow Americans was that first things must be attended to first: When a man had worked and saved his way to success and independence, he could then begin to live a fuller or even quite different life" (Rossiter 1953, 304).

6. C. Joseph Lathrop: "Industry and *frugality* are kindred virtues and similar in their principles and effects. They ought always to accompany each other and go hand in hand, for neither without the other can be a virtue, or answer any valuable purpose to the individual or to society. He that is laborious only that he may have the means of extravagance and profuseness; and he that is parsimonious only that he may live in laziness and indolence, are alike remote from virtue" ("Frugality" [1786], reprinted in Hyneman and Lutz 1983, 1:663).

7. C. Perry Miller: Franklin's plan was "a far remove from the schemes of the normal aspirant for success in the American system, who works in his calling in order eventually to make some sort of splurge, in a mansion, yacht, collection of paintings or of race horses. Franklin put the benefit of mankind in the place of the Puritan's glory of God; thus he could move at ease from Calvinism into the Enlightenment" (Miller 1962, 1:87).

8. C. Thomas Jefferson: "Whenever there are in any country uncultivated lands and unemployed poor, it is clear that the laws of property have been so far extended as to violate natural right. The earth is given as a common stock for man to labor and live on" (Jefferson to James Madison, 28 October 1785, in Jefferson 1984, 841–842).

9. C. Earle D. Ross: "the small independent farmer appeared to him as the basis of social and political security alike in the colonies and in the new nation. The even distribution of wealth and the predominance of small laboring owners largely averted 'those Vices that arise usually from Idleness' [W8:613]. Such social evils as did exist were to be found mainly in the towns" (Ross 1929, 61–62).

10. C. Benjamin Rush: "From the numerous competitions in every branch of business in Europe, success in any pursuit may be looked upon in the same light as a prize in a lottery. But the case is widely different in America. Here there is room enough for every human talent and virtue to expand and flourish" ("Information to Europeans Who Are Disposed to Migrate to the United States" [1790], in Rush 1951, 1:556).

11. C. Benjamin Rush: "Men who are philosophers or poets, without other pursuits, had better end their days in an old country . . . . Painting and sculpture flourish chiefly in wealthy and luxurious countries . . . . To the cultivators of the earth the United States open the first asylum in the world" ("Information to Europeans Who Are Disposed to Migrate to the United States," in Rush 1951, 1:550).

12. In spite of the earlier belief that Franklin had maintained some sort of experimental farm in New Jersey (c. Ross 1929, 55; Carey 1928, 170–171), more recent evidence has proven that this belief was mistaken (c. 3:436; Woodward,1943, 179–189).

13. C. 16:200–201; 17:22–23; 18:32; 19:134–139, 268, 316–317, 323–324; 20:40, 95–97; 24:89; Carey 1928, 168–195; Woodward 1943, 193–197.

14. Here Franklin was influenced by the French Physiocrats. C. 15:181–182; Aldridge 1957, 23–30; Carey 1928, 134–167; Mott and Jorgenson 1936, lxxiv–lxxxi.

15. C. Thomas Jefferson: "Those who labour in the earth are the chosen people of God, if ever he had a chosen people, whose breasts he has made his peculiar deposit for substantial

and genuine virtue . . . . Corruption of morals in the mass of cultivators is a phaenomenon of which no age nor nation has furnished an example" (Jefferson 1984, 290).

16. C. Thomas Jefferson: "While we have land to labour then, let us never wish to see our citizens occupied at a work-bench, or twirling a distaff. Carpenters, masons, smiths, are wanting in husbandry: but, for the general operations of manufacture, let our work-shops remain in Europe" (Jefferson 1984, 291).

17. Karl Marx refers to this passage in *Capital* when he writes: "One of the first economists, after William Petty, to have seen through the nature of value, the famous Franklin . . . is not aware that in measuring the value of everything 'in labour' he makes abstraction from any difference in the kinds of labour exchanged—and thus reduces them all to equal human labour. Yet he states this without knowing it" (Marx 1859 [1977], 142 n. 18; c. Aiken 1966, 378–384).

18. C. Virgle Glenn Wilhite: "value is created by nature rather than by man; however the *magnitude* of the value of food products is indicated by, and equal to, the labor expended in their production; whereas *the amount* of the value of other economic goals is measured by the quantity of labor required to produce the food that is consumed by laborers while they are engaged in the production of these other necessities and conveniences of life" (Wilhite 1958, 298; c. Golladay 1970, 49–50).

19. Franklin certainly engaged in a good deal of activity that would seem to have been useless under this particular formulation: his memoirs, his various hoaxes, his magic squares and circles, his chess playing, his music making, and, of course, his dalliances with women.

20. With regard to the issue of slavery, Franklin moved over the course of his life from an accomplice, who owned slaves and printed advertisements in his *Pennsylvania Gazette* detailing the qualities of others who were available for sale, to the presidency of the Pennsylvania Society for Promoting the Abolition of Slavery (c. Van Doren 1938, 129, 479, 774–775).

21. C. Hoover 1938, 366; Wilhite 1958, 313–319; Williams 1944, 77–91.

22. Franklin did favor some social arrangements that he thought preserved his long-term values. In 1772, for example, he praises an attractive system of elderly housing operating in Holland: "These Institutions seem calculated to *prevent* Poverty which is rather a better thing than *relieving* it. For it keeps always *in the Public Eye* a state of Comfort and Repose in old Age, with Freedom from Care, held forth as an Encouragement to so much Industry and Frugality in Youth as may at least serve to raise the required Sum, (suppose œ50,) that is to intitle a Man or Woman at 50 to a Retreat in these Houses. And in acquiring this Sum, Habits may be acquired that produce such Affluence before that Age arrives as to make the Retreat unnecessary and so never claimed" (19:180).

23. C. Paul K. Conkin: "With enough discipline, with enough hard work, with the requisite habits, with Franklin's good advice drawn from his experience, anyone could rise to the limits of his talents, and many could rise as far as Franklin . . . . In Europe it was different, and [there] Franklin often located the source of misery and poverty in society rather than in deficiencies of character or education . . . the America Franklin knew, [was] an America where it seemed obvious that opportunity abounded, that success was largely a matter of intelligence and determination, and that impoverishment was a matter of stupidity or poor character" (Conkin 1976, 100, 106–107).

24. C. Thomas Jefferson: "By an universal law, indeed, whatever, whether fixed or movable, belongs to all men equally and in common, is the property for the moment of him

who occupies it; but when he relinquishes the occupation, the property goes with it. Stable ownership is the gift of social law, and is given late in the progress of society" (Jefferson to Isaac McPherson, 13 August 1813, in Jefferson 1984, 1291).

Provincialism, Displacement, and Royce's Idea of Community

1. I wish to thank Doug Anderson and William S. Lewis for reading and commenting on an earlier draft of this paper.

2. Royce is not always called a pragmatist, although his most important dialogue partners were Peirce and James. But there are very good reasons for considering him as such, despite his constant friendly wrangling with William James's brand of pragmatism. For an in-depth discussion of Royce's idealistic pragmatism, see Mahowald 1972.

3. In this chapter, the two-volume *Basic Writings of Josiah Royce,* edited by John J. McDermott (Royce 1969), will be referenced with the volume number, followed by the page number.

4. Bruce Kuklick writes, "I am sure that in some sense Royce thought as he did because he was 'an American' and, indeed, 'a Californian.' I am also sure that his personal life influenced his writing. But I have found little evidence which would allow me to interpret his work in these terms; and I would suggest that no one else has such evidence" (Kuklick 1985, 3–4). This is a provocative claim I do not wish to tackle directly in this paper. I will say, however, that Royce himself stresses repeatedly the significance of his California experience—in both negative and positive terms—on his own intellectual formation and ideas. This experience plays a rather large role in his conception of the province, his notion of community, and his political/social/religious/moral philosophy. Furthermore, Royce explicitly argues the case for environmental conditions of a place being crucial in the formation of the character of those who live in that place. (See, for example, *The Pacific Coast: A Psychological Study of the Relations of Climate and Civilization,* in Royce 1969, 1:181–204). Beyond this, I am unsure what would qualify as "evidence" on Kuklick's view.

5. Royce's "post-Peircean insight" theory of interpretation is to be found in Royce 1968, chaps. 11–14. For further discussion, see also Smith 1950 and Corrington 1987. For a brief overview of Peirce and Royce on "Signs, Selves and Interpretation," see Smith 1992, chap. 10.

Does Metaphysics Rest on an Agrarian Foundation?

1. I would like to thank Paul B. Thompson for his extensive comments and criticisms of an earlier draft of this paper. Without his encouragement this paper would never have reached fruition.

2. Also in the *Encyclopedia* article (LW8:21) he writes: "The philosophical tradition of the western world did not originate because of a mere taste for abstract speculation or yet because of pure interest in knowledge divorced from application to conduct."

3. The full passage reads: "In enjoyment of present food and companionship, nature, tradition and social organization have cooperated, thereby supplementing our own endeavors so petty and so feeble without this extraneous reinforcement. Goods are by grace not of ourselves." See also *Experience and Nature,* chapter 2, "Existence as Precarious and as Stable," (LW1:42–68).

4. Dewey argues this point in *Reconstruction in Philosophy* (MW12:83): "To treat the early beliefs and traditions of mankind as if they were attempts at scientific explanation of the world, only erroneous and absurd attempts, is thus to be guilty of a great mistake. The material out of which philosophy finally emerges is irrelevant to science and to explanation."

The Edible Schoolyard

1. Standard references to John Dewey's work are to the critical edition, *The Collected Works of John Dewey,* edited by Jo Ann Boydston (Carbondale and Edwardsville: Southern Illinois University Press, 1969–91), and published as *The Early Works* (EW), *The Middle Works* (MW) and *The Later Works* (LW). These designations are followed by volume and page number.

The Relevance of the Jeffersonian Dream Today

1. Editor's note: C. Wunderlich, this volume, at note 6.

2. The pioneering works on this subject are Weber 1930 (1948); Troeltsch 1958; Tawney 1926; Harkness 1958, see esp. chap. 7.

3. Cited by Fredrick Brown Harris, *The Evening Starr,* editorial page, Washington, D.C., 30 October 1955.

4. In the development of the following summary of Locke's theory of property, the author is especially indebted to two highly original and able interpretations of Locke's theory of property: C. B. MacPherson 1951, 550–566, and Leo Strauss 1953, 234–251.

5. This fact is apparent in Perry's account of the waves of immigrants through the eighteenth century (R. B. Perry 1944, 62–81).

6. To be sure, "the abundance of land" was able to do this only through the assistance of government land policies, which did a fairly effective job of meeting the settlers' minimum request for first opportunities to acquire title to as much land as the usual family could handle with its own labor and the most productive technologies available.

Two on Jefferson's Agrarianism

1. Danbom divides agrarianism into "romantic" and "rational" and places Jefferson in the rational group (Danbom 1991). Neither Griswold (1948) nor Brewster (1963) classify agrarians, but their critiques seem to place Jefferson in the rational group.

2. The substance of Griswold's "Jeffersonian Ideal" appeared originally in "The Agrarian Democracy of Thomas Jefferson," *American Political Science Review* 40, no. 4 (1946): 657–681 (Griswold, 19). The article and book may be considered one work; all quotations are from the book, Griswold 1948.

3. The original version of Brewster, 1963, is contained in *Land Use Policy and Programs in the United States,* edited by Howard Ottoson. A slightly edited and corrected version with the same title is contained in *A Philosopher among Economists* (Brewster 1970), which contains a short biography. Unless otherwise noted, references are to the version in the Ottoson book (Brewster 1963). Another chapter in Ottoson's book appeared later in an anthology of historian Paul Gates's works entitled *The Jeffersonian Dream: Studies in the History of American Land Policy and Development* (Gates 1997).

4. Carl Becker's *The Declaration of Independence* (1922) is attributed by Garry Wills to be the classic presumption of Lockean influence in Jefferson's thinking. With the support of the scholarship of Douglas Adair (1946), Wills provides strong arguments against the Lockean premise. Wills refers several times to Becker's "little book" without citation to title (Wills 1978).

5. Jefferson in a letter from Paris to John Jay, 23 August 1785 (in Koch and Peden 1944, 377). The passage without the italicized portions was cited, for example, in Thompson et al., (1994, 245) in tracing the origins of agrarian populism in America.

6. In the original version of RJDT (Brewster 1963) "causal" was "casual" and the bracketed word [family] inserted before "farming." In the Madden and Brewster version (Brewster 1970) "causal" was corrected but "family" was incorrectly unbracketed. My picky editorial comment goes to the point of whether the meaning of farming was Jefferson's or Brewster's, or, similarly, Griswold's. The plantations of Jefferson and his neighbors seem to stretch almost any definition of "family" farm.

7. In the latter part of his article, Brewster used the terms *dream* and *ideal* interchangeably. The difference does not seem important to the statements.

8. Letter from Jefferson to Hogendorp, Paris, 13 October 1785.

9. John Dewey used Henry George to describe the instrumentalist quality of pragmatism in his foreword to George R. Geiger's *The Philosophy of Henry George*. Dewey said: "Henry George is typically American not only in his career but in the practical bent of his mind, in his desire to *do* something about the phenomena he studied and not to be content with a theoretic study. . . . The 'science' of political economy was to him a body of principles to provide the basis of policies to be executed, measures to be carried out, not just ideas to be intellectually entertained" (Geiger 1933, ix–x). Geiger's book was his Ph.D. dissertation, prepared under the direction of John Dewey. Geiger was, therefore, a contemporary of John Brewster under Dewey at Columbia University. However, Brewster gave no evidence that he shared the Dewey/Geiger interest in Henry George or in land matters as such.

10. Over 60 percent of the private land in the United States (90 percent if forests are included) is in agricultural use (Daugherty 1995, 19).

11. There are a number of excellent books and articles on property, aside from standard legal textbooks, that elucidate property notions in a broad, philosophic context. Two additions of the 1990s are Munzer, *A Theory of Property* (Munzer 1990) and Schultz, *Property, Power, and American Democracy* (Schultz 1992). Another entry is by Ryan, *Property* (Ryan 1987); Alan Ryan more recently has written *John Dewey and the High Tide of American Liberalism* (Ryan,1995). My own version of the property issues is in Wunderlich, *Property In, Taxes On, Agricultural Land* (Wunderlich 1995). For an excellent summary of the takings issue see Duerksen and Christopher, *Takings Law in Plain English* (Duerksen and Christopher 1997).

12. *Nollan* v. *California Coastal Commission* 483 US 825 (1987); *Lucas* v. *South Carolina Coastal Council* 60 USLW 4842 (1992); *Dolan* v. *City of Tigard, Oregon* 62 USLW 4576 (1994).

13. Examples are H.R. 925 in the 104th Congress and H.R. 95 in the 105th. According to Defenders of Property Rights, an advocacy group, twenty-five states have passed or are considering property rights legislation (*State by State Legislative Update, 6/96)*. Since 1991, twenty states have enacted takings legislation. Only Connecticut has not considered takings legislation. See http://www.igc.apc.org/arin/states.

14. New agrarianism perhaps could be divided on the concept of stewardship. The Judeo-Christian concept of stewardship is one of relationship of person to person or people. The responsibility of stewardship rests upon dominion over the earth; *environmentally correct* means doing good things to the environment *for the benefit of present and future people. New agrarianism,* closer to the tradition of Confucian or Buddhist philosophy, would require stewardship of the environment *for all creatures and features of that environment, including people.* For a fuller treatment of this distinction, see Attfield 1991, chap. 3.

Brewster, in his interpretation of agrarianism, drew heavily on Weber's "Protestant work ethic" to explain proficient work attitudes. His focus on the family farm meant that stewardship would be directed at the preservation and betterment of the farm entity, including the resources making up that farm entity. Future generations meant future generations of farm entrepreneurs. Environmental awareness in the context of major cultural changes was beginning toward the close of Brewster's career. The early 1960s divided the old from the new agrarianism.

Bibliography

Abrams, Frank W. 1961. *"The Dangers and Delights of Unlimited Leisure."* Saturday Review, Oct. 14.

Aiken, John R. 1966. "Benjamin Franklin, Karl Marx, and the Labor Theory of Value." *Pennsylvania Magazine of History and Biography* 190 (3): 378–384.

Aiken, William. N.d. "The Goals of Agriculture." In *Agriculture, Change and Human Values,* ed. Richard Haynes and R. Lanier, 29–54. Gainesville: Humanities and Agriculture Program, University of Florida.

Aldridge, Alfred Owen. 1957. *Franklin and His French Contemporaries.* New York: New York University Press.

Allen, Patricia, and Carolyn Sachs. 1992. "The Poverty of Sustainability: An Analysis of the Current Positions." *Agriculture and Human Values* 9 (4): 29–35.

Altieri, Miguel. 1987. *Agroecology: The Scientific Basis of Alternative Agriculture.* Boulder: Westview Press.

Anderson, Paul R., and Max H. Fisch. 1939. *Philosophy in America from the Puritans to James.* New York: D. Appleton Century.

Angoff, Charles. 1931. *A Literary History of the American People: II, 1750–1815.* New York: Knopf.

Anonymous, *Alfred Whitney Griswold, 1906–1963, in Memoriam.* 1964. Stamford, Conn.: Overbrook Press.

The Annals of America. 1968. Vol. 4. Chicago: Encyclopedia Brittanica.

Appleby, Joyce O. 1984. *Capitalism and a New Social Order: The Republican Vision of the 1790s.* New York: New York University Press.

Attfield, Robin. 1991. *The Ethics of Environmental Concern.* 2d ed. Athens: University of Georgia Press.

Avila, Charles. 1983. *Ownership: Early Christian Teaching.* London: Sheed and Ward.

Bailey, Liberty Hyde. 1911. *Outlook to Nature.* New York: Macmillan.

———. 1913. *The Country Life Movement.* New York: Macmillan.

Benson, Jackson J. 1988. *Looking for Steinbeck's Ghost.* Norman: University of Oklahoma Press.

———. 1984. *The True Adventures of John Steinbeck, Writer.* New York: Viking Press.

Benton, Thomas Hart. 1854–56. *Thirty Years' View, or, A History of the American Government for Thirty Years, from 1820 to 1850.* New York: D. Appleton and Co.

Berardi, Gigi M., and Charles C. Geisler, 1984. *The Social Consequences and Challenges of the New Agricultural Technologies.* Boulder: Westview Press.

Berry, Wendell. 1972. *A Continuous Harmony: Essays Cultural and Agricultural.* New York: Harcourt Brace Jovanovich.

———. 1977. *The Unsettling of America.* San Francisco: Sierra Club Books.

———. 1981. *The Gift of the Good Land: Further Essays Cultural and Agricultural.* San Francisco: North Point Press.

———. 1987. *Home Economics.* San Francisco: North Point Press.

———. 1993. *Sex, Economy, Freedom, and Community.* New York: Pantheon.

Berry, Wendell, B. Coleman, and Wes Jackson, eds. 1990. *Meeting the Expectations of the Land: Essays in Sustainable Agriculture and Stewardship.* San Francisco: North Point Press.

Betts, Edwin Morris, ed. 1953. *Thomas Jefferson's Farm Book, with Commentary and Relevant Extracts from Other Writings.* Princeton, N.J.: Princeton University Press.

Blau, Joseph L. 1952. *Men and Movements in American Philosophy.* New York: Prentice Hall.

Boorstin, Daniel. 1964. *The Americans: The Colonial Experience.* New York: Vantage Books.

Borgmann, Albert. 1992. "Cosmopolitanism and Provincialism: On Heidegger's Errors and Insights." *Philosophy Today* 36 (summer).

Borsordi, Ralph. 1972. *Flight from the City: An Experiment in Creative Living on the Land.* New York: Harper and Row.

Bové, Paul. 1988. *Mastering Discourse: The Politics of Intellectual Culture.* Durham, N.C.: Duke University Press.

Bowers, William L. 1974. *The Country Life Movement in America.* Port Washington, N.Y.: Kennikat Press.

Brewster, John M. 1963. "The Relevance of the Jeffersonian Dream Today." In *Land Use Policy and Problems in the United States,* ed. Howard Ottoson, 86–136. Lincoln: University of Nebraska Press.

———. 1970. *A Philosopher among Economists: Selected Works of John M. Brewster,* ed. J. Patrick Madden and David Brewster. Philadelphia: J. T. Murphy Co.

Brewster, John M., and Gene Wunderlich. 1961. "Farm Size, Capital, and Tenure." In *Adjustments in Agriculture—A National Basebook,* ed. Mervin Smith and Carlton Christian, 196–228. Ames: Iowa State University Press.

Bromley, Daniel W. 1989. *Economic Interests and Institutions: The Conceptual Foundations of Public Policy.* Oxford and New York: Basil Blackwell.

Buranelli, Vincent. 1959. "Colonial Philosophy." *William and Mary Quarterly* 16, no. 3 (July).

Burkhardt, Jeffrey. 1985. "Business Ethics: Ideology or Utopia." *Metaphilosophy* 16 (April/July): 118–129.

———. 1988. "Crisis, Argument and Agriculture." *Journal of Agricultural Ethics* 1:123–138.

———. 1998. "The Inevitability of Animal Biotechnology? Ethics and the Scientific Attitude." In *Animal Biotechnology*, ed. A. Holland and A. Johnson. London: Chapman and Hall.

Burroughs, John. 1954. "Thoreau's Wildness." In *Thoreau: A Century of Criticism*, ed. Walter Harding. Dallas: Southern Methodist University Press.

Busch, Lawrence, and William B. Lacy. 1993. *Science, Agriculture, and the Politics of Research*. Boulder: Westview Press.

Busch, Lawrence, William B. Lacy, Jeffrey Burkhardt, and Laura Lacy. 1991. *Plants, Power, and Profit: The Social, Economic, and Ethical Consequences of the New Biotechnologies*. London: Basil Blackwell.

Bushman, Richard L. 1967. *From Puritan to Yankee: Character and the Social Order in Connecticut, 1690–1765*. Cambridge: Harvard University Press.

———. 1992. *King and People in Provincial Massachusetts*. Chapel Hill: University of North Carolina Press.

Bushman, Richard L., ed. 1969. *The Great Awakening*. Chapel Hill: University of North Carolina Press.

Campbell, James. 1987. "Place as Social and Geographical." Paper presented at the Meetings of the Society for Agriculture and Human Values, Orlando, Florida, October.

———. 1990. "Personhood and the Land." *Agriculture and Human Values* 7 (1): 39–43.

———. 1995. *Understanding John Dewey*. Chicago: Open Court.

———. 1998. *Recovering Benjamin Franklin: An Exploration of a Life of Science and Service*. Chicago: Open Court.

Carey, Lewis J. 1928. *Franklin's Economic Views*. Garden City, N.J.: Doubleday, Doran and Co.

Carpenter, Frederick I. 1941. "The Philosophical Joads." *College English* 2 (January). Cited from John Steinbeck, *The Grapes of Wrath: Text and Criticism*, ed. Peter Lisca. New York: Penguin, 1972.

Carson, Rachel. 1962. *Silent Spring*. Boston: Houghton Mifflin.

Casey, Edward S. 1993. *Getting Back into Place: Toward a Renewed Understanding of the Place-World*. Bloomington: Indiana University Press.

Choe, Y. B. 1977. *An Essay on the Idea and Logic of Agricultural Economics*. Ann Arbor, Mich.: University Microfilms International.

Clendenning, John. 1985. *The Life and Thought of Josiah Royce*. Madison: University of Wisconsin Press.

Cohen, Morris R. 1954. *American Thought: A Critical Sketch*. Glencoe, Ill.: Free Press.

Comstock, Gary, ed. 1987. *Is There a Moral Obligation to Save the Family Farm?* Ames: Iowa State University Press.

———. Forthcoming. *Life Science Ethics*. Ames: Iowa State University.

Conkin, Paul K. 1976. *Puritans and Pragmatists: Eight Eminent American Thinkers*. Bloomington: Indiana University Press.

Corrington, Robert S. 1987. *The Community of Interpreters: On the Hermeneutics of Nature and the Bible in the American Philosophical Tradition.* Macon, Ga.: Mercer University Press.

Crèvecoeur, J. Hector St. John de. 1986. *Letters from an American Farmer.* Ed. E. Stone. New York: Penguin Books.

Crosby, Alfred W. 1986. *Ecological Imperialism: The Biological Expansion of Europe, 900–1900.* Cambridge: Cambridge University Press.

Cunha, Euclides da. 1944. *Rebellion in the Backlands.* Trans. and ed. S. Putnam. Chicago: University of Chicago Press.

Danbom, David. 1979. *The Resisted Revolution.* Ames: Iowa State University Press.

———. 1991. "Romantic Agrarianism in Twentieth-Century America." *Agricultural History* 65 (4): 1–11.

Daugherty, Arthur. 1995. "Major Uses of Land in the United States, 1992." *AgEconReport.* U.S. Department of Agriculture, 723.

De Grazia, Sebastian. 1948. *The Political Community: A Study of Anomie.* Chicago: University of Chicago Press.

Dempsey, David. 1958. "Myth of the New Leisure Class." *New York Times Magazine,* Jan. 26.

Dennett, Daniel C. 1995. *Darwin's Dangerous Idea.* New York: Simon and Schuster.

Dewey, John. 1902 [1931]. "Interpretation of the Savage Mind." In *Philosophy and Civilization.* New York: Minton, Balch and Co.

———. 1920 [1948]. *Reconstruction in Philosophy.* Enlarged edition. Boston: Beacon Press.

———. 1927 [1931]. "Philosophy and Civilization." In *Philosophy and Civilization.* New York: Minton, Balch and Co.

———. 1929a. *The Quest for Certainty: A Study of the Relation of Knowledge and Action.* New York: Minton, Balch and Co.

———. 1929b. "Philosophy." In *Research in the Social Sciences: Its Fundamental Methods and Objectives,* ed. Wilson Gee. New York: Macmillan Co.

———. 1934. "Philosophy." In *Encyclopedia of the Social Sciences,* vol. 12. New York: Macmillan.

———. 1938. *Ethics* (revised edition). New York: H. Holt and Co.

———. 1960 [1931]. "Context and Thought." In *Dewey on Experience, Nature, and Freedom: Representative Selections,* ed. Richard J. Bernstein. New York: Liberal Arts Press.

———. 1968. *Problems of Men.* New York: Greenwood Press.

Dewey, John. 1969–91. *The Collected Works of John Dewey,* ed. Jo Ann Boydston. Carbondale and Edwardsville: Southern Illinois University Press.

Diamond, Jared. 1997. *Guns, Germs and Steel: The Fates of Human Societies.* New York: W. W. Norton & Co.

Donaldson, Susan V. 1997. "Gender, Race and Allen Tate's Profession of Letters in the South." In *Haunted Bodies: Gender and Southern Texts,* ed. Anne Goodwyn Jones and Susan V. Donaldson, 492–518. Charlottesville: University Press of Virginia.

Donaldson, Susan V., and Anne Goodwyn Jones. 1997. "Haunted Bodies: Rethinking the South through Gender." In *Haunted Bodies: Gender and Southern Texts,* ed. Anne Goodwyn Jones and Susan V. Donaldson, 1–22. Charlottesville: University Press of Virginia.

Duerksen, Christopher, and Richard Roddewig. 1997. *Takings Law in Plain English.* http://www.igc.apc.org/arin.

Dunn, John. 1969. "The Politics of Locke in England and America in the Eighteenth Century." In *John Locke: Problems and Perspectives,* ed. John W. Yolton. Cambridge: Cambridge University Press.

Eames, S. Morris. 1977. *Pragmatic Naturalism.* Carbondale: Southern Illinois University Press.

Eckstrom, Fannie Hardy. 1954. "Thoreau's Maine Woods." In *Thoreau: A Century of Criticism,* ed. Walter Harding. Dallas: Southern Methodist University.

Eddy, Edward D. 1957. *Colleges for Our Land and Time: The Land-Grant Idea in American Education.* New York: Harper and Brothers.

Edwards, Everett E. 1943. *Jefferson and Agriculture: A Sourcebook.* Agr. History Series No. 7. Washington, D.C.: USDA Bureau of Agricultural Economics.

Ellis, Joseph J. 1997. *American Sphynx: The Character of Thomas Jefferson.* New York: Alfred A. Knopf.

Elster, Jon. 1983. *Sour Grapes: Studies in the Subversion of Rationality.* New York: Cambridge University Press.

Emerson, Ralph Waldo. 1836 [1965]. "Nature," in William H. Gilman, ed. *Selected Writings of Ralph Waldo Emerson.* New York: New American Library.

———. 1841 [1883]. "The Method of Nature." In *Nature, Addresses, and Lectures.* Boston: Houghton, Mifflin and Co.

———. 1844 [1965]. "Experience." In *Selected Writings of Ralph Waldo Emerson,* ed. William H. Gilman. New York: New American Library.

———. 1847 [1883]. "The World-Soul." In *Poems.* Boston: Houghton, Mifflin and Co.

———. 1858 [1942]. "Farming." In *The Selected Writings of Ralph Waldo Emerson,* ed. Brooks Atkinson. New York: Modern Library.

———. 1860 [1965]. "Fate." In *Selected Writings of Ralph Waldo Emerson,* ed. William H. Gilman. New York: New American Library.

———. 1870a [1883]. "Farming." In *Solitude and Society.* Boston: Houghton, Mifflin and Co.

———. 1870b [1904]. "Farming." In *The Complete Works of Ralph Waldo Emerson.* Concord Edition. Boston: Houghton Mifflin.

Fensch, Thomas, ed. 1988. *Conversations with John Steinbeck.* Jackson: University of Mississippi Press.

Finkelman, Paul. 1993. "Jefferson and Slavery: Treason against the Hopes of the World." In *Jeffersonian Legacies,* ed. Peter S. Onuf, 181–221. Charlottesville: University Press of Virginia.

Fox-Genovese, Evelyn. 1976. *The Origins of Physiocracy: Economic Revolution and Social Order in Eighteenth-Century France.* Ithaca, N.Y.: Cornell University Press.

Franklin, Benjamin. 1907. *The Writings of Benjamin Franklin.* Ed. Albert Henry Smyth. New York: MacMillan.

———. 1959–. *The Papers of Benjamin Franklin.* Ed. Leonard W. Labaree, William B. Willcox, and Barbara B. Oberg. New Haven, Conn.: Yale University Press.

———. 1964. *Autobiography.* Ed. Leonard W. Labaree, Ralph L. Ketcham, Helen C. Boatfield, and Helene H. Fineman. New Haven, Conn.: Yale University Press.

Gaither, Gloria. 1992. "John Steinbeck: From the Tidal Pool to the Stars: Connectedness, Is-Thinking, and Breaking Through: A Reconsideration." *Steinbeck Quarterly* 25 (winter/spring).

Gates, Paul. 1997. *The Jeffersonian Dream: Studies in the History of American Land Policy and Development.* Ed. A. and M. Bogue. Albuquerque: University of New Mexico Press.

Goldsmith, Oliver. 1902. *The Deserted Village.* New York: Harper Brothers.

Golladay, V. Dennis. 1970. "The Evolution of Benjamin Franklin's Theory of Value." *Pennsylvania History* 37 (1).

Goodwyn, Lawrence. 1978. *The Populist Moment.* London: Oxford University Press.

Grace, Frank. 1953. *The Concept of Property in Modern Christian Thought.* Urbana: University of Illinois Press.

Greene, Jack P. 1993. "The Intellectual Reconstruction of Virginia in the Age of Jefferson." In *Jeffersonian Legacies,* ed. Peter S. Onuf, 225–253. Charlottesville: University Press of Virginia.

Greene, John C. 1984. *American Science in the Age of Jefferson.* Ames: Iowa State University Press.

Griswold, A. Whitney. 1948. *Farming and Democracy.* New York: Harcourt Brace and Co.

Hacker, Andrew. 1962. "Voice of Ninety Million Americans." *New York Times Magazine,* 4 March, 84.

Hacker, Louis M. 1950. *The Triumph of American Capitalism.* New York: Simon & Schuster.

Hamilton, J. G. de Roulhac, ed. 1926. *The Best Letters of Thomas Jefferson.* Boston and New York: Houghton Mifflin Co.

Hanson, Victor Davis. 1995. *The Other Greeks: The Family Farm and the Agrarian Roots of Western Civilization.* New York: Free Press.

———. 1996. *Fields without Dreams: Defending the Agrarian Idea.* New York: Free Press.

Harding, Walter, ed. 1954. *Thoreau: A Century of Criticism.* Dallas: Southern Methodist University Press.

Hargrove, Eugene C. 1980. "Anglo-American Land Use Attitudes." *Environmental Ethics* 2 (2): 121–148.

Harkness, Georgia. 1958. *John Calvin: The Man and His Ethics.* New York: Abington Press.

Hart, Richard E. 1997. "Steinbeck on Man and Nature: A Philosophical Reflection." In *Steinbeck and the Environment: Interdisciplinary Perspectives/Approaches,* ed. Susan F. Beegel, Susan Shillinglaw, Wesley F. Tiffney, Jr. Tuscaloosa: University of Alabama Press.

Hartz, Louis. 1955. *The Liberal Tradition in America: An Interpretation of American Political Thought since the Revolution.* New York: Harcourt Brace.

Heimert, Alan, and Andrew Delbanco, eds. 1985. *The Puritans in America: A Narrative Anthology.* Cambridge: Harvard University Press.

Hightower, John. 1973. *Hard Tomatoes, Hard Times.* Cambridge, Mass.: Schenkman Publishers.

———. 1976. *Eat Your Heart Out.* New York: Crown Books.

Hofstadter, Richard. 1963. *The Age of Reform.* New York: Knopf.

Hollis, Martin, and Edward J. Nell. 1975. *Rational Economic Man: A Philosophical Critique of Neoclassical Economics.* New York: Cambridge University Press.

Hoover, Herbert. 1938. "On Benjamin Franklin." In *Addresses upon the American Road (1933–1938).* New York: Scribners.

Hospers, John. 1971. *Libertarianism: A Political Philosophy for Tomorrow.* Los Angeles: Nash Publishing.

Howarth, William. 1995. "Land and Word: American Pastoral." In *The Changing American Countryside: Rural People and Places,* ed. Emery N. Castle. Lawrence: University Press of Kansas.

Hume, David. 1993. *Selected Essays.* Oxford: Oxford University Press.

Hurt, R. Douglas. 1994. *American Agriculture, a Brief History.* Ames: Iowa State University Press.

Hyman, Stanley Edgar. 1954. "Henry Thoreau in Our Time." In *Thoreau: A Century of Criticism,* ed. Walter Harding. Dallas: Southern Methodist University.

Hyneman, Charles S., and Donald S. Lutz, eds. 1983. *American Political Writing during the Founding Era.* Indianapolis: Liberty Press.

Inge, M. Thomas, ed. 1969. *Agrarianism in American Literature.* New York: Odyssey Press.

Jackson, Wes. 1985. *New Roots for Agriculture.* Lincoln: University of Nebraska Press.

James, William. 1977. *The Writings of William James.* Ed. John J. McDermott. Chicago: University of Chicago Press.

Jefferson, Thomas. 1984. *Writings.* Ed. Merrill D. Peterson. New York: Literary Classics of the United States/Library of America.

———. 1990. *Public and Private Papers.* New York: Vintage Books/Library of America.

Jones, Anne Goodwyn, and Susan V. Donaldson, eds. 1997. *Haunted Bodies: Gender and Southern Texts.* Charlottesville: University Press of Virginia.

Jones, Bryan. 1982. *The Farming Game.* Lincoln: University of Nebraska Press.

Jones, E. L. 1981. *The European Miracle: Environments, Economies and Geopolitics in the History of Europe and Asia.* Cambridge: Cambridge University Press.

Jones, P. M., and N. Jones. 1977. *Salvation in New England.* Austin: University of Texas Press.

Kant, Immanuel. 1983. *Perpetual Peace and Other Essays.* Indianapolis: Hackett.

Kirkendall, Richard. 1984. "The Central Theme of American Agricultural History." *Agriculture and Human Values* 1 (2): 6–8.

Koch, Adrienne, and William Peden. 1944. *The Life and Selected Writings of Thomas Jefferson.* New York: Modern Library.

Kreyling, Michael. 1998. *Inventing Southern Literature.* Jackson: University of Mississippi Press.

Kuklick, Bruce. 1985. *Josiah Royce: An Intellectual Biography.* Indianapolis: Hackett.

Kunkel, H. O., L. Maw, and C. L. Skaggs, eds. 1996. *Revolutionizing Higher Education in Agriculture: Framework for Change.* Ames: Iowa State University Press.

Langsdorf, Lenore, and A. Smith. 1995. *Recovering Pragmatism's Voice.* Albany: State University of New York.

Lanier, Lyle H. 1930. "A Critique of the Philosophy of Progress." In *I'll Take My Stand: The South and the Agrarian Tradition by Twelve Southerners,* ed. John Crowe Ransom et al., 122–154. Baton Rouge: Louisiana State University Press.

Lawrence, D. H. 1923. *Studies in Classical American Literature.* New York: Viking Press.

Leopold, Aldo. *A Sand County Almanac: And Essays in Conservation from Round River.* New York: Oxford University Press, 1949.

Light, Andrew, and Eric Katz, eds. 1996. *Environmental Pragmatism.* London and New York: Routledge.

Lisca, Peter. 1972. *The Grapes of Wrath: Text and Criticism.* New York: Penguin.

Locke, John. 1690 [1980]. *Second Treatise on Civil Government.* Ed. C. B. MacPherson. Indianapolis: Hackett.

———. 1691 [1824]. "Some Considerations of the Lowering of Interest and Raising the Value of Money." In *The Works of John Locke,* 12th ed., vol. 4 (London: Rivington, pp. 22–23.

MacFayden, J. Tevere. 1984. *Gaining Ground.* New York: Holt, Rinehart and Winston.

MacPherson, C. B. 1951. "Locke on Capitalist Appropriation." *Western Political Quarterly* 4(4): 550–566.

———. 1962. *The Political Theory of Possessive Individualism.* Oxford: Clarendon Press.

Mahowald, Mary Briody. 1972. *An Idealistic Pragmatism: The Development of the Pragmatic Element in the Philosophy of Josiah Royce.* The Hague: Martinus Nijhoff.

Malone, Dumas. 1981. *Jefferson and His Time.* Vol. 6, *The Sage of Monticello.* Boston: Little, Brown.

Marcel, Gabriel. 1956. *Royce's Metaphysics.* Chicago: Henry Regnery Co.

Marx, Karl. 1859 [1977]. *Capital: A Critique of Political Economy.* Vol. 1. Trans. Ben Fowkes. New York: Vintage.

Mayer, A., and J. Mayer. 1974. "Agriculture: The Island Empire." *Daedalus* 103: 83–95.

Mayer, F. 1951. *A History of American Thought*. Dubuque, Iowa: Wm. C. Brown Co.

McDermott, John J. 1985. "Josiah Royce's Philosophy of the Community: Danger of the Detached Individual." In *American Philosophy,* ed. Marcus G. Singer, 153–176. Cambridge: Cambridge University Press.

———. 1986. *Streams of Experience: Reflections on the History and Philosophy of American Culture*. Amherst: University of Massachusetts Press.

McIntosh, James. 1974. *Thoreau as Romantic Naturalist*. Ithaca, N.Y.: Cornell University Press.

McKenna, George, ed. 1974. *American Populism*. New York: G. P. Putnam's Sons.

Miller, Perry. 1939. *The New England Mind: The Seventeenth Century*. Cambridge: Harvard University Press.

———, ed. 1962. *Major Writers of America*. New York: Harcourt, Brace and World.

Montmarquet, James A. 1985. "Philosophical Foundations for Agrarianism." *Agriculture and Human Values* 2 (2): 5–14.

———. 1987. "Agrarianism, Wealth and Economics." *Agriculture and Human Values* 4: 47–52.

———. 1989. *The Idea of Agrarianism: From Hunter-Gatherer to Agrarian Radical in Western Culture*. Moscow: University of Idaho Press.

Moore, Barington, Jr. 1966. *Social Origins of Dictatorship and Democracy: Lord and Peasant in the Making of the Modern World*. Boston: Beacon Press.

Morgan, Edmund S. 1975. *American Slavery American Freedom: The Ordeal of Colonial Virginia*. New York: W. W. Norton & Co.

Morrison, Samuel Eliot, and Henry Steel Commager. 1937. *The Growth of the Republic*. New York: Oxford University. Press.

Mott, Frank Luther, and Chester E. Jorgensen. 1936. Introduction to *Benjamin Franklin: Representative Selections*. New York: American Book Co.

Mumford, Lewis. 1962. "From *The Golden Day*." In *Thoreau: A Collection of Critical Essays,* ed. Sherman Paul. Englewood Cliffs, N.J.: Prentice-Hall.

Munzer, Stephen R. 1990. *A Theory of Property*. Cambridge: Cambridge University Press.

Nash, Roderick. 1967 [1982]. *Wilderness and the American Mind*. New Haven, Conn.: Yale University Press.

National Research Council (NRC). 1989. *Alternative Agriculture*. Washington, D.C.: National Academy Press.

Nikolitch, Radoje. 1962. "Family and Larger Than Family Farms: Their Relative Position in American Agriculture." U.S. Dept. of Agriculture *Agr. Econ. Rpt.* No. 4, Washington, D.C., January 1962.

Nitsch, Ulrich. 1984. "The Cultural Confrontation between Farmers and the Agricultural Advisory Service." *Studies in Communication* (Sweden) 10: 41– 51.

Nordin, D. Sven. 1974. *Rich Harvest*. Jackson: University of Mississippi Press.

O'Brien, Michael. 1988. *Rethinking the South: Essays in Intellectual History*. Baltimore: Johns Hopkins University Press.

Olen, J., and V. Barry, eds. 1999. *Applying Ethics: Text with Readings.* 6th ed. Belmont, Calif.: Wadsworth.

Olexa, Michael. 1995. Personal interview with Michael Olexa, Extension Environmental Specialist, USDA/ University of Florida, conducted by Jeffrey Burkhardt.

O'Neill, Onora. 1986. *Faces of Hunger: An Essay on Poverty, Justice, and Development.* London: G. Allen and Unwin.

Onuf, Peter S. 1993. *Jeffersonian Legacies.* Charlottesville: University Press of Virginia.

Oppenheim, Frank M. 1980. *Royce's Voyage Down Under: A Journey of the Mind.* Lexington: University Press of Kentucky.

Ostrom, Elinor. 1990. *Managing the Commons.* Cambridge: Cambridge University Press.

Ottoson, Howard, ed. 1963. *Land Use Policy and Problems in the United States.* Lincoln: University of Nebraska Press.

Paine, Thomas. 1995. *Collected Writings.* New York: Library of America.

Pangle, Thomas L. 1973. *Montesquieu's Philosophy of Liberalism: A Commentary on "The Spirit of the Laws."* Chicago: University of Chicago Press.

Patterson, Orlando. 1982. *Slavery and Social Death, a Comparative Study.* Cambridge: Harvard University Press.

Paul, Sherman, ed. 1962. *Thoreau: A Collection of Critical Essays.* Englewood Cliffs, N.J.: Prentice-Hall.

Peirce, Charles Sanders. 1992. *The Essential Peirce: Selected Philosophical Writings, Volume 1 (1867-1893).* Ed. Nathan Houser and Christian Kloesel. Bloomington: Indiana University Press.

Perry, Ralph Barton. 1938. *In the Spirit of William James.* New Haven, Conn.: Yale University Press.

———. 1944. *Puritanism and Democracy.* New York: Vanguard Press.

Perry, Richard L., and John Cooper, eds. 1978. *Sources of Our Liberties.* Chicago: American Bar Association.

Peterson, Merrill D. 1970. *Thomas Jefferson and the New Nation.* New York: Oxford University Press.

Pocock, J. G. A. 1975. *The Machiavellian Moment: Florentine Political Thought and the Atlantic Republican Tradition.* Princeton: Princeton University Press.

Railsback, Brian E. 1995. *Parallel Expeditions: Charles Darwin and the Art of John Steinbeck.* Moscow,: University of Idaho Press.

Ransom, John Crowe, et al. 1930 [1977]. *I'll Take My Stand: The South and the Agrarian Tradition by Twelve Southerners.* Baton Rouge: Louisiana State University Press.

Rawls, John. 1971. *A Theory of Justice.* Cambridge, Mass.: Belknap Press.

Redclift, Michael. 1987. *Sustainable Development.* London: Methuen.

Reid, Thomas. 1990. *Practical Ethics, Being Lectures and Papers on Natural Religion.* Princeton: Princeton University Press.

Rhodes, Richard. 1970. *The Inland Ground.* New York: Atheneum.

Robertson, John C. 1987. "Scottish Enlightenment." In *The Blackwell Encyclopaedia of Political Thought*, ed. D. Miller. Oxford: Basil Blackwell.

Rockefeller Foundation. 1982. *Science for Agriculture*. Washington, D.C.: Rockefeller Foundation.

Rosenberg, Charles E. 1976. *No Other Gods*. Baltimore, Md.: Johns Hopkins University Press.

Ross, Earle D. 1929. "Benjamin Franklin as an Eighteenth-Century Agrarian Leader." *Journal of Political Economy* 37 (1).

———. 1969. *Democracy's College: The Land-Grant Movement in the Formative State*. New York: Arno Press.

Rossiter, Clinton. 1953. *Seedtime of the Republic: The Origin of the American Tradition of Political Liberty*. New York: Harcourt, Brace and World.

———, ed. 1961. *The Federalist Papers*. New York: New American Library.

Rousseau, Jean-Jacques. 1762 [1964]. *Émile*. Ed. R. L. Archer. Great Neck, N.Y.: Barron's Educational Series.

Royce, Josiah. 1948. *California, from the Conquest in 1846 to the Second Vigilance Committee in San Francisco: A Study of American Character*. New York: Alfred A. Knopf.

———.1968. *The Problem of Christianity*. Ed. John E. Smith. Chicago: University of Chicago Press.

———. 1969. *Basic Writings of Josiah Royce*. 2 vols. Ed. John J. McDermott. Chicago: University of Chicago Press.

———. 1970. *The Letters of Josiah Royce*. Ed. John Clendenning. Chicago: University of Chicago Press.

Royce, Sarah. 1932. *A Frontier Lady: Recollections of the Gold Rush and Early California*. Ed. Ralph Henry Gabriel. New Haven, Conn.: Yale University Press.

Rush, Benjamin. 1951. *Letters of Benjamin Rush*. Ed. Lyman Henry Butterfield. Princeton, N.J.: Princeton University Press.

Ryan, Alan. 1987. *Property*. Minneapolis: University of Minnesota Press.

———. 1995. *John Dewey and the High Tide sof American Liberalism*. New York: W. W. Norton.

Santayana, George. 1924. *Character and Opinion in the United States*. New York: Charles Scribner's & Sons.

Schama, Simon. 1996. *Landscape and Memory*. New York: Alfred A. Knopf.

Schneider, Herbert Wallace. 1946. *A History of American Philosophy*. New York: Columbia University Press.

Schrader-Frechette, Kristin. 1984. "Agriculture, Property, and Procedural Justice." *Agriculture and Human Values* 1 (3): 15–28.

Schultz, David A. 1992. *Property, Power, and American Democracy*. New Brunswick, N.J.: Transaction Publishers.

Shue, Henry. 1980. *Basic Rights*. Princeton, N.J.: Princeton University Press.

Smith, John E. 1950. *Royce's Social Infinite: The Community of Interpretation*. New York: Liberal Arts Press.

———. 1992. *America's Philosophical Vision.* Chicago: University of Chicago Press.

Steinbeck, John. 1937. *Of Mice and Men.* New York: Viking Penguin.

———. 1937. *The Red Pony.* New York: Viking.

Steinbeck, John, with Edward F. Ricketts Jr. 1951 [1941]. *The Log from the Sea of Cortez.* New York: Penguin.

Stewart, John L. 1965. *The Burden of Time: The Fugitives and the Agrarians.* Princeton, N.J.: Princeton University Press.

Stoller, Leo. 1962. "Thoreau's Doctrine of Simplicity." In *Thoreau: A Collection of Critical Essays,* ed. Sherman Paul. Englewood Cliffs, N.J.: Prentice-Hall.

Strange, Marty. 1988. *Family Farming: A New Economic Vision.* Lincoln: University of Nebraska Press.

Strauss, Leo. 1953. *Natural Right and History.* Chicago: University of Chicago Press.

Stuhr, John J., ed. 1993. *Philosophy and the Reconstruction of Culture: Pragmatic Essays after Dewey.* Albany: State University of New York Press.

Taliaferro, Charles. 1992. "God's Estate." *Journal of Religious Ethics* 20 (1): 69–92.

———. 1994. *Consciousness and the Mind of God.* Cambridge: Cambridge University Press.

———. 1998. *Contemporary Philosophy of Religion.* Oxford: Basil Blackwell.

Tansill, Charles C., ed. 1927. *Documents Illustrative of the Formation of the Union of the American States.* Washington, D.C.: U.S. Government Printing Office.

Tawney, R. H. 1926. *Religion and the Rise of Capitalism.* New York: Harcourt, Brace and Co.

Thesig, C. 1989. "What Is Sustainable Agriculture?" *Land Stewardship Newsletter* 7 (2): 9.

Thompson, E. P. 1971 [1993]. "The Moral Economy of the English Crowd in the Eighteenth Century." In *Customs in Common,* 185–258. New York: Free Press.

Thompson, Paul B. 1986. "The Social Goals of Agriculture." *Agriculture and Human Values* 3: 32–42.

———. 1990. "Agrarianism and the American Philosophical Tradition." *Agriculture and Human Values* 7 (1): 3–8.

———. 1995. *The Spirit of the Soil: Agriculture and Environmental Ethics.* New York: Routledge.

Thompson, Paul B., and Bill A. Stout, eds. 1991. *Beyond the Large Farm: Ethics and Research Goals for Agriculture.* Boulder: Westview Press.

Thompson, Paul B., Robert Matthews, and Eileen van Ravenswaay. 1994. *Ethics, Public Policy, and Agriculture.* New York: Macmillan Co.

Thoreau, Henry David. 1884. *Excursions.* Boston: Houghton, Mifflin and Co.

———. 1937. *Walden and Other Writings of Henry David Thoreau.* Ed. Brooks Atkinson. New York: Random House.

———. 1961. *A Week on the Concord and Merrimack Rivers.* Boston: Houghton, Mifflin and Co.

———. 1966. *Walden and Civil Disobedience.* New York: W. W. Norton and Co.

———. 1984. *The Writings of Henry David Thoreau: Journal.* Vols. 2–3. Ed. Robert Sattelmeyer. Princeton, N.J.: Princeton University Press.

Tilley, Charles. 1992. *Coercion, Capital and European States: 990–1992.* Rev. ed. Oxford: Basil Blackwell.

Tocqueville, Alexis de. 1898. *Democracy in America.* Trans. Henry Reever. New York: Century and Co.

Townsend, Harvey Gates. 1934. *Philosophical Ideas in the United States.* New York: American Book Co.

Troeltsch, Ernst. 1958. *Protestantism and Progress.* Boston: Beacon Press.

U.S. Code Annotated. 1964. Brooklyn: U.S. Government Printing Office.

Van Doren, Carl. 1938. *Benjamin Franklin.* New York: Viking.

Wallerstein, Immanuel. 1974. *The Modern World-System I: Capitalist Agriculture and the Origins of the European World-Economy in the Sixteenth Century.* San Diego, Calif.: Academic Press.

Weber, Max. 1930 [1948]. *The Protestant Ethic and the Spirit of Capitalism.* London: Butler and Tanner Ltd.

Webster, Daniel. 1851. "First Settlement of New England." In the *Works of Daniel Webster.* Boston: C.C. Little and J. Brown.

Welchman, Jennifer. 1995. *Dewey's Ethical Thought.* Ithaca, N.Y.: Cornell University Press.

White, Morton, and Lucia White. 1962. *The Intellectual Versus the City.* New York: Mentor Books.

Whitford, Kathryn, and Philip Whitford. 1954. "Thoreau: Pioneer Ecologist and Conservationist." In *Thoreau: A Century of Criticism,* ed. Walter Harding. Dallas: Southern Methodist University.

Wilhite, Virgle Glenn. 1958. *Founders of American Economic Thought and Policy.* New York: Bookman and Associates.

Williams, B. 1984. "The Impact of Mechanization of Agriculture on the Farm Population of the South." In *The Social Consequences and Challenges of the New Agricultural Technologies,* ed. G. Berardi and C. Geisler. Boulder: Westview Press. Originally published in *Rural Sociology* 4 (1939).

Williams, Howell V. 1944. "Benjamin Franklin and the Poor Laws." *Social Service Review* 18 (1): 77–91.

Williams, Raymond. 1973. *The Country and the City.* Oxford and New York: Oxford University Press.

Wills, Garry. 1979. *Inventing America: Jefferson's Declaration of Independence.* New York: Vintage Books.

———. 1997. "American Adam." *New York Review of Books,* 6 March, 30–33.

Wilson, R. Jackson. 1968. *In Quest of Community: Social Philosophy in the United States, 1860-1920.* New York: John Wiley and Sons.

Wojcik, Jan. 1984. "The American Wisdom Literature of Farming." *Agriculture and Human Values* 1 (4): 26–37.

———. 1989. *The Arguments of Agriculture*. West Lafayette, Ind.: Purdue University Press.

Woodward, Carl R. 1943. "Benjamin Franklin: Adventures in Agriculture." In *Meet Doctor Franklin*. Philadelphia: Franklin Institute.

Works, George A., and Barton Morgan. 1939. *The Land-Grant Colleges*. Washington, D.C.: U.S. Government Printing Office.

Worldwatch Institute. 1997. *State of the World: 1997*. New York: Norton.

Wunderlich, Gene. 1984. "Commentary on Schrader-Frechette." *Agriculture and Human Values* 1 (3): 29–30.

———. 1995. *Property in, Taxes On, Agricultural Land*. LTC Paper 153, Madison, Wisc.: Land Tenure Center.

Contributors

DOUGLAS R. ANDERSON is a teacher of philosophy at the Pennsylvania State University whose primary interests are the history of philosophy and American philosophy. He is author of *Creativity and the Philosophy of C. S. Peirce* (1987) and *Strands of System: An Introduction to the Philosophy of C. S. Peirce* (1995) and is coeditor of *Philosophy in Experience* (with Richard Hart, 1997) and *The Contemporary Vitality of Pragmatism* (with Sandra Rosenthal and Carl Hausman, 1999). Recently he has written several essays engaging the work of Henry Thoreau, Henry Bugbee, and John William Miller.

JOHN M. BREWSTER (1905-1965) completed a Ph.D. in philosophy at Colum-bia University in 1938. His professional career was spent entirely in various economic research offices within the U.S. Department of Agriculture. A posthumous collection of his philosophical essays on agriculture was published under the title *A Philosopher among Economists* (1970).

JEFFREY BURKHARDT is professor of ethics and policy studies in the Food and Resource Economics Department, Institute of Food and Agricultural Sciences (IFAS), University of Florida. Recent publications include "Scientific Values and Moral Education in the Teaching of Science," *Perspectives on Science* (1999), and "The Inevitability of Animal Biotechnology?" in Holland & Johnson, eds., *Animal Biotechnology and Ethics* (1998). He is currently researching the ethical aspects of the growing corporate control over food biotechnology, while desperately trying to preserve the U.S. land grant system from a complete takeover by multinational corporations.

JAMES CAMPBELL is professor of philosophy at the University of Toledo and author of a number of works in the area of American philosophy, including *The Community Reconstructs: The Meaning of Pragmatic Social Thought* (1992), *Understanding John Dewey* (1995), and *Recovering Benjamin Franklin* (1998).

ROBERT S. CORRINGTON teaches systematic theology at Drew University. He is the author of many books on the religious significance of classical American pragmatism, including *The Community of Interpreters* (1995) and *Nature's Self: Our Journey from Origin to Spirit.* His most recent book is *Nature's Religion* (1998).

RICHARD E. HART is Cyrus H. Holley Professor of Applied Ethics at Bloomfield College in New Jersey. He is editor or coeditor of *Ethics and the Environment* (1992), *Philosophy in Experience: American Philosophy in Transition* (1997), and *Plato's Dialogues: The Dialogical Approach* (1997) and is author of some fifty articles and reviews in the fields of ethics, social philosophy, philosophy and literature, American philosophy, and the teaching of philosophy.

LARRY A. HICKMAN is director of the Center for Dewey Studies and professor of philosophy at Southern Illinois University—Carbondale. He is the author of *Modern Theories of Higher Level Predicates* (1980) and *John Dewey's Pragmatic Technology* (1990). He is the editor of *Technology as a Human Affair* (1990), *Reading Dewey: Interpretations for a Postmodern Generation* (1998), *The Essential Dewey* (with Thomas Alexander, 1998), and *The Correspondence of John Dewey, Vol. 1: 1871–1918* (1999).

THOMAS C. HILDE taught and conducted research on sustainability and the philosophy of technology at Texas A&M before moving on to complete his Ph.D. in philosophy at the Pennsylvania State University. His present doctoral research focuses on the pragmatist conception of community, contemporary feminist and postmodern critiques of the idea of community, and the modern reality of displacement.

ARMEN MARSOOBIAN received his doctorate in philosophy from the State University of New York at Stony Brook, where he studied with Justus Buchler and John McDermott, among others. He is (with Kathleen Wallace and Robert Corrington) the editor of *Nature's Perspectives: Prospects for Ordinal Metaphysics.* Marsoobian is professor of philosophy at the University of Southern Connecticut, where he also edits the journal *Metaphilosophy.*

JAMES A. MONTMARQUET is professor of philosophy at Tennessee State University, Nashville, Tennessee. He is the author of *The Idea of Agrarianism* (1989) and *Epistemic Virtue and Doxastic Responsibility* (1993) and numerous articles in the areas of ethics and epistemology.

CHARLES TALIAFERRO, professor of philosophy and member of Environmental Ethics Studies at St. Olaf College, has published papers in *Agriculture and Human Values, Environmental Ethics, A Companion to Environmental Ethics, New Essays in Bioethics,* and elsewhere. He is the author of *Consciousness and the*

Mind of God (1994), *Contemporary Philosophy of Religion* (1998), and coeditor of *sA Companion to Philosophy of Religion* (1997). His current research projects include environmental aesthetics and the role of virtues in philosophy of mind.

PAUL B. THOMPSON holds the Joyce and Edward E. Brewer Chair in Applied Ethics and is a member of the Department of Philosophy at Purdue University. He has authored many books and articles on topics in agriculture and agrarian thought, including *The Spirit of the Soil: Agriculture and Environmental Ethics* (1994). He also directs Purdue University's Center for Animal Well-Being and Productivity in the Agricultural Program.

GENE WUNDERLICH is a consultant on land problems and policies, formerly with the Economic Research Service, U.S. Department of Agriculture. He has written extensively on land tenure, land reform, real estate markets, taxation, information systems, property, ethics and land policy, United States and abroad. He has edited *Land Ownership and Taxation in American Agriculture* and *Agricultural Land Ownership in Transitional Economies.*

Abraham, 81, 82

Index

Abrams, Frank W., 225
Achilles, 150
activism, 238, 240, 241
Adair, Douglas, 313n. 4
Adam, 141, 151
Adams, John Quincy, 137
Adams, R. M., 308n. 5
Adorno, Theodor, 5
aesthetics, 44–46, 48
Africa, 136
Agamemnon, 150
agrarian/agrarianism, 1–5, 12–18, 19–21, 25–36, 41–42, 51–52, 58, 60, 61–64, 69–76, 96, 101, 118, 119, 132, 255, 263–264, 266, 273, 275, 276, 278, 279, 285, 287, 297, 305n. 2, 306n. 4; America and, 4, 14, 51, 103, 104, 113, 115, 265; ancient Greeks, 36–39; community and, 167–170; pragmatist influences on, 6, 26; romanticism and, 3, 42–48, 51, 72, 183
agrarian New Critics. *See* Vanderbilt agrarians
agrarian philosophy, 9, 11, 25, 27, 29, 31, 32–34, 37, 48–50, 72, 134, 138, 291; ancient Greeks and, 36–40
agrarian pragmatism, 77
agrarian society, 14, 29, 46
agribusiness, 279, 286, 291, 301
Agricultural Adjustment Act (AAA), 286
agricultural development, 19, 260
agricultural ethics, 280, 282, 296–303
agricultural experiment stations (AES), 283–286
Agricultural Market Transaction Act, 266
agricultural policy, 18, 69, 75, 236, 243, 244–246, 248–249, 251, 260, 261, 263, 292, 307n. 15, 312n. 6
agricultural practices. *See* farming
agricultural production, 17, 34, 37, 48–49, 60, 69, 70, 265, 290–291, 305n. 2, 306n. 9
agricultural sustainability, 20, 267, 282, 289–290, 294, 296, 301, 302
agricultural technology, 41, 237–238, 243, 247–250, 265, 281, 286, 288
agriculture, 1–3, 11–14, 16–17, 25–27, 29–30, 34–35, 39–40, 41–44, 46, 48, 51, 52–53, 55–56, 57, 59–61, 67–68, 69–70, 72, 74–75, 77, 96, 98, 101, 129, 131, 133, 134, 135, 136, 167–168, 185, 237, 258, 266, 283, 287, 296, 303
Agriculture, Food and Human Values Society, 298–299
Aldridge, Alfred Owen, 309n. 14
Allen, Ethan, 93
Allen, Patricia, 301
Altieri, Miguel, 294
American Agricultural Economics Association (AAEA), 18
American colonialism, 77–79, 81, 88, 89, 93, 97, 98, 307n. 1
American Dream, 214, 240
American Philosophical Society, 106
American philosophy, 1–2, 4, 6, 13–14, 19, 20, 101, 119, 139, 169, 271
American Revolution, 126
American Seed Trade Association, 286

American transcendentalism, 1, 6, 20, 26, 35, 44, 118, 130, 155, 271, 274
Amish, 66
ancient Greeks, 16, 25, 36–40, 188, 193, 27
Anderson, Douglas R., 15, 119, 178, 311n. 1
Anderson, Paul R., 90, 91–93, 95
Anglicanism, 84, 88, 307n. 1; *see also* Church of England
Angoff, Charles, 309n. 4
animal husbandry, 21, 26, 32
Antonius, 122
Apel, Karl-Otto, 6
Apollo, 268
Appleby, Joyce O., 82
applied ethics, 296, 301, 302
Aquinas, St. Thomas, 64
Arbella, 84
Arendt, Hannah, 129
Aristotle, 11, 38, 39, 71, 89, 193, 271, 294, 295
art. *See* aesthetics
Athens, Greece, 38–39
Attfield, Robin, 313n. 14
Augustine, 181
Austin, Benjamin, 128–129
Australia, 171
Avila, Charles, 80

Bailey, Liberty Hyde, 13, 35, 51, 52, 60–64, 69, 72–74, 119, 139, 306n. 10; Country Life Movement and, 62–63, 307nn. 6, 11, 12
Barry, V., 273
Baxter, Richard, 85, 222–223, 225
Becker, Carl, 313n. 4
Benson, Jackson J., 270–271
Bentham, Jeremy, 293
Benton, Thomas Hart, 218
Berardi, Gigi M., 281
Berkeley, Bishop George, 91, 94
Berry, Wendell, 5, 13, 17, 51, 52, 64–69, 74, 119, 280–281, 285, 290, 294, 298; agrarian philosophy and, 20, 64–69, 71, 72, 75, 76, 139, 167–169, 171, 182, 264, 289
Betts, Edwin Morris, 127
Bible, 81, 82, 144, 170
Bierstadt, Albert, 130
biotechnology, 299, 301
Bonaparte, Napoleon, 128
Boorstin, Daniel, 82
Boston, Massachusetts, 101, 105
Bove, Paul, 26, 28, 31, 304n. 1
Bowers, William L., 61, 62, 306nn. 8, 9, 10
Bradford, William, 86
Brattle, Thomas, 89
Brewster, John, 18–19, 119, 254, 257–260, 263, 265, 266, 312nn. 2, 3, 313nn. 6, 7, 313n. 9, 313n. 14
Briemyer, Harold, 119
Bromley, Daniel, 49
Brown, Harold Chapman, 270
Brown, John, 31
Brueghel, Pieter (the Elder), 45
Bryan, William Jennings, 58–59, 306n. 11
Buchanan, Joseph, 90
Buddhist philosophy, 314n. 14
Buffon, Georges-Louis Leclerc, 11, 41, 48, 125–126, 135, 136
Bulkeley, Peter, 85–86
Buranelli, Vincent, 135n. 1
Burk, John Daly, 126
Burkhardt, Jeffrey, 20, 21, 287, 290, 301
Burroughs, John, 156
Busch, Lawrence, 286, 287, 290
Bushman, Richard L., 79, 307n. 2, 308nn. 6, 10, 11

Calvinism, 88, 309n. 7
Cambridge, Massachusetts, 175, 178
Cambridge Platonists, 98, 308n. 11
Campbell, James, 13–14, 119, 134, 168–169, 293
Canada, 123
Canterbury, 84
Carey, Lewis J., 109

Carpenter, Frederick, 271
Carr, Dabney, 122
Carr, Peter, 121–123
Carson Rachel, 266; *Silent Spring*, 264, 298
Cartesian philosophy, 7
Cash, Wilbur J., 31, 304n. 1
Cato, 70
Central Valley, California, 30
Ceres, 56, 159
Cervantes, Miguel de, 150
Charles I (England), 79
Chastellux, marquis de, 135, 136
Choe, Y. B., 281
Christianity, 13, 88, 90, 93, 97, 173, 182
Chronicles, 81
Church, Frederick, 130
Cicero, 70, 122
Civil War, 57, 137, 138, 283
Clawson, Marion, 18
Clendenning, John, 176
Cobbett, William, 66, 71–72, 306n. 4
Colden, Cadwallader, 89, 93, 308n. 1
Cole, G. D. H., 204
Cole, Thomas, 130
College of Emporia, 257
Columbia University, 18, 91, 257, 306n. 8, 313n. 9
Commager, Henry Steel, 238–239
communitarianism, 16
Comstock, Gary, 299, 308n. 11
Concord, Massachussetts, 55, 57, 151, 156
Condorcet, Jean-Antoine, 135
Condillac, Etienne Bonnot de, 90, 135
Confederacy, 283
Confucius/Confucianism, 65, 313n. 14
Congregationalism, 88
Congress, 59, 266, 283, 284, 313n. 13
Conkin, Paul K., 109, 310n. 23
Connecticut, 138, 313n. 13
Constitutional Convention, 217
Cooper, John, 261
Cooper, Thomas, 90
Cornell University, 60
Corrington, Robert S., 14–15, 28, 119, 131, 311n. 5
Cotton, John, 81
Country Life Commission, 59, 60–61, 284–285
Country Life Movement, 13, 59–61, 63–64, 69, 72, 306nn. 7, 8, 9, 10
Country Life Report, 63
Covenantal theology, 84, 307n. 3
Cozens, John Robert, 130
Creel, Richard, 308n. 9
Crevecoeur, J. Hector St. John, 43–45, 46, 51, 61, 75, 127
Cromwell, Alexander, 79
Crosby, Alfred, 50
Cudworth, Ralph, 98
Cunha, Euclides de, 46–47
Cushman, Robert, 86

Dakotas, 166, 186
Danbom, David, 306n. 7
Danford, Merle, 269
Darwinism, 10, 196, 276
Daugherty, Arthur, 313n. 10
Davenport, John, 79
Davidson, Donald, 304n. 1
Davis, Jefferson, 31
DDT, 17, 298
Declaration of Independence, 118, 126, 139, 220, 261
deconstruction, 9, 11
De Grazia, S., 238
Delbanco, Andrew, 81, 82, 84, 86, 96, 97, 307n. 1
Dell, Irve, 308n. 11
democracy, 2, 4, 10, 14, 17–18, 20, 36, 38, 39, 46, 51, 68, 121, 123, 126, 131, 138, 139, 217, 218, 255, 263–264, 265, 284, 291, 303
Dempsey, David, 225
Dennett, Daniel, 10
DeTocqueville, Alexis, 239–240
Dewey, John, 4, 5, 9–11, 16–17, 20, 26, 28–29, 49, 96, 130, 136, 172, 176, 181, 183, 264, 313n. 9; agrarianism and, 190–191, 193, 199–201; agri-

culture and, 190–191, 193, 199–201; democracy and, 10, 204; edible schoolyard and, 196–198, 205; education and, 17, 196–199, 202, 205; ethics and, 291–295, 300, 302; *Experience and Nature,* 186, 191; hunting and, 189–191, 199–201; industrialism and, 202–205; philosophy and, 78, 185–189, 192–194; pragmatism and, 29, 78, 128, 196, 200–201, 273, 274, 291; *Problems of Men,* 78; reconstruction and, 16–17, 192
d'Holbach, Baron, 90, 93–94
Diamond, Jared, 50
Diggers, the, 33, 56, 67
Dionysus, 159–160, 268
Dolan v. *City of Tigard, Oregon,* 313n. 12
Donaldson, Susan, 30, 304n. 1
Donnelly, Ignatius, 58
dualism, 4, 13, 77–78, 88, 90, 93, 94, 96
Dunn, John, 83
Duerkson, Christopher, 313n. 11
Du Pont, Pierre, 262
Dwight, Timothy, 85, 98

Eckstrom, Fannie Hardy, 156
Edwards, Everett E., 255
Edwards, Jonathan, 87, 307n. 1; idealism and, 91–93, 94; *Sinners in the Hands of an Angry God,* 87, 91–93
eighteenth century, 42–48, 80, 87, 88, 134, 137; agrarianism and, 40–42, 50, 126, 129–135, 137, 138, 203; American philosophy and, 14, 90–91, 92, 101; European philosophy and, 25, 42, 118; French philosophy and, 12, 42, 90, 134–135; Puritanism and, 79; Scottish philosophy and, 12, 40, 42, 132
Einstein, Albert, 248
Eliot, Jared, 98, 308n. 10
Eliot, John, 85
Ellis, Joseph J., 118, 125, 126, 129
Elster, John, 302
Emerson, Ralph Waldo, 1, 13, 26, 29, 47, 48, 51, 52–55, 57, 64, 69, 130, 146, 157, 161, 170, 172, 274; agrarianism and, 28, 119, 131, 139, 186; agriculture and, 141–142, 151–152; farming and, 141, 143, 149–152; idealism and, 143, 145, 146; *Nature,* 53–54, 142–144, 146, 148, 151; nature and, 140–152; poetry and, 15, 140, 143–145, 150, 151–152; philosophy and, 15, 52–55, 271
empiricism, 10, 308n. 9
enclosure, 32, 33–34, 42
English parliament, 79
Enlightenment, the, 13, 309n. 7
environmental determinism, 46, 47, 136, 174–175, 178
environmental ethics, 20, 21, 26, 65–66, 71, 296, 297, 299
environmental pragmatism, 20, 21
Ephron, 82
Epictetus, 122
Epistemology, 9–10, 77, 94, 95
Eros, 161
ethics, 20, 77, 94, 302
Euclidean geometry, 267–268
eugenics, 49
Europe, 8, 14, 17, 48, 50, 104, 105, 115, 18, 124, 132, 215, 261
European philosophy, 9, 11
evolution, 10, 49
evolutionary cosmology, 6
experimentalism, 13, 21
extension services, 74, 285, 298, 299

family farming, 18, 209–214, 244–253, 255, 257, 258–259, 264, 265, 285, 287, 288–292, 294, 313nn. 6, 14
farm credit, 19; *see also* Agricultural policy
farming, 3–4, 13, 15, 16–17, 20, 21, 26–27, 28, 30, 32, 33–34, 37–38, 44, 47, 49, 51, 52, 57–58, 59–61, 66, 71, 73–75, 88, 97–98, 129, 130, 135, 167–168, 169, 263–264, 267, 281, 305n. 2

Faulkner, William, 28, 31
Federalist Papers, The, 216, 242–243
feminism, 30
Fensch, Thomas, 270
Ferguson, Adam, 40, 132, 135
feudalism, 215–218
Fichte, J. G., 142
Field, Peter, 308n. 11
Filmer, Sir Robert, 83
Finkelman, Paul, 125
Firdusi, 150
Fisch, Max H., 91–93, John, 92n. 4 95
food, production. *See* agricultural production
Foster, John, 308n. 4
Foucault, Michel, 129
foundationalism, 8, 11, 21
Fox-Genevese, Evelyn, 133
France, 125, 128, 134, 135
Franklin, Benjamin, 11, 14, 21, 93, 101–117, 113, 134; agrarianism and, 101, 119; agriculture and, 13–14, 105, 106–110, 111–113, 117; *Autobiography,* 102; *Poor Richard's Almanack,* 102, 103, 116
freedom, 37, 51, 215, 219, 304n. 1
freeholders, 215, 216–221, 236, 239, 241, 243–245, 260, 286
French natural philosophy, 11, 41–42, 46, 134–135
French Revolution, 134, 239

Gaither, Gloria, 270
Galileo, 89
Gates, Paul, 312n. 3
Geiger, George R., 313n. 9
Geisler, Charles C., 281
Geneva, 84
George, Henry, 313n. 9
German idealism/idealists, 11, 170
German National Socialism, 49
God, 77, 79–98, 141, 144; concepts of, 80–84, 87
Goff, Hiram, 224
Goodwyn, Lawrence, 288
Goldsmith, Oliver, 42–45, 47
Goldstone, Jack, 134
Gospel of John, 98
Gospel of Luke, 81
Gospel of Matthew, 97–98
Grace, Frank, 80
Grange, the, 52, 58–59, 306n. 4, 307n. 12
Great Awakening, the, 1, 87
Great Britain, 79, 128, 131, 134, 135
Great Depression, the, 1, 286
Greek mythology, 58
Greeks. *See* ancient Greeks
Greene, Jack P., 126
Greene, John C., 261
Grice, Paul, 308n. 8
Griswold, A. Whitney, 17–18, 19, 209, 215, 220, 237, 254–256, 259, 260, 263–267, 312nn. 1, 2, 6; *Farming and Democracy,* 17–18, 209, 254–257, 261–262
Gramsci, Antonio, 304n. 1
Guggenheim Memorial Foundation, 255

Habermas, Jurgen, 6
Hacker, Andrew, 219
Hamilton, Alexander, 137, 203
Hanson, Victor Davis, 36–40, 46
Harding, Walter, 156
Harkness, Georgia, 312n. 2
Harrington, James, 42
Hart, Richard E., 19, 278
Hartz, Louis, 82
Harvard University, 89, 174, 284, 306n. 8
Hatch Act, 283, 284
Hegel, G. W. F., 11, 48, 118, 292
Heidegger, Martin, 9, 32, 43, 48, 49
Heimert, Alan, 81, 82, 84, 86, 96, 97, 307n. 1
Helvetius, Claude-Adrian, 90
Henry, George, 33
Henry County, Kentucky, 67
Hesiod, 36, 75
Hickman, Larry A., 17
Hilde, Thomas C., 16

Hightower, Jim, 20, 119, 139, 281, 289–290, 295
Hill, James J., 60
Hindu religion, 146
Hittites, 82
Hobbes, Thomas, 40, 89, 93
Hofstader, Richard, 75, 306n. 5
Hogendorp, 262
Hollis, Martin, 293
Holocaust, the, 49
Homer, 127
Hooker, Thomas, 79, 87
House of Representatives/Senate, 123, 138, 219
Howarth, William, 265
Hucheson, Francis, 40
Hume, David, 40, 41–42, 46, 132
Hurt, Douglas, 96–97
Hyman, Stanley, 154, 158, 270

idealism, 54, 89, 90–93, 94–95
Indians. *See* Native Americans
Industrial Revolution, the, 72, 247–248
industrialization/industrial society, 16, 18, 29, 57, 159, 171–172, 177, 283, 288, 300–301
Inge, Thomas, 3
Internet, the, 170
Ireland, 104

Jackson, Stonewall, 31
Jackson, Wes, 76, 168, 290
James, Henry, 262
James, William, 5, 6, 8–10, 11, 12, 173, 261, 262, 311n. 2
Jamestown, 216
Jefferson, Thomas, 4, 19, 20, 21, 26, 29, 40, 43, 45–46, 51, 53, 56, 59–61, 64, 68, 90, 255–256, 263, 265, 309nn. 8, 15, 310nn. 16, 24, 312n. 2; agrarian philosophy and, 32, 118–139; agrarianism and, 14, 17, 19, 60, 63, 68, 72, 120–139, 159, 186, 199, 202–203, 254, 255–256, 261–263, 266, 303; agriculture and, 120, 124, 126–127, 128, 138, 252–253, 256; correspondence with Jay, 121–124, 128, 256; democracy and, 14, 17, 159, 202, 204, 219–220, 256, 260; education and, 14, 284; farming and, 118, 121, 123, 125, 126, 127, 219–220; freeholders and, 209–214, 216, 219, 244, 257, 258, 284, 285; materialism and, 93; Monticello and, 26, 118, 127; naturalism and, 14, 19, 118; *Notes on the State of Virginia*, 14, 45, 120–121, 122–129, 135, 262; pragmatism and, 118, 261, 263–264; slavery and, 125, 126, 136–137
Jessig, John, 224
Jesus Christ, 81, 84, 85, 98, 144
Johns Hopkins University, 173, 284
Johnson, Edward, 86
Johnson, Samuel, 91–92, 94, 95, 308n. 1
Jones, Anne Goodwyn, 304n. 1
Jones, E. L., 50
Jove, 56, 159
justice/equal opportunity, 241–242, 243, 267

Kant, Immanuel, 132–133, 293
Katz, Eric, 21
Kelley, Oliver Hudson, 57, 59
King's College (New York City), 91; *see also* Columbia University
Kirkendall, Robert, 128
Kirkland, James M., 304n. 1
Koch, Adrienne, 256, 262, 265
Kreyling, Michael, 304n. 1
Kucklick, Bruce, 311n. 4

Lacy, William B., 286, 287, 290
land conservation, 59, 62–63
land grant colleges/universities, 20, 69, 237, 279–291, 295–303
land reform, 1, 33, 35, 72, 266; *see also* agricultural policy; property rights
Langsdorf, Lenore, 77
Lanier, Lyle H., 28

Laud, Bishop, 79
Lathrop, Joseph, 309n. 6
Lawrence, D. H., 309n. 4
Leopold, Aldo, 57, 273
Lewis, C. I., 6
Lewis, William S., 311n. 1
Lewis and Clark Expedition, 138, 263
Lévi-Strauss, Claude, 199
libertarianism, 32, 33–35, 305n. 2; *see also* property rights
Liebig, Justus, 286
Light, Andrew, 21
Lincoln, Abraham, 220
literature. *See* aesthetics
Locke, John, 33, 34, 40, 82–83, 89, 260, 263; agrarian philosophy and, 32; primary and secondary qualities and, 89; property/property rights and, 19, 33, 70, 71, 83, 213, 221, 226–227, 230–235, 236–237, 242, 255, 258; *2nd Treatise of Government,* 33–34, 82–83; state of nature and, 297–330; theory of money and, 232–235
London, England, 103, 105
Louden County, Virginia, 128
Louisiana Purchase, 138, 162
Lucas v. *South Carolina Coastal Council,* 313n. 12
Lutheranism, 88
Lytle, Andrew Nelson, 75, 304n. 1

Machpelah, 82
MacIntyre, Alisdair, 308n. 8
MacKinnon, C. N., 269
MacPherson, C. B., 33
Madison, James, 216, 217, 242
Mahowald, Mary Briody, 311n. 2
Malone, Dumas, 156
Malthus, 53, 55
Mannheim, Karl, 304n. 1
Marcuse, Herbert, 129
Marsoobian, Armen, 16
Marx, Karl, 11, 48, 70, 310n. 17
Massachusetts, 151, 173
materialism, 48, 89–90, 93–95, 241
Mather, Cotton, 85–86, 94, 222
Mayer, A., 281
Mayer, F., 93
Mayer, J., 281
McDermott, John J., 167, 176, 178–179
McIntosh, James, 154–155
McKenna, George, 58
Mead, George Herbert, 6, 18, 257, 263
metaphysics, 16, 77, 94
Middlesex, Massachusetts, 151
Midwest, 30
Mill, John Stuart, 48
Miller, David, 18
Miller, Perry, 85, 86–87, 309n. 7, 307n. 1
Milton, John, 150
Mimms, Edwin, 304n. 1
mind/body problem, 77–78, 88, 90, 93, 94–95, 96
Mirabeau, the marquis de, 42, 133, 134, 135
modernization, 25–27, 139
Montesquieu, 14, 134, 135, 136, 139
Montmarquet, James, 1, 12, 13, 32–34, 35–36, 42–44, 47, 153; agrarianism and, 32; *The Idea of Agrarianism,* 32, 35, 51, 58, 69, 70, 71, 76
Moore, Barrington, 40
morality, 40, 41–42
More, Henry, 94, 98
More, Thomas, 70
Morgan, Barton, 283
Morgan, Edmund S., 136, 139
Morrill Act, 282, 283, 284
Morrison, Samuel Eliot, 238
Moses, 81
Mourt, G., 86
Muir, John, 57
Mumford, Lewis, 157, 162
Munzer, Stephen R., 313n. 11

nationalism, 41, 44–45, 47, 49, 136
Native Americans, 4, 78, 88, 305n. 2, 307n. 3

naturalism, 13, 21, 48, 277, 293; agrarianism and, 20, 72–73, 75
nature, 13, 53–54, 273
Nazism. *See* German National Socialism
neo-agrarianism, 14, 20, 21, 26, 266–267, 288–291, 294–298, 300, 302–303; pragmatism and, 279–280, 296, 300
New Criticism. *See* Vanderbilt agrarians
New Deal, 209
New England, 6, 30, 79, 84, 87, 170–171
New Jersey, 123, 137
New Republic, 270
New World, the, 47, 64, 79, 85, 106, 115, 217, 261, 307n. 1
New York, 43
New York City, 91, 177
Newton, Isaac, 89
Nikolitch, Radoje, 247
nineteenth-century, 8, 44, 96, 237; agrarianism and, 203, 303; philosophy in, 5, 48, 77, 130, 131
Nitsch, Ulrich, 294
Noah, 81, 82
Nollan, Lucas, and Dolan, 266
Nollan v. *California Coastal Commission,* 313n. 12
nominalism, 6
Nordin, D. Sven, 306n. 3
North, the, 29–30
Nozick, Robert, 174

O'Brien, Michael, 304n. 1
Oblinsky, Kira, 308n. 11
Ogilve, George, 33
Ohio University, 269
Oklahoma, 18
Old World, the, 79, 239
Olen, J., 273
Olexa, Michael, 298
O'Niell, Onora, 294
Onuf, Peter, 119
Ostrom, Elinor, 49
Ottoson, Howard W., 18

Paine, Thomas, 83, 93, 98
Panama Canal, 59, 62
Paris, France, 17, 195, 125
pastoralism, 27
Patterson, Orlando, 88
Peden, William, 256, 262, 265
Peirce, Charles Sanders, 1, 5, 6–9, 181, 261, 293, 311nn. 2, 5
Pennsylvania Hospital, 114
Pennsylvania Gazette, 310n. 20
Perry, Ralph Barton, 170, 222, 223, 312n. 5
Perry, Richard L., 261
Peterson, Merrill, 125
Philadelphia, Pennsylvania, 105
Philadelphia Academy, 106*; see also* University of Pennsylvania
philosophes, 41, 45, 135
Physiocrats, 14, 54, 68, 70, 133–134, 139, 309n. 14
Pioneer-Hybrid Seed Co., 286
Plantinga, Alvin, 308nn. 5, 9
Plato, 39, 95, 122, 193
Platonism, 91, 92, 145, 302
Pliny, 157
Plutus, 56, 159
Plymouth Rock, 216
Pocock, J. G. A., 83
poetry. *See aesthetics*
Poor Richard's Almanack. See Franklin, Benjamin
Popper, Karl, 292
Populism/Populist movement, 52, 57, 58–59, 60, 284, 285, 288, 289, 302, 306n. 5, 307n. 12
Pound, Ezra, 49
pragmatism, 1–2, 4, 5–12, 18, 21, 26, 29, 32, 77–78, 88, 118, 139, 261, 263, 267, 271–273, 274–276, 278, 285, 291, 298, 303, 313n. 9; agrarianism in, 11–12, 21, 29, 77, 280
Presbyterianism, 88
Priestly, Joseph, 90
Princeton University, 94
productionism, 286, 291, 300–301

property rights, 13, 33–35, 36–37, 71–72, 83, 217, 225, 226, 235, 266, 267, 305n. 2
Protestant settlers, 6
Protestant work ethic, 258; *see also* proficient work
Protestantism, 88
Psalms, 81
Puritanism, 79–80, 84, 95–96
Pythagoras, 95

Quakers, 88
Quesnay, Francois, 42, 53, 68, 71, 133, 134, 135
Quietism, 238
Quine, W. V. O., 6
Qur'an, 81

Railsback, Brian E., 276
Randolph, Edmund, 126
Ransom, John Crowe, 26, 29, 31, 43, 47, 68, 304n. 1
Rationalism, 312n. 9
Rawls, John, 129, 293
realism, 6
Redclift, Michael, 289
Reformation, 80
Reid, Thomas, 94, 308n. 9
Rhode Island, 91, 123
Rhodes, Richard, 267
Ricardo, David, 53
Ricketts, Ed, 270, 272, 275, 276
risk, 210, 246, 250, 257
Ritter, William Emerson, 270
Robertson, John, 41
Robinson, Howard, 308n. 4
Rockefeller Foundation, 281
Roman Catholicism, 84, 88, 96
romanticism, 3, 12, 42–48, 130
Romans, 113
Rome, Italy, 84
Roosevelt, Franklin Delano, 119
Roosevelt, Theodore, 59, 62, 284
Rorty, Richard, 6, 160
Rosenberg, Charles E., 280
Ross, Earle D., 309n. 9
Rossiter, Clinton, 309n. 5
Rousseau, Jean-Jacques, 53, 130–131, 133, 172; *Emile,* 130–131
Royce, Josiah, 1, 6, 16; agrarianism and, 16, 172; California and, 171, 172–178, 182, 183, 311n. 4; community and, 169–172, 174–184; *The Philosophy of Loyalty,* 167, 174–175, 180, 182; pragmatism and, 171, 180, 311n. 2; *The Problems of Christianity,* 180–181; provincialism and, 170–171, 176–184
Royce, Sarah, 172, 175
rural development, 17, 26
rural life, 9, 29, 60–62, 69, 72, 75, 306n. 7
Rush, Benjamin, 90, 309nn. 10, 11
Russell, Bertrand, 5
Ryan, Alan, 204, 313n. 11

Sachs, Carolyn, 301
Sacramento Valley, 176
San Francisco Chronicle, 270
Santayana, George, 146, 241
Sartre, Jean-Paul, 129
Sauer, Carl, 18
Saxons, 36, 305n. 2
Schama, Simon, 44
Schelling, F. W. J., 130
Schneider, Herbert, 92
scholasticism, 89
Schultz, David A., 313n. 11
science, 7–8, 51, 53, 57, 69, 89, 261, 288
Scotland, 104, 131
Scottish Enlightenment, 40, 94, 131, 132, 133
Searle, John, 308n. 8
Selma, California, 37
Seneca, 122
seventeenth century, 11, 42, 80, 87, 88, 213, 221; England in, 25, 134, 226; puritanism and, 79
Shue, Henry, 34
Sinclair, Sir John, 128S
sixteenth century, 11, 213, 221

slavery, 88, 125, 126, 138, 307n. 3
Smith, A., 77
Smith, Adam, 14, 40, 46, 48, 49, 71, 132, 133, 135, 136, 139; *The Wealth of Nations,* 132
Smith-Lever Act, 283, 284
social contract theory, 40
Social Science Research Council, 255
socialism, 70
Socrates, 39
sophists, 193
South, the, 29–30, 31, 51–52, 88, 138, 249, 304n. 1
Southern Renaissance, 28, 31
Spencer, Herbert, 48
Spengler, Oswald, 48
Stanford University, 270
Steele, Richard, 222
Steinbeck, John, 19–20, 51, 71; agrarianism and, 271, 272; *The Grapes of Wrath,* 20, 269, 271, 274, 278; naturalism and, 20, 273–276, 277; *Of Mice and Men*, 20, 276–277; philosophy and, 269–272, 278; *The Sea of Cortez,* 20, 270, 274, 275
stewardship, 19, 161, 265–266, 313n. 14
Stewart, Dugald, 94
Stewart, John L., 28, 31
stoicism, 193
Stoller, Leo, 157, 159
Stout, Bill A., 290
Strange, Marty, 20, 289–290, 295
subsistence rights, 12, 34–35
Swinburne, Richard, 308n. 5

Taliaferro, Charles, 8, 12, 83, 96
Tansill, Charles C., 217
Tate, Allen, 28, 30, 31, 47, 49, 304n. 1
Tawney, R. H., 222, 312n. 2
Taylor, F. M., 308n. 11
Taylor, John, 50, 52
Taylor, Richard, 159, 308n. 9; technology 57; *see also* agricultural technology
theism, 13, 80, 83, 88, 93, 94, 95, 98, 308n. 9
theology, 79, 81, 84, 88
Theophrastrus, 157
Thompson, E. P., 41
Thompson, Paul B., 1, 2, 5, 12, 14, 40, 167, 254, 264, 272, 286, 290, 296, 311n. 1
Thoreau, Henry David, 1, 11, 13, 15–16, 51, 55–57, 64, 69, 130, 148, 170, 172, 262, 271, 274; agrarianism and, 15, 119, 139, 153–163; as poet, *Walden* and, 55–56, 153–157, 158, 159, 181
Tilly, Charles, 50
Tokyo, 177
Townsend, Harvey, 90–91
transcendentalism. *See* American transcendentalism
Troeltsch, Ernst, 312n. 2
Tufts, James Hyden, 6, 292

United States, 77, 88, 96, 194, 105, 113, 168
United States Department of Agriculture (USDA), 18, 254, 257, 281, 283
University of California at Berkley, 173
University of Chicago, 18, 196, 257, 284, 306n. 8
University of Pennsylvania, 106
University of Texas, 18
University of Virginia, 139
urban agrarians, 60–61, Liberty Hyde, 306n. 10; *see also* Bailey, Liberty Hyde
urban life, 60–62
utilitarianism, 300

Vanderbilt Agrarians/Agrarianism, 12, 27–32, 43, 47, 48, 51, 272, 304n. 1
Vanderbilt University, 26, 27–28, 304n. 1
Virgil, 47, 69; the *Georgics,* 36
Virginia, 123, 124, 126, 136, 139, 255
Virginia Company of London, 97

Wallerstein, Immanuel, 50
Warren, Robert Penn, 28, 31, 304n. 1
Washington, George, 137

Waters, Alice, 195–196, 197–198
Waterville, Maine, 142, 145
Weaver, General James, 57
Weber, Max, 223, 309n. 4, 312n. 2, 313n. 14
Webster, Daniel, 218
Welchman, Jennifer, 292, 293, 302
Whitman, Walt, 274; *Leaves of Grass,* 86
Wilhite, Virgile Glenn, 310nn. 18, 21
Williams, B., 298
Williams, Howell V., 310n. 21
Wills, Gary, 121, 125, 127, 132, 135, 136, 313n. 4.
Wilson, Douglas L., 126
Wilson, Edmund, 28
Winch, Peter, 308n. 8
Winstanley, Gerrard, 33, 56, 67
Winter, Ella, 270
Winthrop, John, 82, 84, 86
Wisconsin State Agricultural Society, 220
Witherspoon, John, 308n. 1
Wittgenstein, Ludwig, 300
Wojcik, Jan, 43–44
Wolterstorff, Nicholas, 93
Wordsworth, William, 67
World War I, 49
World War II, 17, 255, 286
Worldwatch Institute, 288
Wunderlich, Gene, 18–19, 119, 121, 313n. 11

Xenophontis, 122

Yale University, 254
yeoman of Kent, 70
Young, Arthur, 70

Zion, 87